OPTICAL AND FIBER COMMUNICATIONS REPORTS

OPTICAL AND FIBER COMMUNICATIONS REPORTS

The Optical and Fiber Communications Reports (OFCR) book series provides a survey of selected topics at the forefront of research. Each book is a topical collection of contributions from leading research scientists that gives an up-to-date and broad-spectrum overview of various subjects.

The main topics in this expanding field will cover for example:

- specialty fibers (periodic fibers, holey fibers, erbium-doped fibers)
- broadband lasers
- optical switching (MEMS or others)
- polarization and chromatic mode dispersion and compensation
- long-haul transmission
- optical networks (LAN, MAN, WAN)
- protection and restoration
- further topics of contemporary interest.

Including both general information and a highly technical presentation of the results, this series satisfies the needs of experts as well as graduates and researchers starting in the field. Books in this series establish themselves as comprehensive guides and reference texts following the impressive evolution of this area of science and technology.

The editors encourage prospective authors to correspond with them in advance of submitting a manuscript. Submission of manuscripts should be made to one of the editors. See also http://springeronline.com/series/ 4810.

Andrea Galtarossa   Curtis R. Menyuk
Editors

Polarization Mode Dispersion

With 109 Figures

Springer

Andrea Galtarossa
Department of Information Engineering
University of Padova
Via Gradenigo 6/B
35131 Padova
Italy
andrea.galtarossa@unipd.it

Curtis R. Menyuk
Department of Computer Science and Electrical Engineering
University of Maryland Baltimore Country
1000 Hilltop Circle
Baltimore, MD 21250
USA
menyuk@umbc.edu

Library of Congress Control Number: 2005925979

ISBN-10: 0-387-23193-5 Printed on acid-free paper.
ISBN-13: 978-0387-23193-8

Printed in the United States of America. (SPI/EB)

9 8 7 6 5 4 3 2 1

springeronline.com

To Hermann Haus

"For the leader. A Psalm of David. O LORD, Thou hast searched me, and known me."
—Psalm 139

Preface

Andrea Galtarossa and Curtis R. Menyuk

This volume had its origin in conversations we had with Professor Carlo Someda of the Università di Padova in summer 2001. All three of us have had a long-standing interest in polarization mode dispersion and, more generally, polarization effects in optical fibers. We were all impressed by the great increase in interest in this subject that had occurred in the past two years. This subject has many theoretical and experimental subtleties, and we were concerned by the propagation of misconceptions in the scientific literature. The idea of a summer school arose, focused on polarization mode dispersion—or PMD—as it is usually known. Perhaps, it would be an appropriate topic for Lake Como, where summer schools are held every year. Carlo heartily endorsed the idea of a summer school, but suggested that we hold it in Venice. Thus, the idea of holding a summer school on PMD in Venice was born.

The next year was a flurry of activity in which we decided what topics should be taught and who should be asked to teach them. We asked many of the most distinguished researchers in the field to be instructors, and, to our great delight, everyone that we asked was able to accept. At the same time, Carlo was arranging for the Istituto Veneto di Scienze, Lettere, ed Arti to host the summer school and was finding support from commercial sponsors. This financial support, for which we are very grateful, along with the hard work of the local organizing committee and in particular by Marco Santagiustina, allowed us to invite the instructors gratis and to greatly reduce the cost to the participants from what we had originally anticipated.

During this year, the telecommunications bubble burst, and we were worried that not enough paying participants would come to the school. In the end, we need not have worried. Almost every possible seat was filled. Indeed, the school made a small profit, which was expended as prizes for graduate students and non-tenure-track research faculty at universities.

The linchpin of the school was two-hour lectures. The first, which was delivered by Bob Jopson and Lynn Nelson, gave an introduction to the subject of PMD. The second, which was delivered by Henning Bülow and Stéphanie Lanne, discussed the important topic of PMD mitigation; Andrea Galtarossa and Anna Pizzinat presented the model for low-PMD fibers; Nicolas Gisin covered the increasingly important topic of the interaction of PMD with polarization dependent loss. Other topics that were included in the school were: "PMD models," which was covered by Antonio Mecozzi and Mark Shtaif; "Interaction of PMD with nonlinearity and chromatic dispersion," which was covered by Curtis Menyuk; "PMD measurement techniques," which was covered by Paul Williams and by Marco Schiano in two separate lectures; "Spatially resolved measurement of fiber polarization properties," which was covered by Luca Palmieri and Andrea Galtarossa; "PMD impact on optical systems," which was covered by Magnus Karlsson and by Francesco Matera in two separate lectures; "Polarization effects in recirculating loops," which was covered by Brian Marks, Gary Carter, and Yu Sun; "PMD Emulation," which was covered by Alan Willner and Michelle Hauer; and, finally, "Applications of importance sampling to PMD," which was covered by Gino Biondini, Bill Kath, and Sarah Fogal. Dipak Chowdhury worked with Artis and VPI—two producers at that time of commercial software for modeling optical fiber communications systems—to present a lecture that covered numerical modeling of PMD. Additionally, we had lectures on special topics by Hermann Haus, Jim Gordon, Herwig Kogelnik, and Carlo Someda. Finally, we had a poster session, which gave the lecturers the opportunity to learn something from our participants.

The feedback that we received from the participants and the lecturers was overwhelmingly positive. This success was due to the great time and energy that all the instructors put into their lectures. At the summer school and thereafter, we continued to receive the suggestion from many of the lecturers and participants that the summer school lectures would be of interest to a broad audience in the optical fiber communications community.

After asking our lecturers to put so much time and energy into their lecturers, we were a little reluctant to request an additional effort. There was also a concern that the material would have to be updated. However, since the material is largely tutorial in nature, little updating was in fact needed. In the end, all the lecturers provided us with contributions, and the result is the volume that you have before you. We hope that it will be of use to researchers and the students in the field of optical fiber communications who want to have an introduction to PMD. If this volume is as successful as the school, it will be due to the hard work of the contributors.

As a final note, we owe a significant debt to Professor Hermann Haus of MIT. He enthusiastically participated in the summer school, sitting in the front row and posing illuminating questions to the lecturers. It was his suggestion to include poster sessions in the school. He passed away suddenly in May 2003. The loss to our community of one of its most distinguished members is still deeply felt.

Padova, Italy — Andrea Galtarossa
Baltimore, Maryland, USA — Curtis R. Menyuk

Contents

DOI: 10.1007/s10297-004-0001-5
Originally published in J. Opt. Fiber. Commun. Rep. 1, 312–344 (2004)

Introduction to polarization mode dispersion in optical systems

L. E. Nelson[1] and R. M. Jopson[2]

[1]OFS Laboratories, Somerset, NJ USA
[2]Bell Labs, Lucent Technologies
Holmdel, NJ 07733 USA

Abstract. This introduction covers concepts important to the understanding of polarization mode dispersion (PMD), including optical birefringence, mode coupling in long optical fibers, the Principal States Model, and the time and frequency domain behavior of PMD. Other topics addressed include the concatenation rules, bandwidth of the Principal States, PMD statistics and scaling, PMD system impairments, and PMD outage probability calculations.

1. Introduction

Long-haul, fiber-optic transmission systems can be limited by impairments due to polarization mode dispersion (PMD). PMD is caused by the birefringence of optical fiber and the random variation of its orientation along the fiber length. PMD causes different delays for different polarizations, and when the difference in the delays approaches a significant fraction of the bit period, pulse distortion and system penalties occur. Environmental changes including temperature and stress cause the fiber PMD to vary stochastically in time, making PMD particularly difficult to manage or compensate. In addition, although amplifiers or other components such as add-drop multiplexers in an optical system may have constant birefringence, variable polarization rotations between them due to the environment cause these components to add or subtract randomly to or from the PMD of the total system. While improvements in the PMD of optical fiber and optical components have enabled the transmission of ever higher bit-rates, these improvements have been accompanied by requirements of ever more stringent specifications for system PMD.

In this paper, we will review concepts important to the understanding of PMD and its systems impairments. We aim to provide sufficient background for the following, more detailed chapters.

2. Basics of PMD

2.1. *Representation of Polarization*

In free space, light is a transverse wave, where the wave motion is perpendicular to the direction of propagation. Here we choose z as the direction of propagation, such that a continuous wave is described by $E(z,t)e^{j(\omega_0 t-\beta z)}$, where ω_0 is the angular carrier frequency, t is the time, and β is the propagation constant. x, y are the transverse coordinates. The term "polarization" is a vector quantity describing, over the course of an optical cycle, the behavior of the electric field, $E(z,t)$, in the x, y plane centered at time t and propagation distance z. When the phase difference between the x and y components, E_x and E_y, is 0 or an integer multiple of π, the light wave is linearly polarized and the plane of oscillation is constant. When the x and y components have equal amplitude and their phase difference is an odd integer multiple of $\pi/2$, the light is circularly polarized. Coherent light not having linear or circular polarization has elliptical polarization.

The Jones formalism [36] provides a concise representation of polarization using the electric field vector itself. We define $| s\rangle$ as the 2D Jones (column) ket vector,

$$| s\rangle = \begin{pmatrix} s_x \\ s_y \end{pmatrix}, \tag{1}$$

where s_x, s_y are complex quantities. The bra $\langle s |$ indicates the corresponding complex conjugate row vector, i.e., $\langle s | = (s_x^*, s_y^*)$. The bra-ket notation is used to distinguish Jones vectors from Stokes vectors. Our Jones vectors are all of unit magnitude, i.e., $\langle s | s\rangle = s_x^* s_x + s_y^* s_y = 1$, as we assume coherent light except as noted. This concise formalism provides an intuitive understanding of changes in polarization since "lab-space" coordinates are used. It includes optical phase, which is necessary when combining light, but only fully polarized waves can be described by Jones vectors.

The Stokes formalism [75] is an alternative description of polarization and uses four (real) Stokes parameters, which are functions only of observables of the light wave. The polarization state of any light beam (totally, partially, or not polarized) can be described. For coherent light, the Stokes parameters are

$$s_0 = s_x s_x^* + s_y s_y^* \tag{2a}$$
$$s_1 = s_x s_x^* - s_y s_y^* \tag{2b}$$
$$s_2 = s_x s_y^* + s_x^* s_y \tag{2c}$$
$$s_3 = j(s_x s_y^* - s_x^* s_y). \tag{2d}$$

(The broader definition in terms of intensity differences applicable to incoherent light is discussed in section 2.9.) We define $\hat{s} = (s_1, s_2, s_3)$ as a 3D Stokes vector of unit length indicating the polarization of the field and corresponding to $| s\rangle$. By this definition, $s_1 = 1$ for linear polarization aligned with the x-axis; $s_2 = 1$ for linear polarization at 45° to this axis; and $s_3 = 1$ for right-circular polarized light ($s_y = js_x$) conforming to the traditional optics definition. The assignment of either left-circular or right-circular polarization to $s_3 = 1$ is often implicit in PMD literature and has non-trivial consequences in subsequent analyses. We always use the same letter symbols for corresponding Jones and Stokes vectors. Note that a common phase shift of both components of $| s\rangle$ does not change $\hat{s}$.

The locus of Stokes vectors, (s_1, s_2, s_3), representing all possible states of polarization of coherent light forms a unit sphere in Stokes space. This is the Poincaré sphere [64], which provides a unique three-dimensional representation of polarization states (see for example, [18, 33]). Linear polarizations are plotted along the equator; circular polarizations are at the poles. Elliptical polarizations are plotted elsewhere on the surface of the Poincaré sphere. (See Fig. 2.) The Stokes formalism, together with the Poincaré sphere, provides an intuitive understanding of the effects on polarized light of birefringence and hence of PMD. It can describe partially polarized light, but it ignores the optical phase of light.

Partially polarized light is important in a method of PMD monitoring that relies on the degree of polarization (DOP), also discussed in section 2.9. One measure of DOP is the length of the normalized 3D Stokes vector. The DOP of coherent light is unity, whereas the DOP of unpolarized light is zero. DOP is a concept having subtleties not always appreciated. It is necessary to average over time or space to obtain a DOP that differs from unity and the measured DOP will depend on the amount of averaging. In single-mode lightwave systems, there is little spatial averaging, and the time average usually lasts many bit periods.

When light passes through an optical element (e.g., optical fiber, where light is approximately a transverse wave), the input polarization $\hat{s}$ will emerge as a new polarization $\hat{t}$ at the output. The transformation of the polarization can be described using a 3×3 Müller matrix R, through $\hat{t} = R\hat{s}$. Similarly, the Jones vector $| \, s\rangle$ will emerge from the same optical element as $| \, t\rangle$, where the 2×2 unitary transmission matrix T relates output to input via $| \, t\rangle = T \, | \, s\rangle$. The 2×2 Jones matrix U is related to T by $T = e^{-j\phi_0}U$, where $\det U = 1$ determines ϕ_0, the common phase. In the following, we will consistently use this designation of fiber input and output, as shown in Fig. 1.

Fig. 1. Block diagram of optical fiber under test.

2.2. *Birefringence*

PMD has its origins in optical birefringence. Although telecommunications fibers are often called "single mode," even in an ideal circularly symmetric fiber, there are two orthogonally polarized HE_{11} modes. In a perfect fiber, these modes have the same group delay. In reality, fibers have some amount of asymmetry due to imperfections in the manufacturing process and/or mechanical stress on the fiber after manufacture. The asymmetry breaks the degeneracy of the orthogonally polarized HE_{11} modes, resulting in birefringence: a difference in the phase velocities of the two modes. Even very small amounts of birefringence can cause evolution of the polarization state as light propagates through fiber.

Both intrinsic and extrinsic perturbations cause birefringence in optical fibers. The manufacturing process can set up permanent, intrinsic perturbations in the fiber. Form (geometric) birefringence arises due to a noncircular waveguide, whereas stress birefringence is due to forces set up by a noncircular core. When fiber is spooled, cabled, or embedded in the ground, extrinsic perturbations, including lateral stress, bending, or twisting, induce birefringence. These perturbations will change as the fiber's external environment changes.

In a short section of fiber, the birefringence can be considered uniform. The difference between the propagation constants of the slow and fast modes can be expressed as

$$\Delta\beta = \frac{\omega n_s}{c} - \frac{\omega n_f}{c} = \frac{\omega \Delta n}{c}, \tag{3}$$

where ω is the angular optical frequency, c is the speed of light, and $\Delta n = n_s - n_f$ is the differential effective refractive index between the slow (s) and fast (f) modes. Generally, the perturbations discussed above create linear birefringence where there are two linearly polarized waveguide modes whose electric field vectors are aligned with the symmetry axes of the fiber. Circular birefringence results from fiber twist [78].

When an input wave that is linearly polarized at 45° to the birefringent axes is launched into a short fiber, the state of polarization evolves in a cyclic fashion as the light propagates down the fiber, i.e., from linear to elliptical to circular and back through elliptical to a linear state orthogonal to the launch state. Analogously, for a fixed-input polarization state, if the light frequency is varied, the output polarization state from a short length of birefringent fiber will cycle in the same way through the various states. This frequency-domain picture of PMD is illustrated in Fig. 2 for a launch state near the birefringent axis. The output polarization traces a circle on the surface of the Poincaré sphere. Several PMD measurement techniques, including Jones Matrix Eigenanalysis [32], Poincaré Sphere Method [1], and Müller Matrix Method [37] use this frequency-domain picture.

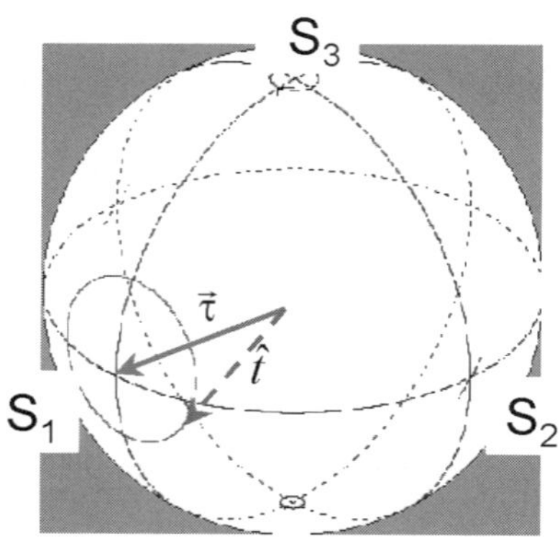

Fig. 2. Frequency-domain behavior of PMD in a short birefringent fiber, where the fiber's birefringent axis $\boldsymbol{\tau}$ is aligned with the S_1 axis.

The differential index, together with the optical wavelength λ, allows us to define a beat length, $L_b = \lambda/\Delta n$, as the propagation distance for which a 2π phase difference accumulates between the two modes or, equivalently, the polarization rotates through a full cycle. Standard telecommunications-type fibers can have beat lengths of $\sim$ 10 m [25], giving $\Delta n \sim 10^{-7}$, which is much smaller than the $\sim 10^{-3}$ index difference between core and cladding.

In the time-domain picture, for a short section of fiber, the differential group delay (DGD), $\Delta\tau$, is defined as the group-delay difference between the slow and fast modes. This $\Delta\tau$ can be found from the frequency derivative of the difference in propagation constants [Eq. (3)]:

$$\frac{\Delta\tau}{L} = \frac{d}{d\omega}\left(\frac{\Delta n\omega}{c}\right) = \frac{\Delta n}{c} + \frac{\omega}{c}\frac{d\Delta n}{d\omega}. \tag{4}$$

This "short-length" or "intrinsic" PMD, $\Delta\tau/L$, is often expressed in units of picoseconds per kilometer of fiber length. The linear length dependence of DGD applies when the birefringence can be considered uniform, as in a short fiber. For the "long-length" PMD regime, the orientation of the fast and slow modes becomes a function of location and the DGD has a square-root-of-length dependence.

Figure 3 shows the time-domain effect of PMD in a short fiber, where a pulse launched with equal power on the two birefringent axes results in two pulses at the output, separated by the DGD, $\Delta\tau$. From Eq. (4) and ignoring the dispersion of Δn, we can then see that the DGD for a single beat length, L_b, is equal to an optical cycle:

$$\Delta\tau_b = L_b\frac{\Delta n}{c} = \frac{\lambda}{c} = \frac{1}{\nu}, \tag{5}$$

which is 5.2 fsec at 1550 nm. The polarization-dependent signal delay method [58] for measuring PMD relies on the time-domain picture of PMD.

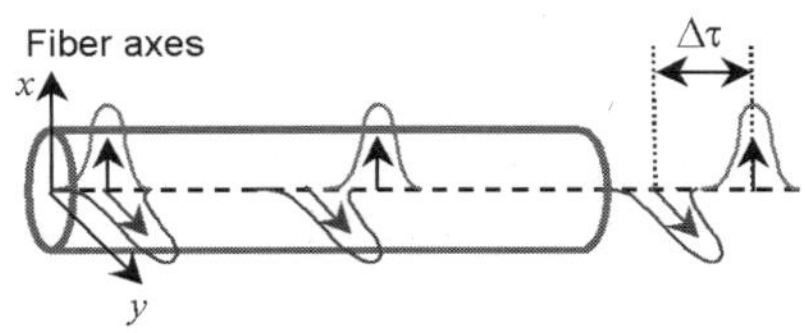

Fig. 3. Time-domain behavior of PMD in a short birefringent fiber.

2.3. *Polarization Mode Coupling*

While DGD in the short-length regime is predictable because the birefringence is inherently additive, fiber lengths in today's terrestrial and submarine transmission systems are 100s or 1000s of km, and the birefringence is no longer additive. Random variations in the axes of the birefringence occur along the fiber length, causing polarization-mode coupling wherein the fast and slow polarization modes from one segment decompose into both the fast and slow modes of the next segment. Polarization-mode coupling results from localized stress during spooling/cabling/deployment, from splices and components, from variations in the fiber drawing process, and from intentional fiber "spinning" during drawing, which induces mode coupling at "meter" lengths [40]. A commonly used model for long fibers is a concatenation of birefringent sections whose birefringence axes (and magnitudes) change randomly along the fiber.

Owing to mode coupling, the birefringence of each section may either add to or subtract from the total birefringence, and therefore the DGD does not accumulate linearly with fiber length. Instead, it has been shown that in long fiber spans, the DGD accumulates as a three-dimensional random-walk, and on average increases with the square root of distance [66,71]. Although mode coupling helps to reduce the DGD of a fiber span, because the mode coupling is determined by the fiber's environment, variations in, for example, external stresses will change the mode coupling and thus the fiber's DGD. Therefore, a statistical approach for PMD must be adopted, as discussed in Section 3.

The categorization of a fiber in the short- or long-length regime is determined by a parameter called the correlation length L_c, also referred to as the coupling length [41]. This parameter describes weak random coupling between two waveguides or the equivalent random coupling between the two polarization modes of a fiber with uniform birefringence subject to random perturbations. One considers the evolution of the polarization modes as a function of length in an ensemble of fibers with statistically equivalent perturbations. While the input polarization is fixed, it is equally probable to observe any polarization state at large lengths. The evolution is characterized by the difference $\langle p_x \rangle - \langle p_y \rangle$ of the ensemble averages of the power in the x and y polarizations. Assuming $\langle p_x \rangle = 1$ and $\langle p_y \rangle = 0$ at the input, this difference evolves from a value of 1 at the input to a value of zero at large lengths. L_c is defined as that length where the power difference has decayed to $\langle p_x \rangle - \langle p_y \rangle = 1/e^2$ (for further detail see [71]). Correlation lengths can be less than 1 m when fiber is spooled (due to large amounts of polarization-mode coupling); conversely, L_c can be $\sim$ 1 km when fiber is cabled. The actual fiber behavior is affected by fiber "spinning" during draw, temperature, spool diameter and tension, cable design, installation conditions, and fiber relaxation.

The correlation length thus defines the two different PMD regimes. When the fiber transmission distance L satisfies $L << L_c$, the fiber is in the short-length regime and the DGD increases linearly with distance. When $L >> L_c$, the fiber is considered to be in the long-length regime and the mean DGD, $\overline{\Delta\tau}$, increases with the square root of distance. Transmission systems are generally in the long-length regime, so fiber PMD is often specified using a PMD coefficient having units of ps/(km)$^{1/2}$. While fibers manufactured today can have mean PMD coefficients less than 0.05 ps/(km)$^{1/2}$, "legacy" fibers installed in the 1980s may exhibit PMD coefficients higher than 0.8 ps/(km)$^{1/2}$ [63].

The statistical theory of PMD [21,81] has provided an elegant expression linking the mean square DGD of the fiber to L_b and L_c:

$$\langle \Delta\tau^2 \rangle = 2\left(\Delta\tau_b \frac{L_c}{L_b}\right)^2 (L/L_c + e^{-L/L_c} - 1). \tag{6}$$

This expression is valid for both regimes and also for the transition region. For $L << L_c$, the above relation simplifies to

$$\sqrt{\langle \Delta\tau^2 \rangle} = \Delta\tau_{\rm rms} = \Delta\tau_b L/L_b, \tag{7}$$

whereas for $L >> L_c$,

$$\Delta\tau_{\rm rms} = (\Delta\tau_b/L_b)\sqrt{2LL_c}, \tag{8}$$

reflecting the length dependence discussed earlier. Single-ended backscattering measurement techniques [16] can determine birefringence (L_b) and the mean square DGD. Then the correlation length (L_c) can be inferred from the fundamental Eq. (6) relating the three quantities.

2.4. *Principal States Model*

The propagation of a pulse through a long length of fiber is very complicated due to random mode coupling and pulse splitting at every change in the local birefringence axes. But a surprising aspect of PMD is that even for long fibers, one can still find two orthogonal-polarization launch states at the fiber input that result in an output pulse that is undistorted to first order. Figure 4 shows the resulting output pulse shapes when a 10-Gb/s, 50% duty-cycle return-to-zero signal was launched with various polarizations through a 48-km fiber with large PMD. The figure shows the output pulse for each of two polarization launches that minimize the bit-error rate (BER) at the fiber output. Note the difference in arrival times, the DGD. Also shown is the output pulse shape for a polarization launch that maximizes the BER at the output. It is apparent that the two launches minimizing the BER result in fairly undistorted pulses, while the pulse from the third launch is significantly broadened. The two least distorted pulses are the fastest and the slowest pulses of all the polarizations launched.

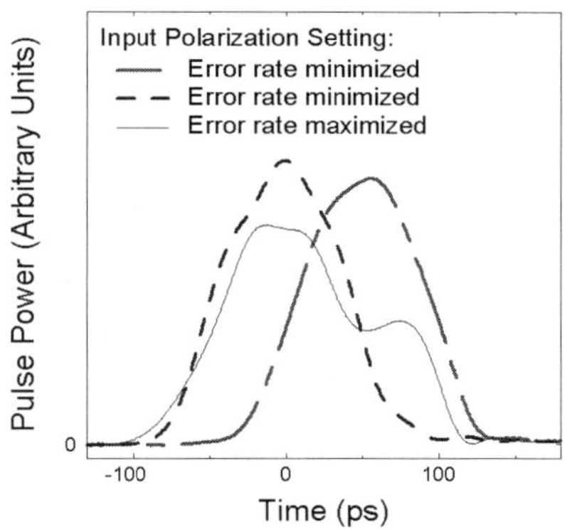

Fig. 4. Output pulse shapes for three polarization launches of a 10Gb/s, 50% duty cycle signal through a 48-km fiber with ~60 ps PMD.

The Principal States Model, originally developed by Poole and Wagner [65] was the first to describe this phenomenon and is still in common use today for the characterization of PMD. The model provides both a time domain and a frequency domain characterization of PMD. Figure 4 illustrates the time-domain picture. The frequency-domain picture allows a very simple definition. It states that, for a length of fiber, there exists for every frequency a special pair of polarization states, called the Principal States of Polarization (PSPs). A PSP is defined as that input polarization for which the output state of polarization is independent of frequency to first order, i.e., over a small frequency range. The Principal States Model applies when the bandwidth of the pulses is small, with a pulse length greater than the PMD-induced DGDs. This is because PMD is intrinsically an interference phenomenon, caused by the coherent addition of the complex amplitudes of the multiplicity of pulses created by the repeated pulse splitting mentioned above.

In the absence of polarization-dependent loss, the PSPs are orthogonal. For each pair of input PSPs, there is a corresponding pair of orthogonal PSPs at the fiber output. The input and output PSPs, designated by the unit Stokes vectors $\hat{p}_s$ and $\hat{p}$, respectively, are related by the fiber's transmission matrix R, via $\hat{p} = R\hat{p}_s$, just as any input polarization is related to a polarization at the fiber output.

2.5. PMD Vector

Using the Principal States Model, PMD can be characterized by the PMD vector:

$$\boldsymbol{\tau} = \Delta\tau\hat{p}, \tag{9}$$

a vector in three-dimensional Stokes space, where the magnitude, $\Delta\tau$, is the DGD. The unit vector $\hat{p}$ points in the direction of the slower PSP, whereas the vector $-\hat{p}$ indicates the orthogonal, faster PSP. In Stokes space, $-\hat{p}$ is $180°$ from $\hat{p}$. Note that the definition used here is in right-circular Stokes space (with S_3 denoting right-circular polarization), whereas the original PMD vector $\boldsymbol{\Omega}$ of Poole et al. [67] was defined in left-circular Stokes space. See [48] for further explanation of the relation between the two.

The PMD vector at the fiber input $\boldsymbol{\tau}_s$ is related to the output PMD vector $\boldsymbol{\tau}$ by $\boldsymbol{\tau} = R\boldsymbol{\tau}_s$. One can then show that the frequency derivative of $\hat{t} = R\hat{s}$ leads directly to the law of infinitesimal rotation

$$\hat{t}_\omega = \frac{d\hat{t}}{d\omega} = \boldsymbol{\tau} \times \hat{t}, \tag{10}$$

where $\boldsymbol{\tau}\times = R_\omega R^T$ and R^T is the transpose of R. Here, the PMD vector describes how, for a fixed input polarization, the output polarization $\hat{t}$ will precess around $\boldsymbol{\tau}$ as the frequency is changed. The direction of $\hat{t}$ relative to $\boldsymbol{\tau}$ determines the angle of precession, whereas the magnitude, $\Delta\tau$, determines the rate at which $\hat{t}$ precesses around $\boldsymbol{\tau}$. For example, if $\hat{s}$ is launched with equal power along the PSPs, $\boldsymbol{\tau} \times \hat{t}$ will have its largest value, and the largest change in the output polarization will occur for a frequency change $\Delta\omega$. The precession has magnitude $\phi = \Delta\tau\Delta\omega$, where ϕ is the rotation angle on the Poincaré sphere. If $\hat{t}$ is aligned with $\pm\boldsymbol{\tau}$, then there is no precession and no change of the output polarization with frequency. This is, of course, the postulate for a PSP. The rotation law, Eq. (10) thus provides a precise mathematical definition of the PSP and of its length, $\Delta\tau$, the DGD. The law also says that there are only two PSPs, corresponding to the two alignments, $\hat{t} = \pm\boldsymbol{\tau}$.

A length of polarization-maintaining fiber (PMF) has a constant PMD vector whose length, the DGD, and direction $\hat{p}$ do not change with frequency. For this simple case, the output vector $\hat{t}$ will trace a circle on the Poincaré sphere as the frequency is varied, as illustrated in Fig. 2. In real fibers, however, both the magnitude and the direction of $\boldsymbol{\tau}$ change with frequency, as shown in Fig. 5.

The rotation law still applies locally in this case, describing $\hat{t}(\omega)$ as a circular arc for a small range of frequencies. In this range, characterized by first-order PMD, the behavior of the real fiber resembles that of PMF. The DGD at an instant in time for such a small frequency range is often called the "instantaneous DGD" to distinguish it from a mean DGD obtained by averaging over time or frequency. The longer-range motion of $\hat{t}(\omega)$ around $\boldsymbol{\tau}(\omega)$ is more complicated, reflecting higher-order PMD.

Whereas the frequency domain provides a continuous wave, single-frequency view of PMD, the time domain involves pulses. This allows an alternative physical interpretation of the DGD parameter, $\Delta\tau$, to the speed of precession identified previously. The

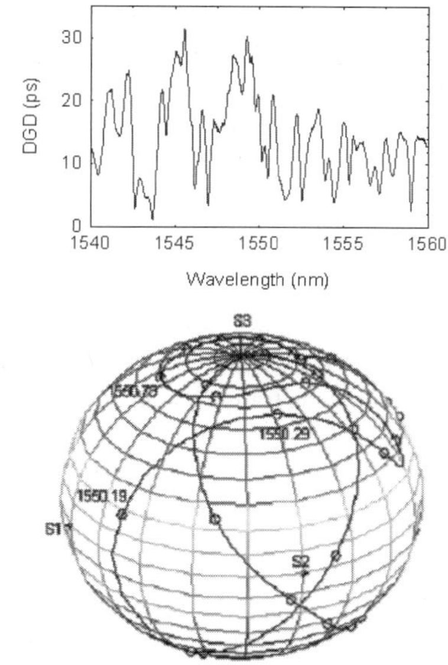

Fig. 5. PMD measurement of a 14.7ps mean DGD fiber. Top: Magnitude of $\boldsymbol{\tau}$. Bottom: Direction of $\boldsymbol{\tau}$ (markers indicate 0.1-nm intervals).

time-domain view uses laboratory coordinates, Jones vectors to characterize polarization, and a 2×2 unitary complex transmission matrix T to relate the input and output Jones vectors, via $\mid t\rangle = T \mid s\rangle$. In this framework, the PSPs are characterized by the unit Jones vectors, $\mid p\rangle$ and $\mid p_-\rangle$, corresponding to the Stokes vectors, $\hat{p}$ and $-\hat{p}$, discussed previously. Using the PSPs as an orthogonal basis set, any input or output polarization can be expressed as the vector sum of two components, each aligned with a PSP. Within the realm of first-order PMD, the output electric field from a fiber with PMD has the form

$$\mathbf{E}_{\text{out}}(t) = a \mid p\rangle E_{\text{in}}(t - \tau_0 - \Delta\tau/2) + b \mid p_-\rangle E_{\text{in}}(t - \tau_0 + \Delta\tau/2), \tag{11}$$

where $E_{\text{in}}(t)$ is the input electric field, a and b are the complex weighting coefficients indicating the field amplitude launched along the slow and fast PSPs, $\mid p\rangle$ and $\mid p_-\rangle$, and τ_0 is the polarization-independent transmission delay. In this formulation, $\Delta\tau$ is identified as the difference in arrival times between the two principal states, explaining its designation as the DGD. It is usually stated in picoseconds (ps). It is apparent from Eq. (11) that PMD can cause pulse broadening due to the DGD, and that there is no pulse broadening when the input is aligned with a PSP, i.e., when a or b is zero. Note that the simple PMD Stokes vector, $\boldsymbol{\tau}$, does not have a vector analog in the laboratory frame where Eq. (11) separates the DGD from the PSP polarizations.

The precise definition of PSPs and DGD in the time domain can be obtained from an expression for polarization-dependent group delay τ_g experienced by a signal launched with polarization $\hat{s}$:

$$\tau_g = \tau_0 + \frac{1}{2}\boldsymbol{\tau} \cdot \hat{s}, \tag{12}$$

where τ_0 is the polarization-independent delay of the fiber (whose wavelength dependence leads to chromatic dispersion). This simple, yet powerful relation was originally derived and further substantiated by analysis based on first moments of the transmitted pulses [28, 42, 53, 73]. This time-domain definition identifies the PSPs as those input polarizations ($\hat{s} = \pm\hat{p}$) that maximize or minimize the signal delay ($\tau_g = \tau_0 \pm \Delta\tau/2$). This definition is equivalent to the earlier frequency-domain definition and is the basis of the polarization-dependent signal delay method for PMD measurement.

The traditional Jones Matrix Eigenanalysis (JME) measurement approach provides yet another equivalent definition bridging the time and frequency domain. Haus [31] observes that the Hermitian appearing in the JME is connected to the energy of the light stored in the fiber. The light in the slow PSP spends more time in the fiber and maximizes the stored energy, whereas transmission along the fast PSP minimizes the stored energy.

The evolution of the PMD vector with fiber length is described by the dynamical equation for PMD [70],

$$\frac{d\boldsymbol{\tau}}{dz} = \frac{d\boldsymbol{\beta}}{d\omega} + \boldsymbol{\beta} \times \boldsymbol{\tau}, \tag{13}$$

relating the PMD vector to the microscopic birefringence. Here z is the position along the fiber. $\boldsymbol{\beta}$ is the three-dimensional, local birefringence vector of the fiber [19] pointing in the direction of the birefringence axis and having magnitude $\Delta\beta$ proportional to Δn (see [28]). This equation is the basis for the statistical theory of PMD [21].

2.6. *Second-order PMD*

Because the fiber PMD vector varies with optical angular frequency ω, a Taylor-series expansion of $\boldsymbol{\tau}(\omega)$ with $\Delta\omega$ about the carrier frequency ω_0 has typically been used for larger signal bandwidths [8, 21, 27],

$$\boldsymbol{\tau}(\omega_0 + \Delta\omega) = \boldsymbol{\tau}(\omega_0) + \boldsymbol{\tau}_\omega(\omega_0)\Delta\omega + \cdots. \tag{14}$$

So-called second-order PMD is then described by the derivative,

$$\boldsymbol{\tau}_\omega = \frac{d\boldsymbol{\tau}}{d\omega} = \Delta\tau_\omega\hat{p} + \Delta\tau\hat{p}_\omega, \tag{15}$$

where the subscript ω indicates differentiation. Second-order PMD thus has two terms. Since $\hat{p}_\omega$, which is not a unit vector, is perpendicular to $\hat{p}$ (i.e., $\hat{p} \cdot \hat{p}_\omega = 0$), the first term on the right-hand side of Eq. (15) is $\boldsymbol{\tau}_{\omega\|}$, the component of $\boldsymbol{\tau}_\omega$ that is parallel to $\boldsymbol{\tau}$, whereas the second term, $\boldsymbol{\tau}_{\omega\perp}$, is the component of $\boldsymbol{\tau}_\omega$ that is perpendicular to $\boldsymbol{\tau}$. Figure 6 shows a vector diagram of the principal parameters and their interrelationships.

The magnitude of the first term, $\Delta\tau_\omega$, is the change of the DGD with wavelength and causes polarization-dependent chromatic dispersion (PCD) [22,68], resulting in polarization-dependent pulse compression and broadening. It can be viewed as a polarization-dependent change in the chromatic dispersion, DL, of the fiber, described by an effective dispersion,

$$(DL)_{\text{eff}} = DL \pm \tau_\lambda. \tag{16}$$

In accordance with the customary dispersion measure, DL, the PCD is defined as,

$$\tau_\lambda = -(\pi c/\lambda^2)\Delta\tau_\omega = \frac{1}{2}\frac{d\Delta\tau}{d\lambda}, \tag{17}$$

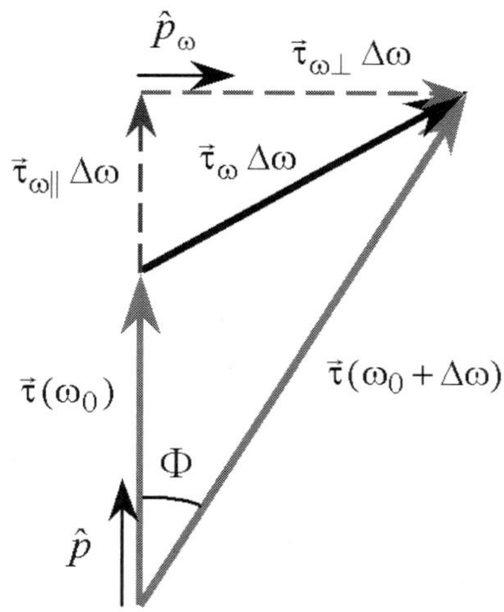

Fig. 6. Schematic diagram of the PMD vector $\boldsymbol{\tau}(\omega)$ and the second-order PMD components showing the change of $\boldsymbol{\tau}(\omega)$ with frequency. Note that $\hat{p}_\omega$ is perpendicular to $\hat{p}$. The angular rotation rate, $d\Phi/d\omega$, of the PMD vector $\boldsymbol{\tau}(\omega)$ with ω is described by $|\ \hat{p}_\omega\ |$.

where c is the velocity of light, λ is the wavelength, and τ_λ is usually expressed in ps/nm. The PCD is proportional to the wavelength derivative of the DGD spectrum. The plus and minus signs in Eq. (16) correspond to alignment with the two PSPs. Note that the magnitudes of $\boldsymbol{\tau}_{\omega||}$ and $\Delta\tau_\omega$ are equal. Figure 7(a) shows the PCD for data depicted in Fig. 5. The DGD data were numerically differentiated to obtain the PCD. It is apparent that PCD causes the effective dispersion to fluctuate rapidly with wavelength.

The second term in Eq. (15), $\Delta\tau\hat{p}_\omega$, describes PSP depolarization, a rotation of the PSPs with frequency. As shown in Fig. 6, the angular rate of rotation, $d\Phi/d\omega = |\ \hat{p}_\omega\ |$, of the PMD vector $\boldsymbol{\tau}(\omega)$ is measured by the magnitude $|\ \hat{p}_\omega\ |$, which we express in ps. Note that $d\Phi/d\nu$[mrad/GHz] $= 2\pi\ |\ \hat{p}_\omega\ |$ [ps], where ν is the optical carrier frequency and $\omega = 2\pi\nu$. We have already seen in Fig. 5 the rapid motion of $\hat{p}$ for the 14.7-ps mean DGD fiber. Figure 7(b) is a plot of $|\ \hat{p}_\omega\ |$ for this same fiber and wavelength range. Pulse distortions caused by depolarization include overshoots and generation of satellite pulses. PSP depolarization can also have a detrimental effect on first-order PMD compensators.

The input and output second-order PMD vectors ($\boldsymbol{\tau}_{s\omega}$ and $\boldsymbol{\tau}_\omega$, respectively) transform the same way as the first-order PMD vector, so that

$$\boldsymbol{\tau}_\omega = R\boldsymbol{\tau}_{s\omega}, \tag{18}$$

where, again, R is the Müller rotation matrix [28]. For the third-order PMD vectors, one can show that

$$\boldsymbol{\tau}_{\omega\omega} = R\boldsymbol{\tau}_{s\omega\omega} + \boldsymbol{\tau} \times \boldsymbol{\tau}_\omega. \tag{19}$$

The statistical theory of second-order PMD [21,22] has provided probability density functions for the various second-order components that have been experimentally confirmed [23, 39], as has their scaling with mean DGD [57]. These results will be outlined

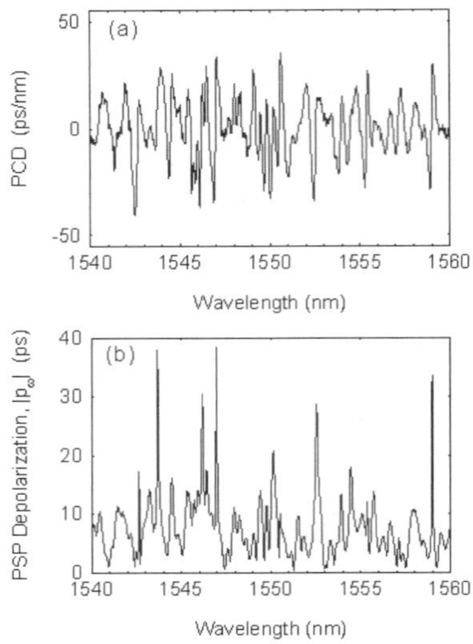

Fig. 7. (a) Plot of the polarization-dependent chromatic dispersion and (b) plot of PSP depolarization, for the same 14.7-ps mean DGD fiber as in Fig. 5.

in Section 3. Higher-order PMD has also been described using other formulations [5, 20, 73] rather than the Taylor-series expansion described above. These formulations attempt to better describe how the PMD vector changes with optical frequency. However, the statistics have not been completely derived for these approaches.

2.7. *The Bandwidth of the Principal States*

The bandwidth of the principal state provides guidance on the change of the PMD vector $\boldsymbol{\tau}(\omega)$ of the fiber with frequency (or wavelength). It is the bandwidth, $\Delta\omega_{\mathrm{PSP}} = 2\pi\Delta\nu_{\mathrm{PSP}}$, or the corresponding wavelength range, $\Delta\lambda_{\mathrm{PSP}}$, over which the PMD vector is reasonably constant. This concept is important for frequency-domain measurement of PMD vectors, where measurements of polarization rotations at two or more frequencies are required. These frequencies have to be confined to the range $\Delta\lambda_{\mathrm{PSP}}$ in order to reduce inaccuracy caused by higher-order PMD. On the other hand, in statistical PMD measurements, the samples of $\boldsymbol{\tau}(\lambda)$ are deemed to be statistically independent if their wavelengths are at least $6\Delta\lambda_{\mathrm{PSP}}$ apart.

While different constants have been reported [3, 5], studies of the accuracy of measurements of PMD provide a good practical estimate for $\Delta\omega_{\mathrm{PSP}}$ given by the relation [37]

$$\Delta\omega_{\mathrm{PSP}} \cdot \overline{\Delta\tau} = \pi/4, \tag{20}$$

where $\overline{\Delta\tau}$ is the mean DGD of the fiber. This implies a frequency band $\Delta\nu_{\mathrm{PSP}} = 1/(8\overline{\Delta\tau})$, or

$$\Delta\nu_{\mathrm{PSP}} = 125\ \mathrm{GHz}/\overline{\Delta\tau}, \tag{21}$$

when $\overline{\Delta\tau}$ is expressed in ps. For wavelengths near 1550 nm, the corresponding wavelength range $\Delta\lambda = \Delta\nu \cdot \lambda^2/c$ can be written in the simple form $\Delta\lambda_{\mathrm{PSP}} = 1\mathrm{nm}/\overline{\Delta\tau}$.

Figure 8 shows measured DGD data for a fiber with a mean DGD $\overline{\Delta\tau} = 40$ ps, shown over a range of 2 nm with 0.1 nm markers. Here, the values of $\Delta\tau$ appear reasonably constant over the calculated $\Delta\lambda_{\text{PSP}}$ of 0.025 nm. Note that the bandwidth of the PSP is consistent with the concept of second-order PMD, $\boldsymbol{\tau}_\omega$, as explained in [48].

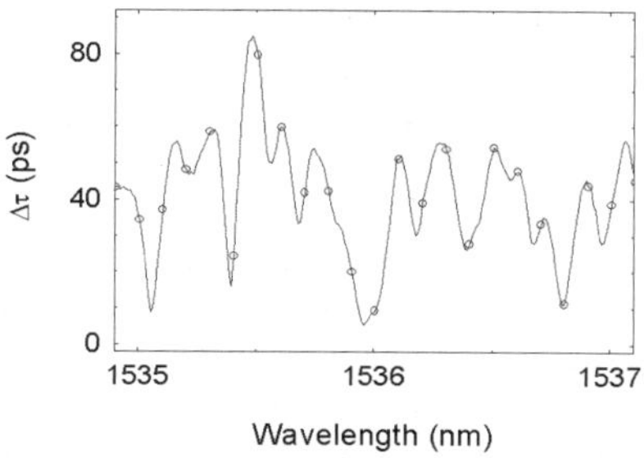

Fig. 8. DGD measurement of a fiber with mean DGD of 40 ps. The measured values of $\Delta\tau$ appear reasonably constant over the bandwidth of the PSP, 0.025-nm. Markers indicate 0.1-nm intervals.

The correlation function of the PMD vectors $\boldsymbol{\tau}(\omega_0)$ and $\boldsymbol{\tau}(\omega_0 + \Delta\omega)$ recently reported by Karlsson and Brentel [43] and Shtaif et al. [74] (and discussed further in Section 3) provides an elegant confirmation and interpretation of the $\Delta\omega_{\text{PSP}}$ concept and its practical implications. Figure 9 shows a normalized plot of this correlation as a function of the frequency separation, $\Delta\nu = \Delta\omega/2\pi$, for a mean DGD of $\overline{\Delta\tau} = 1.25$ ps. For this value, the bandwidth of the PSP is $\Delta\nu_{\text{PSP}} = 100$ GHz. At this frequency spacing the correlation is seen to drop from 1 to 0.89, supporting the idea that $\boldsymbol{\tau}$ is essentially constant over the 100-GHz width. At $6\Delta\nu_{\text{PSP}} = 600$ GHz, the correlation drops to 0.11, indicating that PMD vectors at that frequency spacing are essentially uncorrelated.

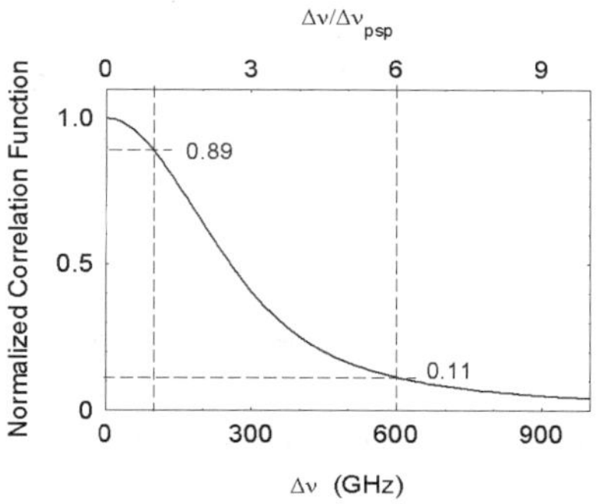

Fig. 9. Plot of the (normalized) correlation function, $\langle \boldsymbol{\tau}(\nu_0) \cdot \boldsymbol{\tau}(\nu_0 + \Delta\nu) \rangle / \langle \Delta\tau^2 \rangle$, as a function of the frequency separation, $\Delta\nu$, for a mean DGD of $\overline{\Delta\tau} = 1.25$ ps and bandwidth of the PSP, $\Delta\nu_{\text{PSP}} = 100$ GHz. The top x axis shows the normalized frequency separation, $\Delta\nu/\Delta\nu_{\text{PSP}}$, allowing general use of the plot.

2.8. Concatenation Rules

The total PMD vector of a series of two or more elements with known PMD vectors can be determined using the simple, but powerful concatenation rules [17, 21, 26, 28, 53, 70]. The concatenation rules have been used in the analysis of how the PMD vector grows with fiber length [17] and for statistical PMD modeling [21]. They are also useful for PMD simulation and in the design of multi-section PMD compensators. Although the concatenation rules have appeared in sum, differential, and integral formulations for both first- and second-order PMD vectors, this section will concentrate on the sum rules. See [28] for the other formulations.

The concatenation rule for first-order PMD is similar to that for transmission-line impedances: To obtain the PMD vector of an assembly, transform the PMD vectors of each individual section to a common reference point and take the vector sum (in three-dimensional Stokes space). This vector can then be transformed to any other location in the system using the known rotation matrices of the different sections. For example, for the two sections shown in Fig.10, the PMD vector, $\boldsymbol{\tau}_m$, at the midpoint is

$$\boldsymbol{\tau}_m = \boldsymbol{\tau}_1 + \boldsymbol{\tau}_{s2} = \boldsymbol{\tau}_1 + R_2^T \boldsymbol{\tau}_2, \tag{22}$$

where all $\boldsymbol{\tau}_i$ and R_i are functions of frequency. To find the total PMD vector at the output, $\boldsymbol{\tau}$, we must then transform it by R_2, so that

$$\boldsymbol{\tau} = R_2 \boldsymbol{\tau}_1 + \boldsymbol{\tau}_2, \tag{23}$$

since $R_2 R_2^T \boldsymbol{\tau}_2 = \boldsymbol{\tau}_2$. The corresponding PMD vector diagram is also shown in Fig. 10. It provides a simple geometrical interpretation of the concatenation.

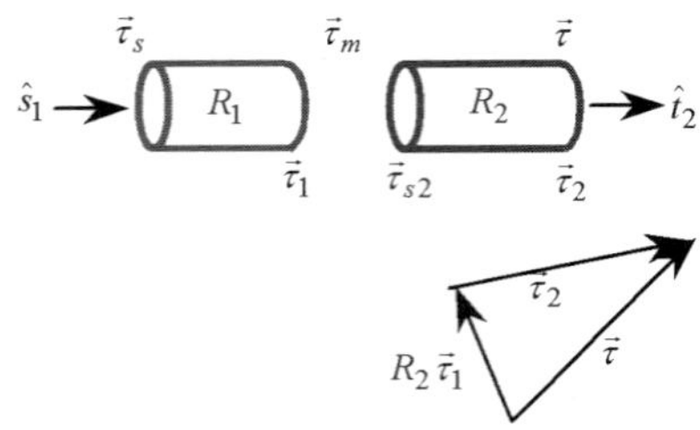

Fig. 10. Diagram of the concatenation of PMD vectors for two sections. The PMD vector $\boldsymbol{\tau}$ at the output is equal to the sum of $\boldsymbol{\tau}_1$ and $\boldsymbol{\tau}_2$, after transforming by R_2.

Equation (23) is the basic concatenation rule. We can use it similarly to find the total PMD vector at the input, $\boldsymbol{\tau}_s$, by the transformation

$$\boldsymbol{\tau}_s = R_1^T \boldsymbol{\tau}_m = R_1^T (\boldsymbol{\tau}_1 + R_2^T \boldsymbol{\tau}_2). \tag{24}$$

The rule can be generalized to multiple sections as well as to differentially small sections. Second-order PMD can also be concatenated. By differentiating Eq. (23) and making the proper substitutions, one can show that

$$\boldsymbol{\tau}_\omega = \boldsymbol{\tau}_2 \times \boldsymbol{\tau} + R_2 \boldsymbol{\tau}_{1\omega} + \boldsymbol{\tau}_{2\omega}. \tag{25}$$

The first- and second-order PMD vectors for many sections can be determined by repeated application of the two section rules in Eqs. (23) and (25). For the fiber in Fig. 11 consisting of m sections, each with known rotation matrix R_n and output PMD vector $\boldsymbol{\tau}_n$, the sum rules of the assembly are for first-order PMD

$$\boldsymbol{\tau} = \sum_{n=1}^{m} R(m, n+1)\boldsymbol{\tau}_n, \tag{26}$$

and for second-order PMD

$$\boldsymbol{\tau}_\omega = \sum_{n=1}^{m} R(m, n+1)\{\boldsymbol{\tau}_{n\omega} + \boldsymbol{\tau}_n \times \boldsymbol{\tau}(n)\}, \tag{27}$$

where we define the rotation matrix of the last $m - n + 1$ sections as $R(m, n) = R_m R_{m-1} \cdots R_n$, where $R(m, m) = R_m$ and $R(m, m + 1)$ is the identity matrix. The differential concatenation rule for PMD shows how $\boldsymbol{\tau}(z)$ changes due to the differential addition of length Δz [28] and is equivalent to Eq. (13), the dynamical PMD equation.

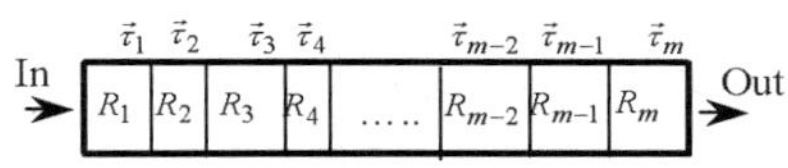

Fig. 11. Concatenation of m sections of PMD, each with known rotation matrix R_n and output PMD vector $\boldsymbol{\tau}_n$. [See Eqs. (26) and (27).]

2.9. *Degree of Polarization*

The concept of the degree of polarization (DOP) characterizes the average polarization state of light over a broad spectral range. For time-dependent signals it is also defined as an average over a specified time period. The definition of the DOP is based on the Stokes parameters routinely measured by a polarimeter-based instrument (containing polarizers and four photodiodes) or DOP monitor.

A coherent sinusoidal optical carrier (or a narrow spectral component) has a well-defined polarization, described by a unit Stokes vector denoted by a lowercase $\hat{s}$. The spectrally averaged Stokes parameters, S_i, are denoted by capital letters, where S_0 is the total intensity of the light. The other three parameters are the differences between the measured intensities of pairs of orthogonal polarizations, where S_1 refers to the vertical/horizontal polarizations, S_2 to the $\pm 45°$ polarizations, and S_3 to right/left circular polarizations. The definition of the DOP is

$$\text{DOP} = \sqrt{S_1^2 + S_2^2 + S_3^2}/S_0. \tag{28}$$

A narrow spectral component then has $\text{DOP} = 1$.

If the filter of the monitor passes several WDM channels (or spectral components), the measured intensity is the sum of all channel intensities for any given polarizer setting. The monitor, therefore, measures Stokes parameters that are the sum of the

Stokes parameters of the individual WDM channels. For the illustrative example of two channels with equal power, one finds

$$\mathrm{DOP} = \sqrt{(1 + \hat{s}_1 \cdot \hat{s}_2)/2} = \cos(\phi/2), \tag{29}$$

where $\hat{s}_1$ and $\hat{s}_2$ are the unit Stokes vectors of the two channels and ϕ is the Stokes angle between the two channels. If the channels have parallel polarizations, the DOP $= 1$; for antiparallel polarizations in Stokes space (orthogonal polarizations in the lab), the DOP $= 0$.

When two channels (or spectral components) having a channel (or spectral) spacing of $\Delta\omega$ are launched with identical polarizations into a fiber with DGD, $\Delta\tau$, the fiber PMD rotates $\hat{s}_2$ relative to $\hat{s}_1$. This reduces the measured DOP at the output. The law of infinitesimal rotation [Eq. (10)] characterizes this relative rotation and leads to the approximate expression

$$\mathrm{DOP} = 1 - \frac{1}{8}(\Delta\omega\Delta\tau \sin\theta)^2, \tag{30}$$

valid for small $\Delta\omega\Delta\tau$. Here θ is the angle between the launched Stokes vector and the PSP of the fiber.

Owing to its dependence on the DGD, $\Delta\tau$, the DOP has been successfully used as a monitor for PMD compensators [72]. Advantages are that the DOP is bit-rate independent and that it can provide high speed, e.g., kHz response time, without requiring electronics operating at frequencies as high as the bit rate.

3. Statistics and Scaling

3.1. Statistics of PMD

PMD differs from many lightwave system properties in that the impairment caused by PMD usually changes stochastically with wavelength and time. This means that one cannot predict the impairment of the system at any particular wavelength and time, but must instead resort to a statistical description. It also means that, in general, one cannot characterize the PMD of a system to arbitrary accuracy, since a measurement will sample only a portion of the statistical ensemble. Most other impairments, such as those due to chromatic dispersion, system attenuation, or self-phase modulation, will have some uncertainty caused by manufacturing tolerances, imprecise wavelength control, aging, and the like. However, in most cases, the probability densities for these impairments will vanish beyond some value, allowing a system to be designed to tolerate the worst-case impairment. Tolerance of worst-case impairment is rarely possible when PMD is a significant source of impairment. The probability densities for PMD in most situations have asymptotic tails extending to unacceptably large impairment. Hence, the system cannot be designed to handle the worst-case PMD impairment and must instead be designed for a specified outage probability. For similar reasons, the goal of PMD compensation cannot be to eliminate the impairment, but rather to reduce the PMD outage probability. To understand and predict system outage probabilities, to design PMD compensators, and to accurately measure PMD in systems, one must understand the statistics of the phenomena associated with PMD.

Interesting statistical characteristics of PMD include the probability densities of first- and higher-order PMD, the scaling of PMD phenomena with changes in the mean DGD, various correlation functions, and characteristics associated with the accuracy of PMD measurements. The first statistical property of PMD to attract interest was the mean DGD [65,1].

Since the DGD usually changes quite slowly with time, it is difficult to obtain an accurate estimate of the mean DGD by repeatedly measuring the DGD at one wavelength. Instead, the DGD is averaged over wavelength and the result is assumed to be equal to the time average. This assumption is of crucial importance. Virtually every measured mean DGD value quoted in the literature was obtained by wavelength averaging. Although some data has been taken covering a broad span of time and wavelength [45, 65], the issue has not been fully explored. Instead, this equality, upon which the accuracy of much experimental PMD work relies, is assumed.

Once the validity of wavelength scanning is assumed, experimental attention can be focused on probability densities. Figure 12 shows the probability density of the DGD for a system having a mean DGD of 1 ps. As shown below, this density and all PMD densities can be easily scaled for a different mean DGD. Measurements, together with simulation results, can be used to establish the veracity of analytic derivations of probability densities and the validity of the PMD models. The densities are then combined with an understanding of PMD-induced system penalties (see Section 4) to predict system outage probability. Powerful scaling laws emerge from the densities and their derivation. With these laws, one can obtain extensive understanding of the statistical properties of the PMD in a particular system through the measurement of a single system characteristic, the mean DGD.

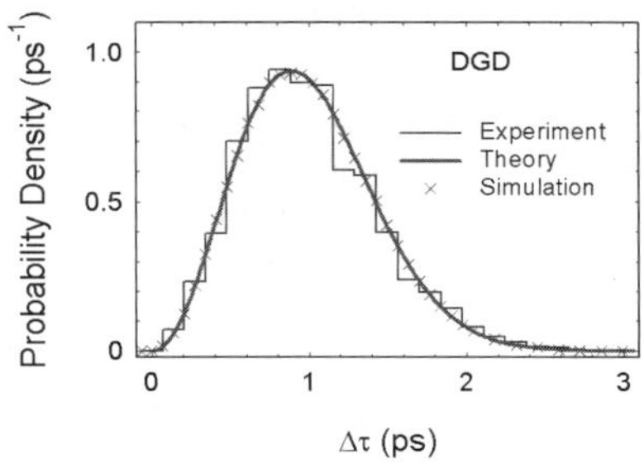

Fig. 12. Maxwellian probability density of DGD for a fiber having 1-ps mean DGD.

3.2. Probability Densities

The probability density of the DGD was the first PMD density to attract attention [17, 70]. For fully random PMD, the problem can be reduced to that of the probability density of the magnitude of the sum of 3-dimensional vectors having random orientation and length. The result is the well-known Maxwellian distribution shown in Fig. 12. The simulation used a quarter million instances of 1200 randomly oriented linearly birefringent plates. The measured data was scaled from a 120-nm scan of a fiber having a mean DGD of 14.7 ps. Using the bandwidth of the principal states discussed above in section 2.7, we expect the 6000-point scan to contain only about 300 statistically independent measurements.

Replotting with a logarithmic vertical scale yields Fig. 13, which starts to reveal the range of density that is more important for the estimation of outage probabilities. The deviations of the simulation values for small probabilities are caused by the small number of simulations achieving the higher values of DGD. The experimental values demonstrate what appears to be a systematic deviation from the expected curve for the high values of DGD. Since only two statistically independent points are involved in the deviation, it is likely a statistical fluctuation. The smoothness of the deviation arises from the data being 20-fold over sampled.

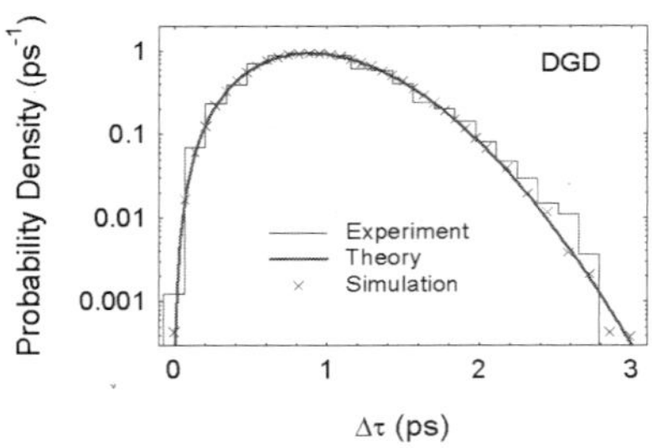

Fig. 13. Maxwellian probability density of DGD for a fiber having mean DGD of 1 ps.

Figure 14 shows the density for the second-order quantity, $\hat{p}_\omega$, which characterizes the depolarization of the PSPs. (As mentioned below, $\hat{p}_\omega$ scales as a first-order quantity.) To obtain the measured data for Fig. 14, the Müller Matrix Method was first used to obtain the spectrum of $\boldsymbol{\tau}$ used for Figs. 12 and 13. From this, $\boldsymbol{\tau}_\omega$ was obtained by numerical differentiation and $\hat{p}_\omega$ obtained by vector algebra. Figures 13 and 14 illustrate the difficulty of verifying densities through measurement or simulation. The interesting regions of the densities are the asymptotic tails containing an integrated probability of 10^{-4} to 10^{-7}. The 300 statistically independent measurement points do not illuminate this region at all and the 250,000-point simulation barely reaches it. One approach to the problem of simulating the important, but improbable regions of the densities is importance sampling, a detailed description of which appears elsewhere [4] and in this tome.

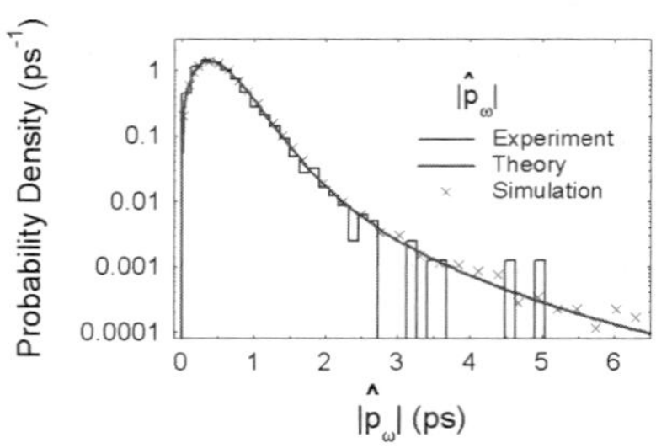

Fig. 14. Probability density for the depolarization of the principal states of polarization for a fiber with mean DGD of 1 ps.

Analytic expressions for probability densities have been obtained for first- and second-order PMD [14, 17, 21, 22, 23, 39, 57, 70]. In particular, densities for both the magnitude and an individual component of first-order PMD are known. For second-order PMD, they are known for the magnitude and a component of the vector as well as the component of the second-order vector that is aligned with the first order vector (the PCD component). The densities are known in integral form for the component of the second-order vector that is perpendicular to the first order vector and for this component normalized by the DGD (the PSP depolarization). Once the densities are known, they can be used to obtain the mean and mean square (or second moment or variance) of each of the variables of interest.

Appendix A summarizes the results. For simplicity, τ rather than $\overline{\Delta\tau}$ is used in the appendix to represent the mean DGD. The estimation of system outage probabilities often requires complementary cumulative probability distributions, which are

$$CC(y) = \int_y^\infty dx\, p_X(x),$$

where $p_X(x)$ could be one of the densities in Appendix A. These integrals provide the probability of exceeding some value, y, which is typically the threshold for unacceptable impairment. Figure 15 shows the complementary cumulative probability distribution together with the cumulative probability distribution for the DGD, once again using a mean DGD of 1 ps. The markers show results from the simulation described previously. The deviation of the simulation from theoretical predictions for $\Delta\tau > 3.1$ ps is likely a result of the small number of realizations (five) that generated a DGD larger than 3.1 ps.

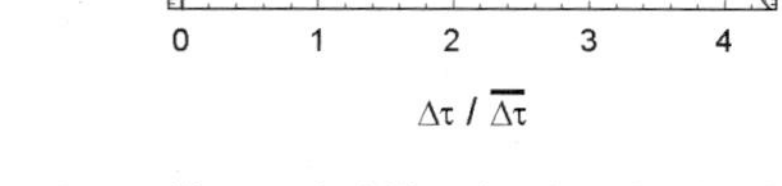

Fig. 15. Integrals of the Maxwellian probability density of DGD for a mean DGD of 1 ps.

4. System Effects of PMD

Fiber PMD can cause a variety of impairments in optical fiber transmission systems. For a single digital channel, intersymbol interference (ISI) impairment is caused by the differential group delay, $\Delta\tau$, between the two pulses propagating in the fiber when the input polarization, $\hat{s}$, of the signal does not match the PSP of the fiber, $\hat{p}$. In this first-order PMD effect the fractional powers launched into the PSPs are $\gamma = \langle s \mid p\rangle\langle p \mid s\rangle = (1 + \hat{p}\cdot\hat{s})/2$ and $(1-\gamma) = (1 - \hat{p}\cdot\hat{s})/2$. For larger signal bandwidths system impairments can occur due to second- and higher-order PMD, particularly when these PMD components combine with chromatic fiber dispersion or signal chirp.

Power penalties and system outage in digital systems due to first-order PMD are described in the succeeding sections. We also mention the effects of PMD on multi-channel systems. PMD can cause severe impairments in analog systems; the reader is referred to [71] and to [14] for information on this topic. While the combined effects of PMD and polarization dependent loss is now an active area of investigation [30, 34, 82], the following sections assume absence of PDL.

4.1. Power Penalties due to First-order PMD

In the first-order picture, PMD splits the input signal entering the fiber into two orthogonally polarized components that are delayed by $\Delta\tau$ relative to each other during transmission. The impairment caused by this effect can be expressed as a power penalty ϵ of the form [69]

$$\epsilon(\mathrm{dB}) = (A/T^2)\Delta\tau^2\gamma(1-\gamma) = A(\Delta\tau/2T)^2 \sin^2\theta, \tag{31}$$

where the penalty, expressed in dB, is assumed to be small. Here, T is the bit interval, $0 \leq \gamma \leq 1$ is the power-splitting ratio, and θ is the angle between the input polarization, $\hat{s}$, and the input PSP, $\hat{p}$. The DGD value, $\Delta\tau$, appearing in Eq. (31) is the "instantaneous" DGD value, assumed to be constant during the penalty measurement. The $\gamma(1-\gamma)$ dependence shown in this expression has been verified by experiment in [47]. The dimensionless A-parameter depends on pulse shape, modulation format, and specific receiver characteristics such as the detailed response of the electrical filter and whether optical or thermal noise predominates. For pin receivers the reported values for A range from 10 to 40 for non-return-to-zero (NRZ) modulation format and from 20 to 40 for return-to-zero (RZ) modulation format, whereas the A ranges for optically preamplified receivers are 10 to 70 for NRZ and 10 to 40 for RZ.

Jopson et al. [38] report the penalty measurements, emulations, and simulations for optical preamplifiers shown in Fig. 16. For these specific receiver types, the data for NRZ transmission show a good fit to the penalty formula (31) for an A-parameter of about 60 to 70. For RZ transmission the measured A-parameters range from about 15 to 25. Sunnerud et al. [76] use a different approach, reporting simulations for PMD-induced RZ and NRZ system degradation.

At this time the penalty formula [Eq. (31)] should be regarded as a semi-empirical rule. The receiver parameter, A, in particular, should be determined by simulation and experiment [83]. However, the general nature of the formula, particularly the dependence of the penalty on the DGD and the direction of the launch polarization relative to the PSP, is in good agreement with analysis of the moments of the received signal [28, 42, 73].

As reflected in Eq. (31) the power penalty, ϵ, changes with the DGD and with the launch penalty factor, g, given by

$$g = \sin^2\theta = [1-(\hat{p}\cdot\hat{s})^2] = (\hat{p}\times\hat{s})^2. \tag{32}$$

This factor depends on the alignment between the Stokes vectors, $\hat{s}$, of the input polarization and the instantaneous PSP, $\hat{p}$. We have $g = 0$ for perfect alignment, where there is no power penalty, and a maximum of $g = 1$ when the Stokes vectors $\hat{s}$ and

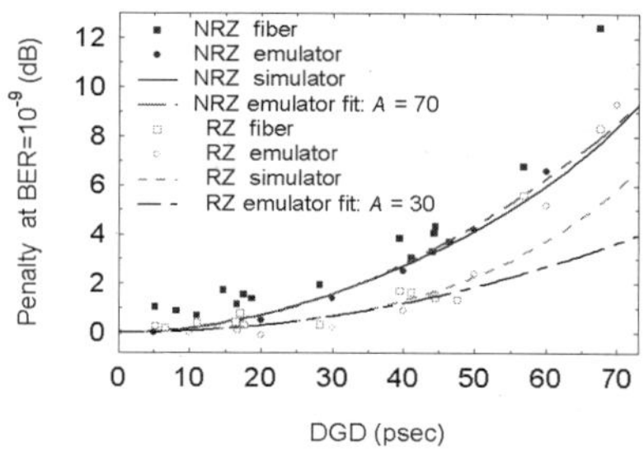

Fig. 16. NRZ and RZ penalties for measurements of real fiber and a first-order PMD emulator, as well as for simulations and optically preamplified receivers plotted as a function of instantaneous DGD. Worst-case polarization launch was used in all cases. Emulator fits use Eq. (31).

$\hat{p}$ are perpendicular (i.e., when equal powers are launched on the PSPs). The latter is the worst-case launch resulting in the maximum power penalty, and is often used in receiver penalty experiments such as those shown in Fig. 16.

4.2. System Outage Due to PMD

Outage specifications for optical fiber transmission systems depend on the application. Usually the power penalty contributions of PMD are required to be less than 1 dB for all but a specified cumulative probability, often a fraction of time ranging from 10^{-4} and 10^{-8}. This translates to average cumulative outages ranging from fractions of a second to 60 minutes per year.

To specify this limit one requires knowledge of the probability density, $p_\epsilon(\epsilon)$, for the occurrence of a power penalty ϵ. The penalty formula (31) shows that ϵ is proportional to the product of the launch penalty g and a DGD term, $\Delta\tau^2$. The probability density of g for uniform distribution over the Poincaré sphere is known (see for example, [48]). The density of $\Delta\tau$ is Maxwellian (Appendix A) from which the density of the $A(\Delta\tau/2T)^2$ can be derived. From these densities, $p_\epsilon(\epsilon)$ can be deduced using standard probability theory for the density of a product [71].

However, there is another, more direct, approach shedding further light on the mechanisms of PMD penalty statistics [48]. In view of

$$\Delta\tau^2(\hat{p}\times\hat{s})^2 = (\boldsymbol{\tau}\times\hat{s})^2 = (\boldsymbol{\tau}_\perp\times\hat{s})^2 = \Delta\tau_\perp^2,$$

we can rewrite the first-order penalty formula as

$$\epsilon = (A/4T^2)\Delta\tau_\perp^2, \qquad (33)$$

where $\boldsymbol{\tau}_\perp$ is the component of $\boldsymbol{\tau}$ perpendicular to $\hat{s}$, and $\Delta\tau_\perp$ is its magnitude. The penalty ϵ is caused only by the components of the PMD vector $\boldsymbol{\tau}$ that are perpendicular to $\hat{s}$. The statistics of the two components of $\boldsymbol{\tau}_\perp$ are described by two independent Gaussians. The magnitude $\Delta\tau_\perp$, therefore, follows a Rayleigh distribution, with the probability density

$$p_{\Delta\tau_\perp}(x) = \frac{x}{\sigma^2}\cdot e^{-x^2/2\sigma^2}, \qquad (34)$$

where

$$\sigma = \frac{1}{2}\sqrt{\frac{\pi}{2}}\overline{\Delta\tau}, \tag{35}$$

and $\overline{\Delta\tau}$ is the mean DGD. The mean penalty parameter is, therefore,

$$\bar{\epsilon} = \int_0^\infty dx\ \epsilon(x) p_{\Delta\tau_\perp}(x) = A\sigma^2/2T^2 = (\pi A/16T^2)\overline{\Delta\tau}^2. \tag{36}$$

One can now transform the Rayleigh density for $\Delta\tau_\perp$ to the density for ϵ, and obtain the known exponential (Boltzmann) distribution

$$p_\epsilon(\epsilon) = p_{\Delta\tau_\perp}(\Delta\tau_\perp(\epsilon))\frac{d\Delta\tau_\perp}{d\epsilon} = (1/\bar{\epsilon})e^{-\epsilon/\bar{\epsilon}}. \tag{37}$$

Poole and Nagel [71] have tested the exponential density of Eq. (37) and found good agreement between the simulated data and the predictions.

From Eq. (37) follows the outage probability, P_{out}, i.e., the probability for the penalty to exceed N dB,

$$P_{\text{out}} = P_{\epsilon \geq N} = \int_N^\infty d\epsilon \cdot p_\epsilon(\epsilon) = e^{-N/\bar{\epsilon}}, \tag{38}$$

where the penalty limit, N, is usually specified as 1 or 2 dB. Thus, the mean penalty associated with the specification of an outage probability, P_{out}, is

$$\bar{\epsilon} = N/\ln(1/P_{\text{out}}). \tag{39}$$

Inserting Eq. (36) we can formulate this condition as a requirement for the mean DGD of the fiber

$$\overline{\Delta\tau}/T = 4\sqrt{N}/\sqrt{\pi A \ln(1/P_{\text{out}})}, \tag{40}$$

where A is the receiver parameter discussed earlier. Figure 17 shows a plot of this requirement for three values of the parameter A/N covering a range of values encountered in NRZ and RZ systems. Note that for an outage probability of 10^{-7} (3 sec/yr), a typical RZ system $(A = 30, N = 1)$ requires a $\overline{\Delta\tau}/T$ ratio of 10% whereas the corresponding typical NRZ system $(A = 70, N = 1)$ requires $\overline{\Delta\tau}/T$ to be less than 7% (assuming receiver noise dominated by optical noise). We emphasize that A values are strongly dependent on pulse shape and receiver characteristics and must be determined for each specific system.

4.3. *Second-order PMD Impairments*

The concept and definition of second- (and higher-) order PMD, discussed in Section 2, suggest that its effects on system performance increase with the frequency separation, $\Delta\omega$, from the carrier $(\omega = \omega_0)$. No significant second-order effects are expected as long as the bandwidth of the signal, $\Delta\nu$, is smaller than the bandwidth of the PSP, $\Delta\nu_{\text{PSP}}$ (see Section 2.7). The PMD outage criteria discussed earlier specify the allowed ratio of mean DGD to bit interval as approximately $\overline{\Delta\tau}/T \leq 0.1$. As the signal bandwidth is roughly equal to $1/T$, and $\overline{\Delta\tau}$ is related to ν_{PSP} via Eq. (21), the outage criterion is roughly equivalent to

$$\Delta\nu \leq \Delta\nu_{\text{PSP}}. \tag{41}$$

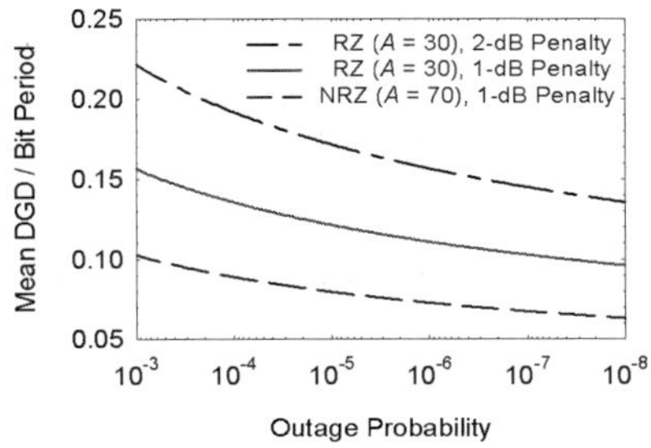

Fig. 17. Plot of the allowed mean DGD/bit period $(\overline{\Delta\tau}/T)$ as a function of the outage probability for RZ and NRZ transmission.

Together, the PSP-bandwidth statement and the outage specification imply that second-order penalties are negligible as long as the first-order outage specification is met. This important observation has been confirmed repeatedly by experiments and simulations [7, 8, 27]. Restated in plain language this means there is no need to worry about second-order PMD as long as there is no need for first-order PMD compensation. However, when the outage specification for $\overline{\Delta\tau}$ is not met, there is a need for PMD compensation as well as a need for concern about second-order PMD impairments.

The systems impacts of second-order PMD have been studied by a number of different groups [5, 7, 8, 24, 27, 50, 57, 68]. The simplest second-order impairment is the polarization-dependent chromatic dispersion (PCD) discussed in Section 2. The PCD increases or decreases the effective chromatic dispersion of the transmission system [Eq. (16)], depending on the input state of polarization. However, statistically, the PCD is a relatively minor component of the second-order PMD. The dominant impairment mechanism is caused by the PSP depolarization term that is proportional to $|\hat{p}_\omega|$. Pulse distortions due to second-order PMD can be visualized as the interplay between six different pulse replicas leading to pulse overshoots and satellite pulses.

Statistical analyses and numerical simulations of system impairments indicate a strong interaction between chromatic dispersion in the fiber and second-order PMD in general [5, 62]. Similarly, it has been found that chirp in the transmitted signal has a significant impact on second-order PMD impairments [8].

4.4. PMD Impairments in WDM Systems

PMD can also cause impairments in WDM systems because it changes the relative polarization of the different wavelength channels as they propagate along the optical fiber. As discussed in Section 2.5, the PMD vector describes how, for a fixed input polarization, the output polarization $\hat{t}$ will precess around $\boldsymbol{\tau}$ as the frequency is changed. For worst-case launch, the precession angle is $\phi = \Delta\tau\Delta\omega$, where $\Delta\omega$ is the angular frequency separation. Thus, in WDM systems where adjacent wavelength channels are launched with orthogonal polarizations to suppress nonlinear impairments such as cross-phase modulation (XPM) or four-wave mixing (FWM), PMD destroys the orthogonality of these polarizations [29, 48, 49]. Another example is a system where polarization multiplexing is used for close packing of WDM channels in order to achieve high bandwidth efficiency. Here, PMD induces coherent cross-talk between multiplexed channels leading to system impairments [35, 59, 60].

PMD is not the only effect that can destroy the initial alignment of signal polarizations of two or more different frequencies (or channels) in the fiber. Cross phase modulation depends on the relative polarizations of the channels and causes two Stokes vectors to precess around one another [54]. This "power-dependent PMD" or polarization scattering accelerates the scrambling or depolarization of the WDM channels beyond the rate caused by PMD alone. Experiments have shown that the polarization of a channel can change on a bit-to-bit time scale, making compensation very difficult [15, 46, 55].

5. Leaving the Laboratory

5.1. Reduction of System PMD

The simplest and potentially least costly way of reducing PMD impairment is to deploy fiber having low PMD. For example, fiber routes having a mean DGD less than 2% of the current bit rate would have sufficiently low PMD for current systems, as well as next-generation systems, assuming that bit rates will increase by a factor four. Since significant birefringence can arise from less than 1% core ellipticity and the associated stress, much work in fiber manufacturing has focused on reducing deviations in the circularity of the fiber preform. Much of this effort is hidden behind factory walls. It is widely known that the deviations from circular symmetry of preforms produced by the various soot-deposition techniques can differ, so some manufacturing techniques require greater care in preform preparation than others.

Early advances in fiber manufacturing techniques [2, 12, 13, 61, 79] initially led to lower intrinsic fiber birefringence. As further reduction in the intrinsic fiber birefringence became increasingly difficult, attention shifted to mitigating it. One method used to reduce the effect of intrinsic fiber birefringence is spinning. Spinning the fiber during draw [40, 51, 52] decreases the coupling length of the fiber and thus reduces its PMD. It has enabled successively lower PMD in each new generation of fiber. These efforts devoted to the reduction of fiber PMD have been rewarded with great success.

Field measurements suggest that some production fiber manufactured in the late 1980s had PMD well beyond 1 ps/(km)$^{1/2}$ [63]. It is currently possible to purchase large quantities of fiber having PMD less than 0.1 ps/(km)$^{1/2}$ on the spool and considerably less than that when cabled.

A parameter used to specify fiber PMD is the "link design value" (LDV), which defines the maximum value of the PMD coefficient in terms of the probability Q for links with at least N concatenated sections, where typically $Q = 10^{-4}$ and $N \geq 20$ (IEC SC 86A/WG1). The LDV serves as a statistical upper bound for the PMD coefficient of the concatenated fibers comprising an optical cable link.

This specification allows for small variations of the PMD coefficient from section to section. Some manufacturers have recently specified LDVs as low as 0.04 ps/(km)$^{1/2}$. Referring to Fig. 17 and allowing a 1-dB penalty for 30 min/yr, a 40-Gb/s NRZ ($A = 70$) signal could be transmitted over 2900 km of fiber with LDV of 0.04 ps/(km)$^{1/2}$, compared to 460 km of fiber with LDV of 0.1 ps/(km)$^{1/2}$. As the bit rate and system reach continue to increase, fibers produced today will not be "future proof," and manufacturers will continue to search for methods to reduce the PMD of cabled fibers.

5.2. *PMD of Fiber + PMD of Components*

The total PMD of a system is determined by all the components through which the signal travels, including not only the transmission fiber but also the amplifiers, dispersion-compensating modules, and multiplexers/demultiplexers for add/drop. Figure 18 shows a plot of the PMD limited transmission distance for 40 Gb/s RZ transmission with and without component PMD. This calculation assumes an A parameter equal to 30 and an allowed penalty of 1-dB for not more than 3 min/yr [from Eq. (40)]. If the component DGD associated with each fiber span is about 0.4 ps, then for 100-km fiber spans with LDV of 0.04 ps/(km)$^{1/2}$, the mean fiber DGD and mean component DGD will be the same. Thus the total distance over which a 40-Gb/s, RZ signal could be transmitted before being limited by PMD is reduced by a factor of 2, i.e., from 4000 to 2000 km.

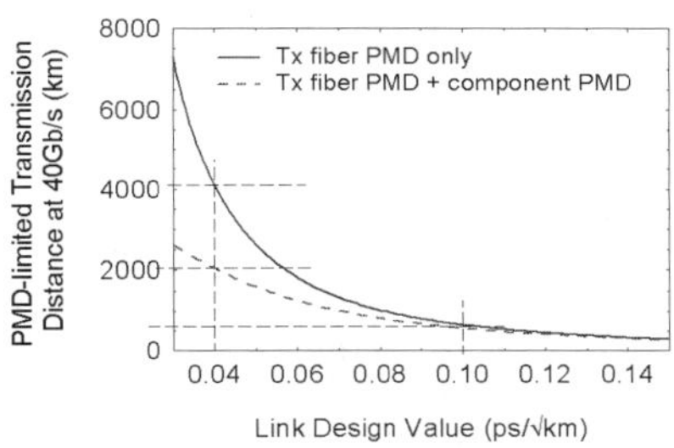

Fig. 18. Effect of fiber and component PMD on transmission distance for a 40 Gb/s RZ system, assuming 0.4 ps component PMD per 100-km span, $A = 30$ for RZ, and an allowed penalty of 1 dB for not more than 3 minutes per year.

5.3. *Time Rate of Change of PMD*

The required speed of PMD compensators depends on the time rate of change of the PMD in installed fiber links. A number of groups have made long-term PMD measurements, and the consensus is that temperature changes in embedded fibers are slow in general and cause slow PMD variations, i.e., on the time scale of hours [6, 10, 27, 44, 45, 77]. In fact, the DGD versus wavelength spectrum for an embedded cable can be quite stable over intervals as long as several weeks [27]. A detailed, long-term study with simultaneous measurements of two fibers in the same embedded cable showed that drift averaged over wavelength was 96% correlated between the two fibers for both the DGD and PSPs [44, 45]. A strong correlation between changes in the DGD and PSP was observed, and the study also confirmed that the rate of temporal change of the PMD increased with the cable length and the mean DGD.

In another comprehensive study, Nagel et al. [56] measured the DGD of a 41-ps mean DGD embedded fiber every 5 to 10 minutes over a 70-day period. All observed large amplitude, rapid DGD fluctuations were caused by human activity. They found that the measured DGD values at a given wavelength did, in fact, follow a Maxwellian distribution over long times, with average correlation times of 19 hours for DGD and 5 hours for PSPs, indicating that the PSPs are changing more rapidly. From their measurements, they predicted that the DGD would exceed three times the mean DGD at an average rate of once every 3.5 years, for a duration of 13 minutes.

In addition to slow temperature variations in embedded cable, human operation in huts, mechanical vibrations, or wind for aerial fibers can cause state of polarization (SOP) variations on the Poincaré sphere of up to 50 revolutions per second [9]. Bülow et al. measured the fastest PMD fluctuations to take place in a timescale of 6 to 13 msec on a 52-km, 7.3-ps mean DGD embedded cable, presumably due to moving of the fiber pigtails in the central office. Measurements of aerial cables have also been reported, where larger strain and temperature changes can occur. In one study where the interferometric technique was used to measure the $\sim$ 9-ps mean DGD (over a 70-nm wavelength range) of a 96-km aerial cable, temperature alone caused mean DGD fluctuations on a time scale of about 5 minutes [10]. Another study measured the upper limit of SOP changes to be 1.8 seconds [80] on a 12.7-ps mean DGD aerial fiber, seemingly a contradiction to the faster PMD fluctuations measured by Bülow et al. [9]. Further investigation of these fast perturbations to PMD and SOP is still warranted.

6. Conclusion

While much progress has been made in understanding PMD, the search for a practical, inexpensive method for PMD compensation is difficult. The first strategy for reducing impairment caused by PMD is to use terminal and in-line equipment that has low PMD. When this measure does not suffice, one can avoid PMD impairment by laying new fiber having sufficiently low PMD. Electrical or optical compensation offers the potential of reducing impairment to acceptable levels for some systems. Since fiber spans differ in their mean PMD, one approach is to compensate some fiber spans and abandon spans containing too much PMD. Whatever the method developed to mitigate PMD impairment in current systems, PMD is likely to remain an important impairment in the years to come.

Appendix A

$$\text{Density } p_X(x) \geq 0, \qquad \int_{-\infty}^{\infty} dx\, p_X(x) = 1$$

X	$p_X(x)$	$EX \triangleq \int_{-\infty}^{\infty} dx\, x p_X(x)$	$EX^2 \triangleq \int_{-\infty}^{\infty} dx\, x^2 p_X(x)$
τ_i	$\frac{2}{\pi\tau} e^{-(2x/\tau)^2/\pi}$, Gaussian	0	$\frac{\pi}{8}\tau^2 = \sigma^2$
$\mid \boldsymbol{\tau} \mid \triangleq \Delta\tau$	$\frac{8}{\pi^2\tau}\left(\frac{2x}{\tau}\right)^2 e^{-(2x/\tau)^2/\pi}$; $x \geq 0$, 0 ; $x < 0$ Maxwellian	τ	$\frac{3\pi}{8}\tau^2 = 3\sigma^2$
$\tau_{\omega i}$	$\frac{4}{\pi\tau^2}\text{sech}(4x/\tau^2)$, soliton amplitude	0	$\left(\frac{\pi}{8}\right)^2 \tau^4 = \sigma^4$
$\mid \boldsymbol{\tau}_\omega \mid$	$\frac{8}{\pi\tau^2}\frac{4x}{\tau^2}\tanh(4x/\tau^2)\text{sech}(4x/\tau^2)$; $x \geq 0$, 0 ; $x < 0$	$\frac{2G}{\pi}\tau^2$	$3\left(\frac{\pi}{8}\right)^2 \tau^4 = 3\sigma^4$
$\mid \boldsymbol{\tau} \mid_\omega \triangleq \Delta\tau_\omega$	$\frac{2}{\tau^2}\text{sech}^2(4x/\tau^2)$, soliton intensity	0	$\frac{1}{3}\left(\frac{\pi}{8}\right)^2 \tau^4 = \frac{1}{3}\sigma^4$
$\mid \boldsymbol{\tau}_{\omega\perp} \mid$	$x\left(\frac{8}{\pi\tau^2}\right)^2 \int_0^\infty d\alpha\, J_0\left(\frac{8\alpha x}{\pi\tau^2}\right)\text{sech}(\alpha)(\alpha\tanh\alpha)^{1/2}$		$\frac{8}{3}\left(\frac{\pi}{8}\right)^2 \tau^4 = \frac{8}{3}\sigma^4$
$\mid \hat{p}_\omega \mid$	$x\left(\frac{24}{\pi\tau^2}\right)\int_0^\infty d\beta \frac{\sinh^{3/2}\beta}{\sqrt{\beta}\cosh^{5/2}\beta}\, {}_1F_1\left(\frac{5}{2};1;-\frac{x^2 \cdot 4\beta}{\pi\tau^2}\tanh\beta\right)$		$\frac{4}{3}\frac{\pi}{8}\tau^2 = \frac{4}{3}\sigma^4$

Probability densities of a component of first-order PMD (τ_i), the magnitude of first-order PMD ($\mid \boldsymbol{\tau} \mid$), a component of second-order PMD ($\tau_{\omega i}$), the magnitude of second-order PMD ($\mid \boldsymbol{\tau}_\omega \mid$), the second-order PMD component associated with PCD ($\mid \boldsymbol{\tau} \mid_\omega = (\Delta\tau)_\omega$), the perpendicular second-order component ($\mid \boldsymbol{\tau}_{\omega\perp} \mid$), and the depolarization ($\mid \hat{p}_\omega \mid$), all scaled to the mean DGD, $\overline{\Delta\tau}$, denoted here by τ. Catalan's constant, G, is 0.915965..., J_0 is the zero-order cylindrical Bessel function, and ${}_1F_1$ is the standard hypergeometric function

References

[1] Andresciani, D., F. Curti, F. Matera, and B. Daino, "Measurement of the group-delay difference between the principal states of polarization on a low-birefringence terrestrial fiber cable," Opt. Lett., **12**, 844–846 (1987).

[2] Barlow, A. J., J. J. Ramskov-Hansen, and D. N. Payner, "Birefringence and polarization mode-dispersion in spun single-mode fibers," Appl. Opt., **20**, 2962–2968 (1981).

[3] Betti, S., F. Curti, B. Daino, G. De Marchis, E. Iannone, and F. Matera, "Evolution of the bandwidth of the principal states of polarization in single-mode fibers," Opt. Lett., **16**, 467–469 (1991).

[4] Biondini, G. and W. L. Kath, "Applications of Importance Sampling to PMD Problems," Proc. Venice Summer School on Polarization Mode Dispersion, 2002.

[5] Bruyere, F., "Impact of first- and second-order PMD in optical digital transmission systems," Opt. Fiber Technol., **2**, 269–280 (1996).

[6] Bülow, H. and G. Veith, "Temporal dynamics of error-rate degradation induced by polarization mode dispersion fluctuation of a field fiber link," Proc. European Conference on Optical Communication, ECOC'97. 1, 115–118, 1997.

[7] Bülow, H., "Analysis of system outage induced by second order PMD in the presence of chromatic dispersion," Proc. European Conference on Optical Communication, ECOC'98. Paper WdC5, 1998.

[8] Bülow, H., "System outage probability due to first- and second-order PMD," IEEE Photon. Technol. Lett., **10**, 696–698 (1998).

[9] Bülow, H., W. Baumert, H. Schmuck, F. Mohr, T. Schulz, F. Küppers, and W. Weiershausen, "Measurement of the maximum speed of PMD fluctuation in installed field fiber," Optical Fiber Communications Conference, OFC'99. Technical Digest W:83–85, 1999.

[10] Cameron, J., X. Bao, and J. Stears, "Time evolution of polarization mode dispersion for aerial and buried cables," Proc. Optical Fiber Comm. Conf., OFC'98. Paper WM51:249–250,1998.

[11] Cameron, J., L. Chen, X. Bao, and J. Stears, "Time evolution of polarization mode dispersion in optical fibers," IEEE Photon. Technol. Lett., **10**, 1265–1267 (1998).

[12] Chiang, K. S, "Conditions for obtaining zero polarization-mode dispersion in elliptical-core fibres," Electron. Lett., **21**, 592–593 (1985).

[13] Chiang, K. S, "Linearly birefringent fibres with zero polarization-mode dispersion," Electron. Lett., **21**, 916–917 (1985).

[14] Ciprut, P., B. Gisin, N. Gisin, R. Passy, J. P. Von der Weid, F. Prieto, and C. Zimmer, "Second-order PMD: impact on analog and digital transmissions," IEEE J. Lightwave Technol., **16**, 757–771 (1998).

[15] Collings, B. C. and L. Boivin, "Nonlinear polarization evolution induced by cross phase modulation and its impact on transmission systems," IEEE Photon. Technol. Lett. , **12**, 1582–1584 (2000).

[16] Corsi, F., A. Galtarossa, L. Palmieri, M. Schiano, and T. Tambosso, "Continuous-wave backreflection measurement of polarization mode dispersion," IEEE Photon. Technol. Lett., **11**, 451– 453 (1999).

[17] Curti, F., B. Daino, G. De Marchis, and F. Matera, "Statistical treatment of the evolution of the principal states of polarization in single-mode fibers," IEEE J. Lightwave Technol., **LT-8**, 1162–1166 (1990).

[18] Derickson, D., *Fiber Optic Test and Measurement* (Prentice Hall, New Jersey, 1998).

[19] Eickhoff, W., Y. Yen, and R. Ulrich, "Wavelength dependence of birefringence in single-mode fiber," Appl. Opt., **20**, 3428–3435 (1981).

[20] Eyal, A., W. Marshall, M. Tur, and A. Yariv, "Representation of second-order PMD," Elect. Lett. **35**, 1658–1659 (1999).

[21] Foschini, G. J. and C. D. Poole, "Statistical theory of polarization dispersion in single mode fibers," IEEE J. Lightwave Technol., **LT-9**, 1439–1456 (1991).

[22] Foschini, G. J., R. Jopson, L. Nelson and H. Kogelnik, "The statistics of PMD-induced chromatic fiber dispersion," IEEE J. Lightwave Technol., **17**, 1560–1565 (1999).

[23] Foschini, G. J., L. E. Nelson, R. M. Jopson, and H. Kogelnik, "Probability densities of second-order polarization mode dispersion including polarization dependent chromatic fiber dispersion," IEEE Photon. Technol. Lett., **12**, 293–295 (2000).

[24] Francia, C., F. Bruyere, D. Penninckx, and M. Chbat, "PMD second-order effects on pulse propagation in single-mode fibers," IEEE Photon. Technol. Lett., **10**, 1739–1741 (1998).

[25] Galtarossa, A., L. Palmieri, A. Pizzinat, M. Schiano, and T. Tambosso, "Measurement of local beat length and differential group delay in installed single-mode fibers," IEEE J. Lightwave Technol. **18**, 1389–1394 (2000).

[26] Gisin, N. and J. P. Pellaux, "Polarization mode dispersion: Time versus frequency domains," Opt Commun., **89**, 316–323 (1992).

[27] Gleeson, L., E. Sikora, and M. J. O'Mahoney, "Experimental and numerical investigation into the penalties induced by second order polarization mode dispersion at 10 Gb/s," Proc. European Conference on Optical Communication, ECOC'97. 15–18, 1997.

[28] Gordon, J. P. and H. Kogelnik, "PMD Fundamentals: Polarization mode dispersion in optical fibers," Proc. Natl. Acad. Sci. USA., **97**, 4541–4550 (2000).

[29] Hansryd, J., H. Sunnerud, P. A. Andrekson, and M. Karlsson, "Impact of PMD on four-wave-mixing-induced crosstalk in WDM systems," IEEE Photon. Technol. Lett. , **12**, 1261–1263 (2000).

[30] Haunstein, H. F. and H. M. Kallert, "Influence of PMD on the performance of optical transmission systems in the presence of PDL," Proc. Optical Fiber Communication Conference, OFC'01. Paper WT4, 2001.

[31] Haus, H. A., "Group velocity, energy, and polarization mode dispersion," J. Opt. Soc. Am. B, **16**, 1863–1867 (1999).

[32] Heffner, B. L., "Automated measurement of polarization mode dispersion using Jones matrix eigenanalysis," IEEE Photon. Technol. Lett., **4**, 1066–1069 (1992).

[33] Huard, S., *Polarization of Light* (Wiley, 1997).

[34] Huttner, B., C. Geiser, and N. Gisin, "Polarization-induced distortions in optical fiber networks with polarization-mode dispersion and polarization-dependent loss," J. Select. Topics Quant. Electron., **6**, 317–329 (2000).

[35] Ito, T., K. Fukuchi, K. Sekiya, D. Ogasahara, R. Ohhira, and T. Ono, "6.4 Tb/s (160 × 40 Gb/s) WDM transmission experiment with 0.8 bit/s/Hz spectral efficiency," Proc. of European Conference on Optical Communication. Paper PD1.1, 2000.

[36] Jones, R. C., "New calculus for the treatment of optical systems. I. Description and discussion of the calculus," J. Opt. Soc. Am., **31**, 488 (1941).

[37] Jopson, R., L. Nelson, and H. Kogelnik, "Measurement of second-order PMD vectors in optical fibers," IEEE Photon. Technol. Lett., **11**, 1153–1155 (1999).

[38] Jopson, R., L. Nelson, G. J. Pendock, and A. H. Gnauck, "Polarization-mode dispersion impairment in return-to-zero and nonreturn-to-zero systems," Proc. Optical Fiber Communication Conference, OFC'99. Paper WE3, 1999.

[39] Jopson, R., L. E. Nelson, H. Kogelnik, and G. J. Foschini, "Probability densities of depolarization associated with second-order PMD in optical fibers," Proc. Optical Fiber Communication Conference, OFC'01. Paper ThA4, 2001.

[40] Judy, A. F., "Improved PMD stability in optical fibers and cables," Proc. of the 43rd International Wire & Cable Symposium, 658–664, 1994.

[41] Kaminow, I. P., "Polarization in optical fibers," IEEE J. Quant. Electron., **QE-17**, 15–22 (1981).

[42] Karlsson, M., "Polarization mode dispersion-induced pulse broadening in optical fibers," Opt. Lett., **23**, 688–690 (1998).

[43] Karlsson, M. and J. Brentel, "Autocorrelation function of the polarization-mode dispersion vector," Opt. Lett., **24**, 939–941 (1999).

[44] Karlsson, M., J. Brentel, and P. Andrekson, "Simultaneous long-term measurement of PMD on two installed fibers," Proc. European Conference on Optical Communication, ECOC'99. 2:12, 1999.

[45] Karlsson, M., J. Brentel, and P. Andrekson, "Long-term measurement of PMD and polarization drift in installed fibers," IEEE J. Lightwave Technol., **18**, 941–951 (2000).

[46] Khosravani, R., Y. Xie, L-S. Yan, Y. W. Song, A. E. Willner, and C. R. Menyuk, "Limitations to first-order PMD compensation in WDM systems due to XPM-induced PSP changes," Proc. Optical Fiber Communication Conference, OFC'01. Paper WAA5, 2001.

[47] Kim, C. H., E. C. Burrows, and R. M. Jopson, "Polarization dependence of polarization mode dispersion penalties," private communication, 2001

[48] Kogelnik, H., R. M. Jopson, and L. E. Nelson, "Polarization Mode Dispersion," In *Optical Fiber Telecommunications IVB*, I. P. Kaminow and T. Li, eds. (Academic Press, CA, 2000), pp. 725–861.

[49] Kovsh, D. I., S. G. Evangelides, Jr., and A. N. Pilipetskii, "The impact of PMD on nonlinear interchannel crosstalk in DWDM transoceanic systems," Proc. Optical Fiber Communication Conference, OFC'01. Paper WT1, 2001.

[50] Leppla, R. and W. Weiershausen, "A new point of view regarding higher order PMD pulse distortion: Pulse spreading interpreted as multi path propagation," Proc. European Conference on Optical Communication, ECOC 2000. 3:211–212 (2000).

[51] Li, M. J. and D. A. Nolan, "Fiber spin designs for producing fibers with low polarization mode dispersion," Opt. Lett., **23**, 1659–1661 (1998).

[52] Li, M. J., A. F. Evans, D. W. Allen, and D. A. Nolan, "Effects of lateral load and external twist on PMD of spun and unspun fibers," Proc. European Conference on Optical Communication, ECOC 1999. 2:62–63, 1999.

[53] Mollenauer, L. F. and J. P. Gordon, "Birefringence-mediated timing jitter in soliton transmission," Opt. Lett., **19**, 375–377 (1994).

[54] Mollenauer, L. F., J P. Gordon, and F. Heismann, "Polarization scattering by soliton-soliton Collisions," Opt. Lett., **20**, 2060–2062 (1995).

[55] Möller, L., L. Boivin, S. Chandrasekhar, and L. L. Buhl, "Impact of cross-phase modulation on PMD compensation," Proc. IEEE LEOS Annual Meeting 2000. Paper PD1.2, 2000.

[56] Nagel, J. A., M. W. Chbat, L. D. Garrett, J. P. Soigné, N. A. Weaver, B. M. Desthieux, H. Bülow, A. R. McCormick, and R. M. Derosier, "Long-term PMD mitigation at 10 Gb/s and time dynamics over high-PMD installed fiber," Proc. European Conference on Optical Communication, ECOC 2000. 2:31, 2000.

[57] Nelson, L., R. Jopson, H. Kogelnik, and G. Foschini, "Measurement of depolarization and scaling associated with second-order PMD in optical fibers," IEEE Photon. Technol. Lett., **11**, 1614–1616 (1999).

[58] Nelson, L., R. Jopson, H. Kogelnik, and J. P. Gordon, "Measurement of polarization mode dispersion vectors using the polarization-dependent signal delay method," Opt. Express., **6**, 158–167 (2000).

[59] Nelson, L. E. and H. Kogelnik, "Coherent crosstalk impairments in polarization multiplexed transmission due to polarization mode dispersion," Opt. Express., **7**, 350–361 (2000).

[60] Nelson, L. E., T. N. Nielsen, and H. Kogelnik, "Observation of PMD-induced coherent crosstalk in polarization-multiplexed transmission," IEEE Photon. Technol. Lett., **13**, 738–740 (2001).

[61] Norman, S. R., D. N. Payne, M. J. Adams, and A. M. Smith, "Fabrication of single-mode fibres exhibiting extremely low polarization birefringence," Elect. Lett., **24**, 309–311 (1979).

[62] Penninckx, D. and F. Bruyere, "Impact of the statistics of second-order PMD on system performance," Optical Fiber Communication Conference, OFC '98, Technical Digest. 340–342, 1998.

[63] Peters, J., A. Dori, and F. Kapron, "Bellcore's fiber measurement audit of existing cable plant for use with high bandwidth systems," National Fiber Optics Engineers Conference, NFOEC'97, 19–30, 1997.

[64] Poincaré, H., *Théorie Mathématique de la Lumière* (Guathiers-Villars, Paris, 1892), Vol. 2, Chap. 12.

[65] Poole, C. D. and R. E. Wagner, "Phenomenological approach to polarization dispersion in long single-mode fibres, " Elect. Lett., **22**, 1029–1030 (1986).

[66] Poole, C. D., "Statistical treatment of polarization dispersion in single-mode fiber," Opt. Lett. **13**, 687–689 (1988).

[67] Poole, C. D., N. S. Bergano, R. W. Wagner, and H. J. Schulte, "Polarization dispersion and principal states in a 147-km undersea lightwave cable," IEEE J. Lightwave Technol., **LT-6**, 1185–1190 (1988).

[68] Poole, C. D. and C. R. Giles, "Polarization-dependent pulse compression and broadening due to polarization dispersion in dispersion-shifted fiber," Opt. Lett., **13**, 155–157 (1988).

[69] Poole, C. D., R. W. Tkach, A. R. Chraplyvy, and D. A. Fishman, "Fading in lightwave systems due to polarization-mode dispersion," IEEE Photon. Technol. Lett., **3**, 68–70 (1991).

[70] Poole, C. D., J. H. Winters, and J. A. Nagel, "Dynamical equation for polarization dispersion," Opt. Lett., **16**, 372–374 (1991).

[71] Poole, C. D. and J. A. Nagel, "Polarization effects in lightwave systems," In *Optical Fiber Telecommunications IIIA*, I. P. Kaminow and T. L. Koch, eds. (Academic Press, CA, 1997). pp. 114–161.

[72] Roy, F., C. Francia, F. Bruyere, and D. Penninckx, "A simple dynamic polarization mode dispersion compensator," Optical Fiber Communication Conference, OFC'99. Technical Digest. 1:275–278, 1999.

[73] Shieh, W, "Principal states of polarization for an optical pulse," IEEE Photon. Technol. Lett. **11**, 677–679 (1999).

[74] Shtaif, M., A. Mecozzi, and J. Nagel, "Mean-square magnitude of all orders of PMD and the relation with the bandwidth of the principal states," IEEE Photon. Technol. Lett., **12**, 53–55 (2000).

[75] Stokes, G. G., "On the composition and resolution of streams of polarized light from different sources," Trans. Cambridge Phil. Soc., **9**, 399 (1852).

[76] Sunnerud, H., M. Karlsson, and P. A. Andrekson, "A comparison between NRZ and RZ data formats with respect to PMD-induced system degradation," Proc. Optical Fiber Communication Conference, OFC'01. Paper WT3, 2001.

[77] Takahasi, T., T. Imai, and M. Aiki, "Time evolution of polarization mode dispersion in 120 km installed optical submarine cable," Elect. Lett., **20**, 1605–1606 (1993).

[78] Ulrich, R. and A. Simon, "Polarization optics of twisted single-mode fibers," Appl. Opt., **18**, 2241– 2251 (1979).

[79] Vengsarkar, A. M., A. H. Moesle, L. G. Cohen, and W. L. Mammel, "Polarization mode dispersion in dispersion shifted fibers: An exact analysis," Proc. Optical Fiber Communication Conference, OFC'93. Paper ThJ7, 1993.

[80] Waddy, D., P. Lu, L. Chen, and X. Bao, "The measurement of fast state of polarization changes in aerial fiber," Proc. Optical Fiber Communication Conference, OFC'01. Paper ThA3, 2001.

[81] Wai, P. K. A and C. R. Menyuk, "Polarization mode dispersion, decorrelation, and diffusion in optical fibers with randomly varying birefringence," IEEE J. Lightwave Technol., **14**, 148–157 (1996).

[82] Wang, D. and C. R. Menyuk, "Calculation of penalties due to polarization effects in a long-haul WDM system using a Stokes parameter model," IEEE J. Lightwave Technol., **19**, 487–494 (2001).

[83] Winzer P. J., H. Kogelnik, C. H. Kim, H. Kim, R. M. Jopson, L. E. Nelson, "Effect of receiver design on PMD outage for RZ and NRZ," Proc. Optical Fiber Communication Conference, OFC'02. Paper TuI1, 2002.

DOI: 10.1007/s10297-004-0002-4
Originally published in J. Opt. Fiber. Commun. Rep. 1, 248–265 (2004)

Modelling of polarization mode dispersion in optical communications systems

Mark Shtaif[1] and Antonio Mecozzi[2]

[1] School of Electrical Engineering
Dept. of Physical Electron.
Tel-Aviv University
Tel-Aviv, Israel
Email: shtaif@eng.tau.ac.il

[2] Department of Electrical Engineering
University of L'Aquila
L'Aquila, Italy
and
Istituto Nazionale di Fisica della Materia
Italy
Email: amecozzi@ing.univaq.it

Abstract. With the rapid increase in the data rates transmitted over optical systems, as well as with the recent extension of terrestrial systems to ultra-long haul reach, polarization mode dispersion (PMD) has become one of the most important and interesting limitations to system performance. This phenomenon originates from mechanical and geometrical distortions that break the cylindrical symmetry of optical fibers and create birefringence. It is the random variations of the local birefringence along the propagation axis of the optical fiber that create the rich and complicated bulk of phenomena that is attributed to PMD. The detailed statistical properties of the local birefringence and its dependence on position are only important as long as the overall system length is comparable with the correlation length of the birefringence in the fiber. In typical systems, however, the latter is smaller by more than three orders of magnitude so that the specific properties of the local birefringence become irrelevant. Instead, the fiber can be viewed as a concatenation of a large number of statistically independent birefringent sections characterized only by the mean square value of their birefringence. This model has been used extensively in the study of PMD and its predictions have been demonstrated to be in excellent agreement with experimental results. This approach opens the door to the world of stochastic calculus, which offers many convenient tools for studying the PMD problem. In this article we review the modelling of PMD and discuss the properties of this phenomenon as a stochastic process. We explain the use of stochastic calculus for the analysis of PMD and describe the derivation of the frequency autocorrelation functions of the PMD vector, its modulus and the principal states. Those quantities are then related to commonly used parameters such as the bandwidth of the first order PMD approximation, the bandwidth of the principal states and to the accuracy of PMD measurements.

1. What is PMD?

When the cylindrical symmetry of optical fibers is disturbed as a result of mechanical stress, or geometrical distortions, the transmission properties of the fiber become polarization dependent. If the violation of cylindrical symmetry is constant along the fiber, it is possible to identify two orthogonal states of polarization that are characterized by slightly different refractive indices. We will call these states of polarization the eigenpolarizations of the birefringent fiber. When an arbitrarily polarized optical pulse is injected into such a fiber, its components along the two eigenpolarizations are propagated at different group velocities such that the pulse at the output consists of two identically shaped and orthogonally polarized replicas of the original pulse, that are merely delayed relative to each other (see Fig. 1a). One can imagine that such a distortion can be detrimental to optical communications systems where the injected pulses carry actual data. Nevertheless, this case of constant birefringence constitutes, as we shall see, a very degenerate form of PMD, which although potentially detrimental in its raw effect, can be easily avoided. For example, one could consider launching the signal into only one of the eigenpolarizations in which case it will propagate without distortions. A more realistic scenario is that where the birefringence changes both in its strength (the difference between the refractive indices) and orientation (the eigenpolarizations) along the propagation axis of the optical fiber. To picture what happens in this situation we illustrate in Fig. 1b the case in which the birefringence has a stepwise constant behavior. In other words, the fiber consists of many short sections of constant birefringence. In this case the signal splits into two delayed replicas after every individual section such that a very large number of replicas is present at the link output. If instead of an impulse we consider an input pulse of some finite bandwidth, it smears in the process of propagation and becomes distorted in a manner that cannot be easily characterized in the general case. This is in fact the true effect of PMD. It is interesting that when the bandwidth of the pulse is sufficiently narrow, the distortion that it experiences can be described in terms that are equivalent to the case of constant birefringence. In other words, one could replace the link with a fiber of constant birefringence with some well defined delay and eigenpolarizations that would distort the injected signal in a similar manner. The approximate description of PMD in terms of equivalent birefringence is the so called first order PMD approximation. Generally speaking, it results from the first order expansion of the fiber transmission matrix with respect to frequency, but its justification and range of validity will be discussed in what follows.

A formal treatment of PMD requires some mathematical concepts the first of which is the description of polarizations in Stokes space and their representation on the Poincaré sphere. We will now very briefly review the use of these concepts without going into any of the gory details which are very well covered by the known literature

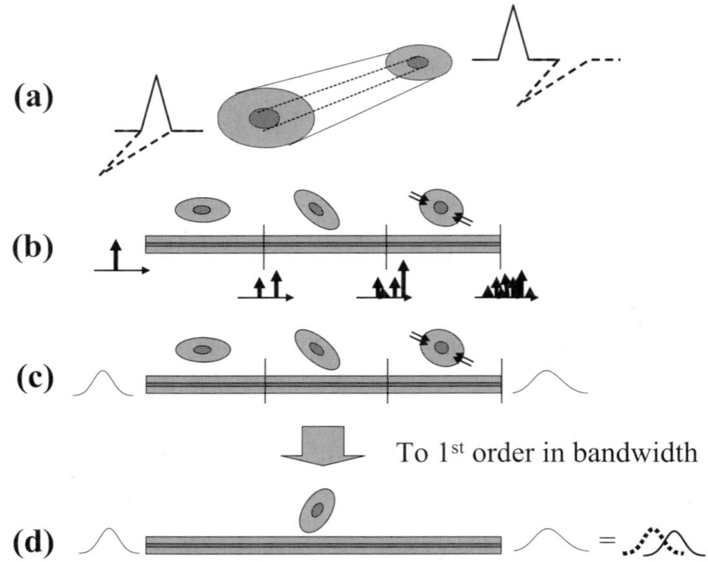

Fig. 1. Constant birefringence introduces a delay between two orthogonal replicas of the transmitted pulse (a). When the fiber consists of many sections with different birefringence, the transmitted pulse splits many times along the propagation and significant distortions to the transmitted waveform may follow as a result of PMD (b). When the injected pulse is sufficiently narrowband (c) the effect of a fiber with changing birefringence becomes equivalent to first order to the effect of constant birefringence (d).

on polarizations [2],[3],[4]. As is well known, an optical signal at any frequency and position along the fiber $\boldsymbol{E}(\omega, z)$ can be represented by a vector $\boldsymbol{S}(\omega, z)$ in a three dimensional space called the Stokes space. Unlike the two components of $\boldsymbol{E}(\omega, z)$ that are complex, the three components of $\boldsymbol{S}(\omega, z)$ are real, thereby allowing a simple pictorial representation. The vector $\boldsymbol{S}(\omega, z)$ contains all the information in $\boldsymbol{E}(\omega, z)$ except for the optical phase of $\boldsymbol{E}(\omega, z)$ that cannot be extracted from $\boldsymbol{S}(\omega, z)$.[3] The length of the Stokes vector is equal to the optical power $S(\omega, z) = |\boldsymbol{E}(\omega, z)|^2$ and its orientation $\hat{\boldsymbol{S}}(\omega, z)$ represents the polarization state of $\boldsymbol{E}(\omega, z)$. The tip of $\hat{\boldsymbol{S}}(\omega, z)$ resides on the unit sphere in Stokes space, which is known as the Poincaré sphere, and the various points on its surface uniquely represent the polarization state of $\boldsymbol{E}(\omega, z)$, as is illustrated in Fig. 2. Notice particularly, that orthogonal polarization states are represented by antipodal points on the sphere. The Stokes space representation has many convenient properties. For example, the effect of constant birefringence is represented in Stokes space by the precession of the vector $\boldsymbol{S}(\omega, z)$ about some fixed axis, as is illustrated in Fig. 3a. It is not difficult to become convinced that the orientation of that axis must coincide with the eigenpolarizations of the birefringent fiber as they

[3] Since the vector $\boldsymbol{E}(\omega, z)$ has two complex components it consists of 4 degrees of freedom. One of them, the mutual phase of the two components, is lost when shifting to the three-dimensional Stokes space representation.

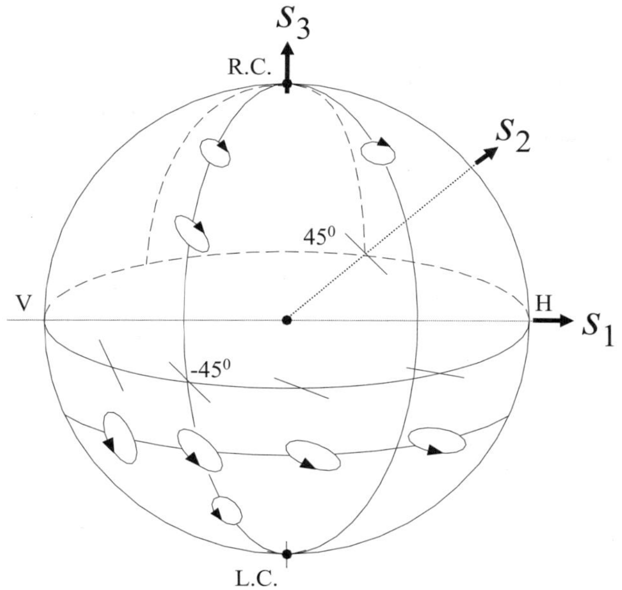

Fig. 2. The relation between the actual polarizations and the position on the Poincaré sphere. The abbreviations H, V, R.C. and L.C. stand for horizontal, vertical, right-circular and left-circular, respectively. After [4].

are represented on the Poincaré sphere. It is convenient to represent birefringence by a vector $\boldsymbol{\beta}$ in Stokes space whose orientation coincides with the slow (or fast in the formalism of [5]) eigenpolarization state and whose magnitude is equal to the difference between the wave vectors of the two eigenpolarizations. Neglecting the dispersion of the birefringence, one may write $|\boldsymbol{\beta}| = \Delta n\omega/c$ with Δn being the difference in refractive indices and with c being the speed of light. The phase difference acquired between the eigenpolarizations when propagating through a distance z is equal to $|\boldsymbol{\beta}|z$ and it is also equal to the angle that the vector $\boldsymbol{S}$ rotates about the birefringence vector $\boldsymbol{\beta}$. It is now instructive to consider the way in which a change in the optical frequency of the launched signal affects the polarization state of the signal at the output of a fiber with constant birefringence. Having neglected the dispersion of the birefringence, the dependence of the output polarization on frequency is identical to its dependence on position because the rotation angle $|\boldsymbol{\beta}|z = \Delta n\omega z/c$ is symmetric with respect to its dependencies on frequency ω and position z. In the following we will refer to the delay vector $\boldsymbol{\beta}'(\omega, z) = \partial\boldsymbol{\beta}(\omega, z)/\partial\omega$ that is equal to the time delay between the two eigenpolarizations per unit of fiber length. The evolution of the output polarization in response to a continuous change in the optical frequency is illustrated in Fig. 3b.

The spectral evolution of the output state of polarization becomes more complicated when the birefringence vector changes along the propagation axis of the fiber. Once again we consider for simplicity the stepwise constant evolution of the birefringence vector that corresponds to the concatenation of many constant birefringence sections with different birefringence vectors. This situation is illustrated in Fig. 4a.

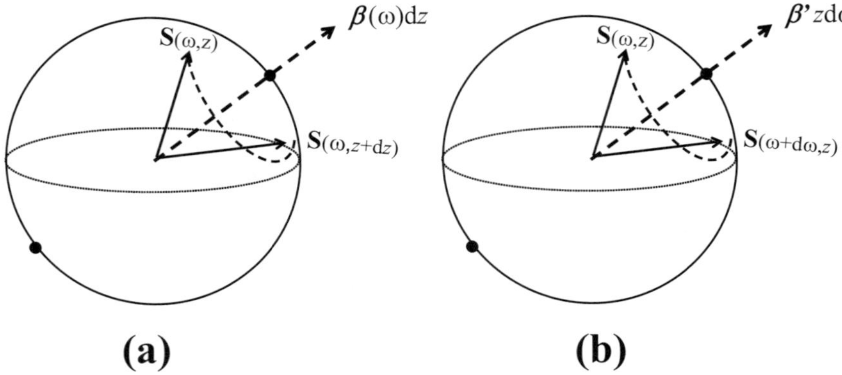

Fig. 3. The effects of changing position (a) along a constant birefringence fiber and of changing the optical frequency (b) are both reflected through a rotation of the polarization around the birefringence vector in Stokes space.

Figures 4b and 4c describe the evolution of the polarization state at two different optical frequencies when the input state of polarization is kept constant. In the process of passing through each one of the constant birefringence sections the state of polarization rotates around the corresponding birefringence vectors and draws an arch on the

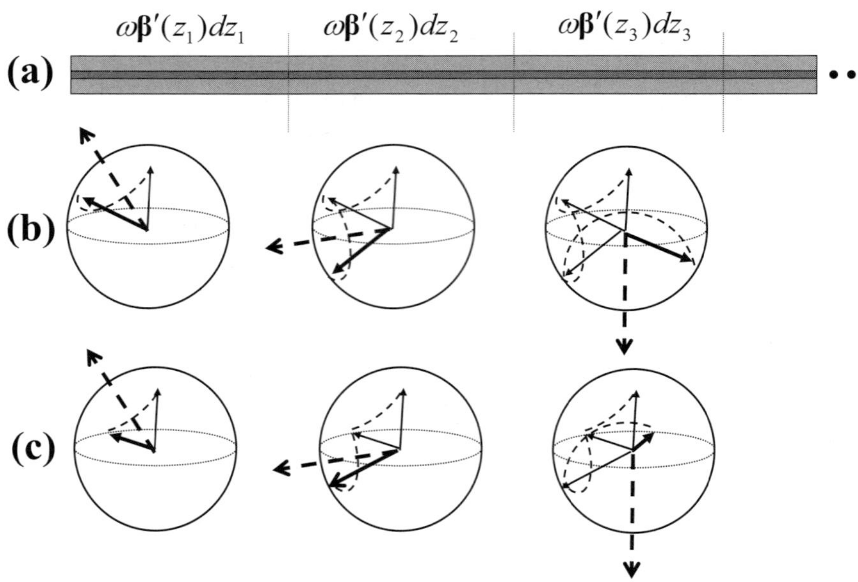

Fig. 4. (a) is an example of a case where the birefringence changes with position along the link. (b) and (c) show what happens to the polarization Stokes vector of the signal when passing through the individual sections. The optical frequency in (c) is slightly lower than in (b) such that all the arches are proportionally shorter. The figure shows that there is no simple relation between the output polarization states that correspond to different frequencies in a complex fiber.

Poincaré sphere. The lengths of the arches are different at the two different frequencies and as a result, the dependence of the output state of polarization on frequency is rather complex. When the number of birefringent sections becomes large the output state of polarization draws a seemingly random curve on the surface of the Poincaré sphere in response to a continuous change in the optical frequency, as is illustrated in Fig. 5. Notice, that locally, in the immediate vicinity of any optical frequency ω_0 the evolution of the curve can be described as precession[4] about some axis $\boldsymbol{\tau}(\omega_0)$. It is customary to set the length of this vector $\boldsymbol{\tau}(\omega_0)$ to be equal to the rotation angle of the output state of polarization per unit of frequency in the vicinity of ω_0. When defined in this way $\boldsymbol{\tau}(\omega)$

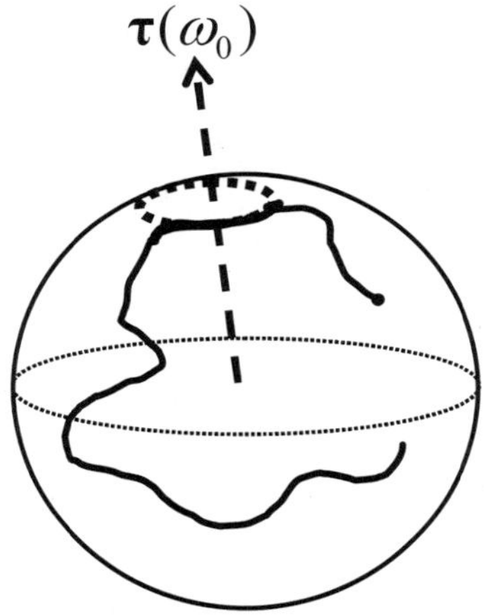

Fig. 5. An illustration of a trajectory that the polarization state at a given position z along the link draws on the Poincaré sphere. Locally the variation of the state of polarization with frequency can be described as rotation around the PMD vector $\boldsymbol{\tau}$.

is called the PMD vector, its orientation $\hat{\boldsymbol{\tau}}(\omega)$ defines the so-called principal states of polarization (PSP)[5] and its magnitude $\tau(\omega_0) = |\boldsymbol{\tau}(\omega)|$ is the differential group delay (DGD). The physical meaning of $\boldsymbol{\tau}(\omega)$ can be understood through analogy with the birefringence vector. Specifically, if we consider a narrowband optical signal such that within its bandwidth the frequency dependence of $\boldsymbol{\tau}$ can be neglected, then the frequency components of that signal lie on an arch that is part of a circle surrounding the vector $\boldsymbol{\tau}(\omega_0)$ with ω_0 being the central optical frequency. In this case the dependence on frequency is identical to what would result from propagation in a constant birefringence fiber whose birefringence vector satisfies $\boldsymbol{\beta}'z = \boldsymbol{\tau}(\omega_0)$. It is therefore clear by analogy that the output pulse will consist of two orthogonally polarized replicas of the original pulse that are delayed relative to each other. The polarizations states of these two replicas coincide with the PSPs and the delay between them is the DGD of the fiber.

[4] Notice that since the effect of the entire fiber link consists of a concatenation of many infinitesimal rotation operators, one can be certain that indeed the local effect can be described as a rotation around $\boldsymbol{\tau}(\omega_0)$, regardless of what the polarization at ω_0 is.

[5] More accurately, the principal states are the two polarizations states that are parallel and antiparallel to $\hat{\boldsymbol{\tau}}$. It is customary to define the orientation of $\hat{\boldsymbol{\tau}}$ such that the rotation of $\boldsymbol{S}$ around it follows the right screw convention. In this case $\hat{\boldsymbol{\tau}}$ coincides with the slow PSP.

With this we justified the first order PMD approximation that we initially introduced earlier. We have also seen that this approximation remains valid as long as the PMD vector can be considered constant within the bandwidth of the transmitted signal. This statement will be further quantified later. To account for the spectral dependence of the PMD vector it is customary to represent it in the form of a Taylor expansion

$$\boldsymbol{\tau}(\omega, z) = \boldsymbol{\tau}(\omega_0, z) + (\omega - \omega_0)\boldsymbol{\tau}'(\omega_0, z) + \frac{1}{2}(\omega - \omega_0)^2\boldsymbol{\tau}''(\omega_0, z) + \dots, \quad (1)$$

where the primes over $\boldsymbol{\tau}$ denote derivatives with respect to frequency. The term $\boldsymbol{\tau}(\omega_0, z)$ represents the first-order PMD approximation since it neglects the spectral variation of PMD altogether. The frequency derivatives of $\boldsymbol{\tau}$ represent the so-called high orders of PMD, where it has become customary to associate the nth frequency derivative of the PMD vector with the $n + 1$ PMD order [6]. Other, alternative, definitions of high orders of PMD have also been proposed recently [7, 8].

2. The PMD Equations

The above description can be mathematically formulated in the following way:

$$\boldsymbol{S}(\omega, z + \mathrm{d}z) = \mathsf{R}(\boldsymbol{\beta}(\omega, z)\mathrm{d}z)\boldsymbol{S}(\omega, z) \quad (2)$$
$$\boldsymbol{S}(\omega + \mathrm{d}\omega, z) = \mathsf{R}(\boldsymbol{\tau}(\omega, z)\mathrm{d}\omega)\boldsymbol{S}(\omega, z), \quad (3)$$

where $\mathsf{R}(\boldsymbol{V})$ is a rotation operator that rotates the vector that it acts upon by an angle $|\boldsymbol{V}|$ around the axis $\hat{\boldsymbol{V}}$. This operator can also be represented in an exponential operator form as $\mathsf{R}(\boldsymbol{V}) = \exp(\boldsymbol{V}\times)$, with $(\boldsymbol{V}\times)$ representing the operation of a cross product. When the rotation operators are expanded to first order with respect to their argument ($\mathsf{R}(\boldsymbol{\beta}\mathrm{d}z) \simeq 1 + \mathrm{d}z\boldsymbol{\beta}\times$ and $\mathsf{R}(\boldsymbol{\tau}\mathrm{d}\omega) \simeq 1 + \mathrm{d}\omega\boldsymbol{\tau}\times$) the more familiar form of expressions (2) and (3) follows [9, 10, 11]:

$$\frac{\partial \boldsymbol{S}}{\partial z} = \boldsymbol{\beta} \times \boldsymbol{S} \quad (4)$$
$$\frac{\partial \boldsymbol{S}}{\partial \omega} = \boldsymbol{\tau} \times \boldsymbol{S}. \quad (5)$$

Furthermore, by combining Eqs. (2) and (3) and using the algebra of rotations, it can also be shown (see appendix A and [11, 12]) that the evolution of $\boldsymbol{\tau}(\omega, z)$ is described by

$$\boldsymbol{\tau}(\omega, z + \mathrm{d}z) = \mathsf{R}(\boldsymbol{\beta}(\omega, z)\mathrm{d}z)\boldsymbol{\tau}(\omega, z) + \boldsymbol{\beta}'(\omega, z)\mathrm{d}z, \quad (6)$$

where we recall that $\boldsymbol{\beta}'$ is the delay vector, which is equal to the frequency derivative of $\boldsymbol{\beta}(\omega, z)$. We should stress that while the vector $\boldsymbol{\tau}(\omega, z)$ in Eq. (6) reflects the PMD that is accumulated between the beginning of the link and point z, the vectors $\boldsymbol{\beta}(\omega, z)\mathrm{d}z$ and $\boldsymbol{\beta}'(\omega, z)\mathrm{d}z$ represent the birefringence and delay, respectively, of the section of fiber that starts at z and ends at $z + \mathrm{d}z$. This clarification becomes important when the birefringence vector is characterized by abrupt changes along the link as in the case where the fiber is constructed from many sections of constant birefringence. Equation (6) can also be simplified if the rotation operator is expanded to first order in $\mathrm{d}z$, but let us defer this simplification to a later stage, for reasons that will become clear shortly.

The details of the evolution of the birefringence vector $\boldsymbol{\beta}(\omega, z)$ along the fiber and its statistical properties have so far been completely ignored in our discussion. The study of the spatial variation of the birefringence is, in fact, a subject of continuing research [13]. Nevertheless, in most cases of practical interest the detailed statistics of the birefringence has little significance in predicting experimental results. At least, this appears to be the case with long optical links where the overall length of the link is significantly larger than the correlation length of the birefringence, which is the length-scale over which the birefringence vector looses its dependence on the initial conditions. Measurements of the correlation length [13] seem to indicate that it is within hundreds of meters, whereas the length of optical communications links is usually of the order of hundreds to thousands of kilometers - three orders of magnitude larger. Under such conditions it appears that the statistical details of $\boldsymbol{\beta}(z)$ loose their relevance and only the mean square value of $\boldsymbol{\beta}(z)$ determines the effect on signal propagation. This is in fact the limit in which $\boldsymbol{\beta}(z)$ can be safely approximated as white Gaussian noise, meaning that $\langle\boldsymbol{\beta}(z)\cdot\boldsymbol{\beta}(z')\rangle \propto \delta(z-z')$ and $\langle\boldsymbol{\beta}(z)\times\boldsymbol{\beta}(z')\rangle = 0$, where the angular brackets denote ensemble averaging. In this situation the fiber-optic link can be legitimately described in terms of a large number of constant birefringence sections with statistically independent vectors of birefringence. Thus, we have now justified the picture that we have previously introduced merely for illustration purposes. Notice that the white noise approximation applies only to the dependence of the birefringence vector on position. Its dependence on frequency can be effectively approximated as

$$\boldsymbol{\beta}(\omega, z) = \boldsymbol{\beta}(\omega_0, z) + (\omega - \omega_0)\boldsymbol{\beta}'(\omega_0, z). \tag{7}$$

The frequency independent part $\boldsymbol{\beta}(\omega_0, z)$ can be safely ignored since it has no effect on the waveform distortions that PMD generates in optical communications systems. Formally, this term can be removed from all equations by a simple change of variables as can be found in [14]. We will also simplify the notation further by setting $\omega_0 = 0$, such that ω represents the offset from the central frequency of the signal. Finally, instead of referring to the birefringence vector $\boldsymbol{\beta}$, or to the delay vector $\boldsymbol{\beta}'$ that are infinitely wild, it is more convenient to define $\mathrm{d}\boldsymbol{W}(z) = \boldsymbol{\beta}'\mathrm{d}z$ such that $\mathrm{d}\boldsymbol{W}(z)$ is a standard three dimensional Brownian motion whose most relevant property is

$$\langle \mathrm{d}W_i(z)\mathrm{d}W_j(z')\rangle = \frac{1}{3}\gamma_0^2\delta(z-z')\delta_{i,j}\mathrm{d}z\mathrm{d}z' + r(\mathrm{d}z), \tag{8}$$

where $\mathrm{d}W_i$ and $\mathrm{d}W_j$ are the i and j components of $\mathrm{d}\boldsymbol{W}$ (i, j are equal to 1,2 or 3), γ_0 is a proportionality coefficient that will later be tied to the mean square DGD value and $r(\mathrm{d}z)$ is a zero mean remainder whose dependence on $\mathrm{d}z$ satisfies $\lim_{\mathrm{d}z\to 0} r(\mathrm{d}z)/\mathrm{d}z = 0$ in the mean square sense. Notice that using equation (8) one can show that the mean value of $\langle \mathrm{d}\boldsymbol{W}(z)\cdot\mathrm{d}\boldsymbol{W}(z)\rangle$ is $\gamma_0^2\mathrm{d}z$. Recall that as we explained following Eq. (6), $\mathrm{d}\boldsymbol{W}(z)$ corresponds to the delay vector of the fiber section between z and $z+\mathrm{d}z$, implying that $\mathrm{d}\boldsymbol{W}(z)$ and $\boldsymbol{\tau}(z)$ are statistically independent quantities.

We are now ready to transform Eq. (6) to a more usable form. This is done by expanding the rotation operator $\mathsf{R}(\boldsymbol{\omega}\beta'\mathrm{d}z) = \exp(\omega\mathrm{d}\boldsymbol{W}\times)$ to first order in $\mathrm{d}z$. Notice, however that in view of (8), first order in $\mathrm{d}z$ implies second order in $\mathrm{d}\boldsymbol{W}$, namely $\exp(\omega\mathrm{d}\boldsymbol{W}\times) \simeq 1 + (\omega\mathrm{d}\boldsymbol{W}\times) + 1/2(\omega\mathrm{d}\boldsymbol{W}\times)(\omega\mathrm{d}\boldsymbol{W}\times)$, which reduces after some algebraic manipulation to $1 + \omega\mathrm{d}\boldsymbol{W}\times - \gamma_0^2(\omega^2/3)\mathrm{d}z$. This allows us to rewrite Eq. (6) in the more convenient form [6]:

$$\mathrm{d}\boldsymbol{\tau} = \omega \mathrm{d}\boldsymbol{W} \times \boldsymbol{\tau} + \mathrm{d}\boldsymbol{W} - \frac{\omega^2}{3}\gamma_0^2 \boldsymbol{\tau} \mathrm{d}z. \tag{9}$$

It is important to stress that the rightmost term in (9) came up because of the fact that we approximate the birefringence of the fiber as a white noise process[6]. In the absence of that approximation this term disappears and Eq. (9) reduces to the familiar form introduced by Poole in 1991 [10]. The proportionality coefficient γ_0 can now be easily tied to the physical parameters of the system. Setting the frequency to be $\omega = 0$ we obtain $\boldsymbol{\tau} = \boldsymbol{W}$ as can be easily seen from Eq. (9). The mean square DGD is then $\langle \tau^2 \rangle = \gamma_0^2 z$, implying that γ_0^2 is simply the mean square DGD per unit of fiber length. Furthermore we find that $\boldsymbol{\tau}(\omega, z)$ is a Gaussian vector and therefore its magnitude $\tau = |\boldsymbol{\tau}|$ has a Maxwellian distribution [15] such that the probability density function of τ is given by

$$P(\tau) = \sqrt{\frac{54}{\pi}} \frac{\tau^2}{\langle \tau^2 \rangle^{3/2}} \exp\left(-\frac{3\tau^2}{2\langle \tau^2 \rangle} \right). \tag{10}$$

Our choice of $\omega = 0$ does not limit the generality of the result owing to the stationarity of PMD with respect to frequency. This can be learned from the same transformation that allowed the removal of the frequency independent term $\boldsymbol{\beta}(\omega_0, z)$ of Eq. (7) [14].

Finally, by normalizing the distance, the frequency and the modulus of the PMD vector in the following way [16]:

$$z_N = z/L\,, \tag{11}$$

$$\omega_N = \omega\sqrt{\langle \tau^2 \rangle}\,, \tag{12}$$

$$\boldsymbol{\tau}_N = \boldsymbol{\tau}/\sqrt{\langle \tau^2 \rangle}\,, \tag{13}$$

we arrive at a universal equation for the evolution of PMD

$$\mathrm{d}\boldsymbol{\tau}_N = \omega_N \mathrm{d}\boldsymbol{W}_N \times \boldsymbol{\tau}_N + \mathrm{d}\boldsymbol{W}_N - \frac{\omega_N^2}{3}\boldsymbol{\tau}_N \mathrm{d}z_N, . \tag{14}$$

where $\mathrm{d}\boldsymbol{W}_N(z)$ is a standard Brownian motion in three dimensions, satisfying Eq. (8) with $\gamma_0 = 1$. This equation is powerful because of the fact that it has no adjustable parameters. When it is solved (whether numerically or analytically), its solution can be scaled rigorously to any fiber, based solely on the mean square value of the total DGD of that fiber. In other words this normalized equation implies that *any* statistical property of PMD is determined completely by the mean DGD of the fiber. In the following we will only consider the normalized quantities in our analysis, but the subscript N will be omitted in order to simplify the notation. Notice also that Eq. (14) (as well as Eq. (9) for that matter) is a standard stochastic differential equation expressed in the Ito form [14, 17]. Using this formulation opens the door to the rich toolbox of stochastic calculus which we will use in the sections that follow.

[6] In the terminology of stochastic calculus Eq. (9) is called an Ito stochastic differential equation [17] and the right-most term is called the Ito correction term that takes into account the peculiarities of white noise. One could use an alternative, Stratonovich formulation in which that term does not exist, but the terms $\mathrm{d}\boldsymbol{W}(z)$ and $\boldsymbol{\tau}(z)$ are no longer independent. The Stratonovich formulation ends up being more complicated and less natural in implementations.

3. The Statistics of the Frequency Dependence of PMD and Its Related Quantities

The spectral behavior of PMD is of utmost importance in determining the effect of PMD in optical communications systems. To have a full understanding of the spectral dependence, one needs to obtain a complete description of PMD as a stochastic process in frequency, a goal which has so far been unaccomplishable. Instead, we will ask and answer somewhat less ambitious questions concerning the spectral dynamics of PMD. The first thing that we will want to know is how much the PMD vector changes on average in response to a given change in the optical frequency. The answer to this question will give us an idea concerning the bandwidth in which the first order PMD approximation remains valid. We will then ask more specific questions related to the dependence of the PSPs and the DGD on the optical frequency. The answer to the latter question will later be used for quantifying the accuracy of PMD measurements. Other, more advanced questions that we shall answer relate to the conditional moments of the PMD vector at one frequency when the PMD at some given other optical frequency is known.

An important quantity that answers the above questions to some extent is the frequency autocorrelation function [18, 19]. We may think of several such functions that are of interest. The first is the autocorrelation function of the PMD vector itself and it is defined as $\langle \boldsymbol{\tau}(\omega) \cdot \boldsymbol{\tau}(\omega') \rangle$ where we have left the dependence on z implicit. Owing to the stationarity of PMD with respect to frequency, we expect this function to depend only on the frequency difference $\Delta\omega = \omega' - \omega$. This function gives us an idea on how much the optical frequency can be varied before the PMD vector looses memory of its initial value. The disadvantage of this function is that it treats the PMD vector as a whole and does not differentiate between its orientation (the PSPs) and its modulus (the DGD). The dependence of these individual quantities on the optical frequency can be characterized by relating to two additional autocorrelation functions; one for the DGD $\langle \tau(\omega)\tau(\omega') \rangle$ and one for the PSPs $\langle \hat{\boldsymbol{\tau}}(\omega) \cdot \hat{\boldsymbol{\tau}}(\omega') \rangle$ [16]. The width of all three of these quantities when expressed as a function of $\Delta\omega$ will provide us with an idea for the bandwidth in which the first order PMD approximation remains valid.

We will now briefly describe the extraction of these quantities. We start with the vector autocorrelation function $\langle \boldsymbol{\tau}(\omega) \cdot \boldsymbol{\tau}(\omega') \rangle$ which can be obtained from a differential equation that is constructed as follows,

$$\mathrm{d}\langle \boldsymbol{\tau}(\omega) \cdot \boldsymbol{\tau}(\omega') \rangle = \langle \boldsymbol{\tau}(\omega) \cdot \mathrm{d}\boldsymbol{\tau}(\omega') \rangle + \langle \mathrm{d}\boldsymbol{\tau}(\omega) \cdot \boldsymbol{\tau}(\omega') \rangle + \langle \mathrm{d}\boldsymbol{\tau}(\omega) \cdot \mathrm{d}\boldsymbol{\tau}(\omega') \rangle \,, \quad (15)$$

where we may use Eq. (14) in order to express the differentials $\mathrm{d}\boldsymbol{\tau}(\omega)$ and $\mathrm{d}\boldsymbol{\tau}(\omega')$ which can be substituted into Eq. (15). If we recall that the average of terms that contain $\mathrm{d}\boldsymbol{W}$ (in an odd power) is 0 and that $\langle \mathrm{d}\boldsymbol{W} \cdot \mathrm{d}\boldsymbol{W} \rangle = \mathrm{d}z$ we obtain the equation

$$\frac{\mathrm{d}\langle \boldsymbol{\tau}(\omega) \cdot \boldsymbol{\tau}(\omega') \rangle}{\mathrm{d}z} = 1 - \frac{1}{3}\Delta\omega^2 \langle \boldsymbol{\tau}(\omega) \cdot \boldsymbol{\tau}(\omega') \rangle \,. \quad (16)$$

Equation (16) can be easily solved to yield the following autocorrelation function at the end of the link ($z = 1$)

$$\langle \boldsymbol{\tau}(\omega) \cdot \boldsymbol{\tau}(\omega') \rangle = \frac{3}{\Delta\omega^2} \left(1 - \mathrm{e}^{-\Delta\omega^2/3} \right) \,, \quad (17)$$

a result that has been originally obtained in [18] and [19]. We should note that the above derivation becomes particularly simple if one exploits the stationarity of PMD with respect to frequency and sets in Eq. (15) $\omega = 0$ and $\omega' = \Delta\omega$. The autocorrelation function (17) is plotted in Fig. 6. The autocorrelation bandwidth of the

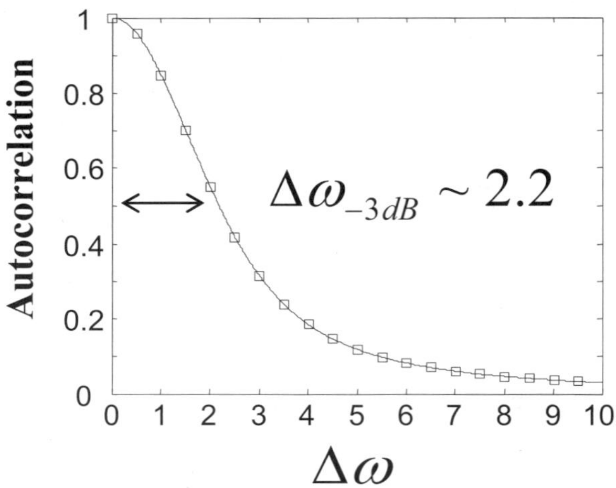

Fig. 6. The autocorrelation function of the PMD vector in normalized units.

PMD vector, which we define by the point where the autocorrelation function reduces to half of its maximal value is equal approximately $\Delta\omega_c = 2.2$. In real world units and in actual frequency (as opposed to angular frequency) this corresponds to $0.35/\sqrt{\langle\tau^2\rangle}$, or equivalently to $0.32/\langle\tau\rangle$ where $\langle\tau\rangle$ is the mean DGD and it is equal to $\langle\tau\rangle = \sqrt{8/(3\pi)\langle\tau^2\rangle}$ as is dictated by the relation between the first two moments of a Maxwellian random variable. The frequency autocorrelation function can also be used to derive the variances and covariances of the various PMD orders that are defined based on the Taylor expansion of $\boldsymbol{\tau}(\omega)$ that we described above. By taking the corresponding derivatives of $\langle\boldsymbol{\tau}(\omega)\cdot\boldsymbol{\tau}(\omega')\rangle$ with respect to ω and ω' one obtains

$$\langle\boldsymbol{\tau}^{(n-k)}\cdot\boldsymbol{\tau}^{(n+k)}\rangle = \frac{(2n)!}{3^n(n+1)!}\langle\tau^2\rangle^{n+1}, \tag{18}$$

$$\langle\boldsymbol{\tau}^{(n-k)}\cdot\boldsymbol{\tau}^{(n+k+1)}\rangle = 0, \tag{19}$$

where $\boldsymbol{\tau}^{(j)}$ denotes the jth derivative of $\boldsymbol{\tau}$ with respect to frequency, or the $j-1$ order of PMD. Setting $k = 0$, we find that the righthand side of Eq. (18) is the variance of the $n-1$ order of PMD. In a similar calculation to the one leading to Eq. (17) we can obtain a rigorous expression for the frequency autocorrelation of the square modulus of the PMD vector (or the square DGD) [16],

$$\langle\tau^2(\omega)\tau^2(\omega')\rangle = \langle\tau^2\rangle^2 + \frac{4\langle\tau^2\rangle}{\Delta\omega^2} - \frac{12}{\Delta\omega^4}\left[1 - e^{-\Delta\omega^2/3}\right], \tag{20}$$

which is plotted together with the results of computer simulations in Fig. 7. The width at half maximum of this function is given by $\Delta\omega_c = 2.8$. Unfortunately the same tech-

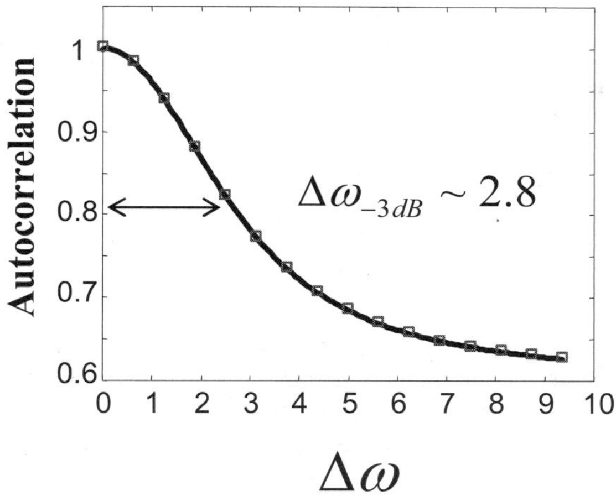

Fig. 7. Autocorrelation function of the square DGD [equation (20)] as well as results of a Monte Carlo simulation of a fiber with 1000 independent sections (squares).

nique does not allow the rigorous analytical derivation of the autocorrelation functions of the DGD and of the PSPs. These quantities can however be numerically extracted from Eq. (14) and since, as we have already explained, this equation is universal, the results can be scaled to any fiber based on its mean DGD value. The results for the autocorrelation functions of the DGD and the PSPs appear in Figure 8 together with a comparison with the autocorrelation functions obtained from experimental data. The correlation bandwidths are $\Delta\omega_c = 2.2$ for the PSPs and $\Delta\omega_c = 2.8$ for the DGD. The two quantities decorrelate at a fairly similar rate, contrary perhaps to some of the common intuition. Nevertheless, since the autocorrelation function of the DGD only changes within 15% of its maximal value, its contribution to the decorrelation of the entire PMD vector is secondary. As a result the autocorrelation function of the PSPs is very similar to the autocorrelation function of the PMD vector (17) as can be observed in Fig. 9. Notice that the autocorrelation function of the PSPs is slightly narrower than the vector autocorrelation function. This is opposite to what would be the case if the DGD and the PSPs were statistically independent, where $\langle \boldsymbol{\tau}(\omega) \cdot \boldsymbol{\tau}(\omega')\rangle$ would be equal to $\langle \tau(\omega)\tau(\omega')\rangle\langle \hat{\boldsymbol{\tau}}(\omega) \cdot \hat{\boldsymbol{\tau}}(\omega')\rangle$. It is interesting that an excellent analytical approximation for the autocorrelation function of the DGD can also be obtained by comparison between the autocorrelation functions $\langle \tau(\omega)\tau(\omega')\rangle$ and $\langle \tau^2(\omega)\tau^2(\omega')\rangle$. In Fig. 10 we plot these two functions after subtracting their values at $\Delta\omega \to \infty$ and normalizing their peaks to unity. Notice the great resemblance between the two functions that is evident both on a linear and a logarithmic scale. This suggests that the autocorrelation of the DGD can be very accurately approximated by the expression

$$\langle \tau(\omega)\tau(\omega')\rangle \simeq \frac{3\pi - 8}{\pi}\left[\frac{2}{\Delta\omega^2} - \frac{6}{\Delta\omega^4}\left(1 - e^{-\Delta\omega^2/3}\right)\right] + \frac{8}{3\pi}\langle \tau^2\rangle\,, \tag{21}$$

which is obtained from Eq. (20).

Finally, it is also interesting to mention without derivation a number of more advances quantities that reflect the spectral behavior of PMD as a stochastic process in frequency. Two such quantities are the mean and the mean square value of PMD at one frequency when the value of the PMD vector at some other, offset optical frequency is known. These results are presented below in our normalized units [20]

$$\langle \boldsymbol{\tau}(\omega) | \boldsymbol{\tau}(\omega') = \boldsymbol{\tau}_0 \rangle = \frac{3\boldsymbol{\tau}_0}{\Delta\omega^2}\left[1 - \exp\left(-\frac{\Delta\omega^2}{3}\right)\right], \tag{22}$$

$$\langle \tau^2(\omega) | \boldsymbol{\tau}(\omega') = \boldsymbol{\tau}_0 \rangle = 1 + \left(\tau_0^2 - 1\right)\frac{6}{\Delta\omega^4}\left[\Delta\omega^2 - 3 + 3\exp\left(-\frac{\Delta\omega^2}{3}\right)\right]. \tag{23}$$

Notice that expression (22) has exactly the same functional dependence on optical frequency as does the autocorrelation function of the PMD vector (17). The evolution of the PMD from its value at ω' (which is $\boldsymbol{\tau}_0$) to its unconditional value (which is 0) at large ω, is therefore characterized only by the autocorrelation bandwidth. A property that is common to both conditional moments of PMD is that the functional dependence on the frequency offset is independent of $\boldsymbol{\tau}_0$. This feature is obvious in the case of the first moment (22) and it can be easily observed in the case of Eq. (23) by expressing the quantity $\left(\langle \tau^2(\omega)|\boldsymbol{\tau}_0\rangle - \langle \tau^2 \rangle\right) / \left(\tau_0^2 - \langle \tau^2 \rangle\right)$ and showing that it does not depend on $\boldsymbol{\tau}_0$. To visualize the consequences of this property, consider for example a WDM optical communications system in which one of the channels suffers from an outage due to PMD. One may think that the rate at which the PMD vector de-correlates may be affected by the fact that the DGD at the optical frequency of the impaired channel is very large. The conditional moments (22) and (23) clearly refute this stipulation. Perhaps somewhat contrary to what may be considered intuitive, expressions (22) and (23) indicate that the rate at which the PMD vector loses memory of its value does not depend on whether its modulus, the DGD, is significantly different from its average. It is perhaps relevant to discuss the value that the results (22) and (23) add to the PMD vector autocorrelation function (17) whose form is nearly identical to that of equation (22). The autocorrelation function of the DGD describes the way in which the PMD

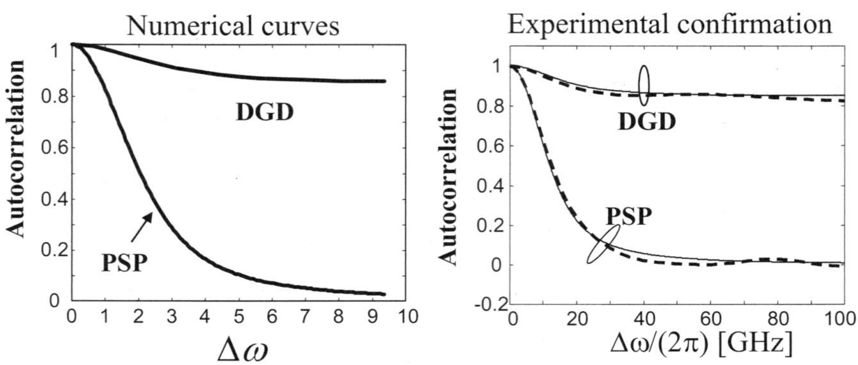

Fig. 8. On the left are the numerically extracted autocorrelation functions for the DGD and the PSPs in normalized units. On the right figure the numerical curves are compared with the results in a simulated fiber with mean DGD of 27ps (after [16]). The oscillations in the experimental curves result from a limited measurement bandwidth and they can be reproduced in simulations [19]

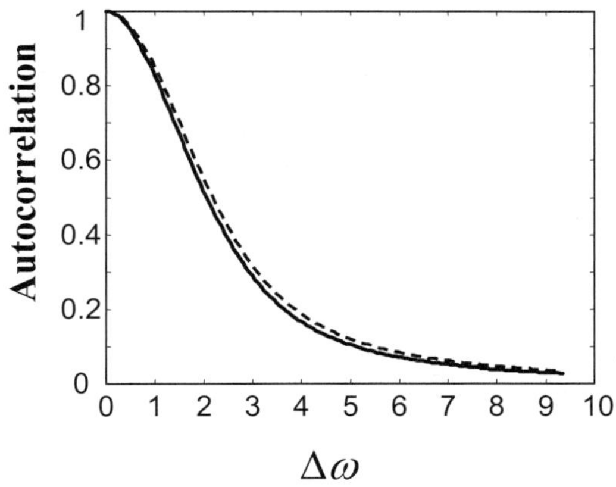

Fig. 9. Autocorrelation functions of the PSPs (solid) and of the PMD vector (17) (dashed).

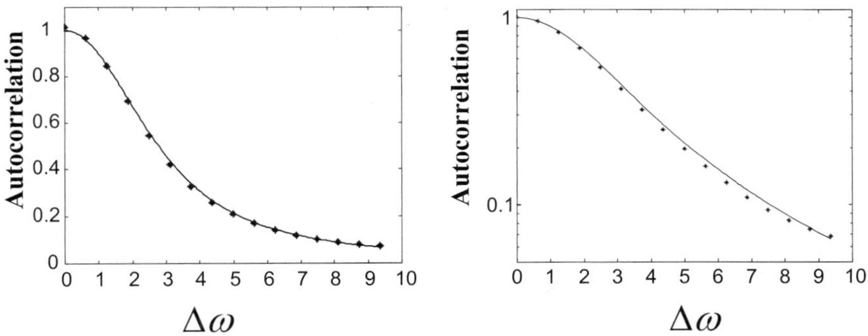

Fig. 10. The normalized autocorrelation functions of the DGD (dots) and of the square DGD (solid curve). The left-hand figure is plotted on linear scale and the right-hand figure is plotted on a logarithmic scale.

vector loses its memory on average, independent of whether its value at some particular frequency is known. It can in fact be obtained from the conditional first moment (22) through the relation $\langle \boldsymbol{\tau}_0 \cdot \boldsymbol{\tau}_\omega \rangle = \langle \boldsymbol{\tau}_0 \cdot \langle \boldsymbol{\tau}_\omega | \boldsymbol{\tau}_0 \rangle \rangle$, whereas the opposite is of course not true.

4. Assessment of the Accuracy of PMD Measurements

We have already emphasized that the only parameter determining the statistical properties of PMD in long optical fibers is the mean DGD. In this section we discuss the accuracy of its measurements [16],[21]. The experimental assessment of the mean DGD is based on the spectral averaging of the DGD that is measured in a broad range of frequencies. In the limit of very large bandwidth, this procedure yields a value that converges to the ensemble average of the DGD[7], which we have previously denoted by $\langle\tau\rangle$. Yet, when the measured bandwidth is limited, the measured frequency average is a random variable whose mean is the desired quantity $\langle\tau\rangle$ and whose standard deviation reflects the accuracy of the measurement. Using the autocorrelation function for the DGD, which we presented in the previous section, we may now obtain an explicit expression for the accuracy of the measurement. Formally, the measured frequency average is given by

$$\bar{\tau}(B) = \frac{1}{B}\int_{-B/2}^{B/2} \mathrm{d}\omega\tau(\omega)\,. \tag{24}$$

The mean of $\bar{\tau}$ is obviously equal to $\langle\tau\rangle$ and its variance is given by

$$\begin{aligned} \mathrm{Var}(\bar{\tau}(B)) &= \langle\bar{\tau}^2(B)\rangle - \langle\tau\rangle^2 \\ &= \frac{1}{B^2}\int_{-B/2}^{B/2}\mathrm{d}\omega\int_{-B/2}^{B/2}\mathrm{d}\omega'\langle\tau(\omega)\tau(\omega')\rangle - \langle\tau\rangle^2\,. \end{aligned} \tag{25}$$

Then, using Eq. (21) for $\langle\tau(\omega)\tau(\omega')\rangle$ one can obtain a lengthy, but analytical solution that is plotted in Fig. 11. Its width at half maximum is 10.4, approximately 4 times larger than the correlation bandwidth of the DGD. If we concentrate on the important limit of such measurements, where $B \gg 1$ then expression (25) approaches $\mathrm{Var}(\bar{\tau}(B)) \simeq 1.23/B$, or $\mathrm{Var}(\bar{\tau}(B)) \simeq 1.34\langle\tau\rangle/B$ in real units. This asymptotic solution is shown by the dashed curve in Figure 11 and it becomes an excellent approximation to the complete solution when $B > 30$.

An alternative method for extracting the mean DGD is to measure the frequency average of the square of the DGD and then use the relation $\langle\tau\rangle = \sqrt{8\langle\tau^2\rangle/(3\pi)}$. The measured quantity in this case is

$$\bar{\tau}^2(B) = \frac{1}{B}\int_{-B/2}^{B/2}\mathrm{d}\omega\tau^2(\omega) \tag{26}$$

and its variance can be expressed in a way similar to equation (25), except that the autocorrelation function of the square DGD (20) needs to be placed inside the integral. The asymptotic solution in this case is $\sqrt{0.5\pi(8/3)^3}/B$, or in real units $(\pi^2/\sqrt{2})\langle\tau\rangle^3/B$. If we use this value to obtain the variance of the mean DGD that we extract with this method we obtain $\mathrm{Var}[\sqrt{8\bar{\tau}^2(B)/(3\pi)}] \simeq 1.26\langle\tau\rangle/B$, slightly better than what we obtain from the frequency averaging of τ. A point to notice with the second method for measuring the mean DGD is that the obtained result is slightly biased on average with respect to the mean DGD value [22], meaning that that the mean of the measured average is not equal to the mean DGD. Nevertheless, the difference between the two values in the relevant range of parameters ($B \gg 1$) is approximately $\langle\tau\rangle - \langle\bar{\tau}(B)\rangle \simeq 0.6/B$ implying an insignificant estimation error. Figure 11 shows that the measured bandwidth should always be large enough (relative to $\langle\tau\rangle^{-1}$) in order to ensure the accuracy of the result. When the mean DGD of the link is small, as is the case in many modern communications systems operating on new fiber, an accurate measurement of PMD often requires an unrealistically broad measurement bandwidth and the development of alternative methods that optimize measurement accuracy becomes highly relevant [22].

---

[7] This is because $\langle\tau(\omega)\tau(\omega')\rangle \to \langle\tau\rangle^2$ when $\Delta\omega \to \infty$, which ensures that $\tau(\omega)$ is mean ergodic with respect to frequency.

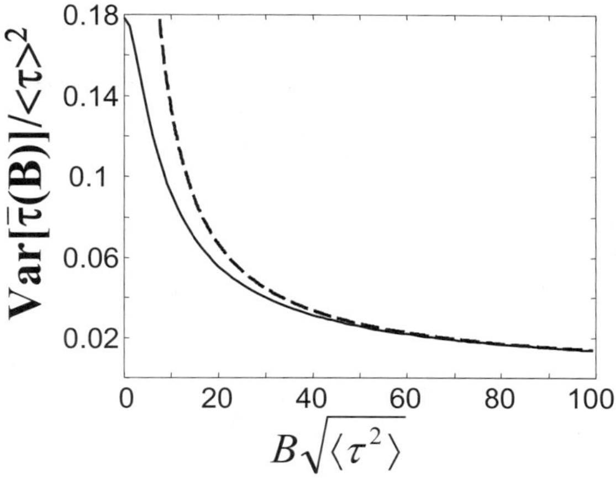

Fig. 11. Variance of the measured mean DGD when it is frequency averaged over an angular frequency bandwidth B and divided by the square of the mean DGD. The dashed curve shows the asymptotic dependence Var$(\bar{\tau}(B)) \simeq 1.34\langle\tau\rangle/B$.

5. Final Note

It is well recognized that PMD is one of the most significant limiting mechanisms in optical communications and as such it has been the subject of very extensive research. In this article we assumed the most broadly used model for PMD, where the fiber is described as a concatenation of a large number of statistically independent constant birefringence sections. Over the years this model has proven itself as extremely robust, and predictions that have been made on its basis were found to be in excellent agreement with experimental results. Nevertheless, as the field of optical communications constantly evolves, it may become necessary to consider situations that require a more refined modelling of the polarization properties of optical fibers. For example, this may be the case when the fiber link is known to have a small number of high PMD contributors that are placed within an otherwise very low PMD fiber. Studies of the evolution of PMD in time may also require a more careful description of the fiber. So far, however, it has been extremely difficult to refine the modelling of optical fibers in a way that is both consistent with the experimental reality and manageable in the mathematical sense. The search for such models will remain an active topic in the future research of PMD.

A. The Derivation of Eq. (6)

Equation (2) can be used to express $\boldsymbol{S}_{\omega+\mathrm{d}\omega,z+\mathrm{d}z}$ in the following way

$$\boldsymbol{S}_{\omega+\mathrm{d}\omega,z+\mathrm{d}z} = \mathsf{R}\left[\boldsymbol{\beta}_{\omega+\mathrm{d}\omega,z}\mathrm{d}z\right]\boldsymbol{S}_{\omega+\mathrm{d}\omega,z}, \tag{27}$$

where in order to shorten the notation we use indices to denote the dependence of the various parameters on z and ω [for example $\boldsymbol{S}_{z,\omega} = \boldsymbol{S}(z,\omega)$ and so on]. Using Eq. (3) we get

$$\boldsymbol{S}_{\omega+\mathrm{d}\omega,z+\mathrm{d}z} = \mathsf{R}\left[\boldsymbol{\beta}_{\omega+\mathrm{d}\omega,z}\mathrm{d}z\right]\mathsf{R}\left[\boldsymbol{\tau}_{\omega,z}\mathrm{d}\omega\right]\mathsf{R}^{-1}\left[\boldsymbol{\beta}_{\omega,z}\mathrm{d}z\right]\boldsymbol{S}_{\omega,z+\mathrm{d}z}. \tag{28}$$

Expanding $\boldsymbol{\beta}_{\omega+\mathrm{d}\omega,z} \simeq \boldsymbol{\beta}_{\omega,z} + \boldsymbol{\beta}'_{\omega,z}\mathrm{d}\omega$ and concentrating on the cases where $\boldsymbol{\beta}'(\omega,z)$ is parallel to $\boldsymbol{\beta}(\omega,z)$, we may rewrite the above equation in the form

$$\boldsymbol{S}_{\omega+\mathrm{d}\omega,z+\mathrm{d}z} = \mathsf{R}\left[\boldsymbol{\beta}_{\omega+\mathrm{d}\omega,z}\mathrm{d}z\right]\mathsf{R}\left[\boldsymbol{\tau}_{\omega,z}\mathrm{d}\omega\right]\mathsf{R}^{-1}\left[\boldsymbol{\beta}_{\omega+\mathrm{d}\omega,z}\mathrm{d}z\right]\mathsf{R}\left[\boldsymbol{\beta}'_z\mathrm{d}\omega\mathrm{d}z\right]\boldsymbol{S}_{\omega,z+\mathrm{d}z}, \tag{29}$$

which is equivalent to

$$\boldsymbol{S}_{\omega+\mathrm{d}\omega,z+\mathrm{d}z} = \mathsf{R}\left[\mathsf{R}\left[\boldsymbol{\beta}_{\omega+\mathrm{d}\omega,z}\mathrm{d}z\right]\boldsymbol{\tau}_{\omega,z}\mathrm{d}\omega\right]\mathsf{R}\left[\boldsymbol{\beta}'_z\mathrm{d}\omega\mathrm{d}z\right]\boldsymbol{S}_{\omega,z+\mathrm{d}z}, \tag{30}$$

where we have used the relation $\mathsf{R}(\boldsymbol{A})\mathsf{R}(\boldsymbol{B})\mathsf{R}^{-1}(\boldsymbol{A}) = \mathsf{R}(\mathsf{R}(\boldsymbol{A})\boldsymbol{B})$, that corresponds to the rotation of the entire space by the operator $\mathsf{R}(\boldsymbol{A})$. To first order in $\mathrm{d}\omega$, Eq. (30) is equivalent to

$$\boldsymbol{S}_{\omega+\mathrm{d}\omega,z+\mathrm{d}z} = \mathsf{R}\left[\mathsf{R}\left[\boldsymbol{\beta}_{\omega+\mathrm{d}\omega,z}\mathrm{d}z\right]\boldsymbol{\tau}_{\omega,z}\mathrm{d}\omega + \boldsymbol{\beta}'_z\mathrm{d}\omega\mathrm{d}z\right]\boldsymbol{S}_{\omega,z+\mathrm{d}z}. \tag{31}$$

On the other hand, we can relate $\boldsymbol{S}_{\omega+\mathrm{d}\omega,z+\mathrm{d}z}$ with $\boldsymbol{S}_{\omega,z+\mathrm{d}z}$ with the help of Eq. (3), obtaining

$$\boldsymbol{S}_{\omega+\mathrm{d}\omega,z+\mathrm{d}z} = \mathsf{R}\left[\boldsymbol{\tau}_{\omega,z+\mathrm{d}z}\mathrm{d}\omega\right]\boldsymbol{S}_{\omega,z+\mathrm{d}z}. \tag{32}$$

Comparison between Eqs. (31) and (32) yields Eq. (6).

References

1. H. Kogelnik, R. M. Jopson and L.E. Nelson, Polarization mode dispersion. In: *Optical Fiber Telecommunications IVb: Systems and Impairments*, edited by I.P. Kaminow and T. Li (Academic Press, San Diego, 2002), Chap. 15.
2. M. Born, E. Wolf, *Principles of Optics*, 6th Ed. (Pergamon, New York, 1986)
3. R.M.A. Azzam, N. M. Bashara, *Ellipsometry and Polarized Light* (NHPL, Elsevier, New York, 1977)
4. D.S. Kliger, J. W. Lewis, C.E. Randall, *Polarized Light in Optics and Spectroscopy* (Academic Press, New York, 1990)
5. C. D. Poole, R. E. Wagner, Electron. Lett., **22**, 1029 (1986)
6. G. J. Foschini, C. D. Poole, J. Lightwave Technol., **9**, 1439 (1991)
7. A. Eyal, W. K. Marshall, A. Yariv, M. Tur, Electron. Lett., **19**, 1658 (1999)
8. H. Kogelnik, J. Opt. Fiber Commun. Rep. **1**, 107–122 (2004) DOI: 10.1007/s10297-004-0005-1
9. R. Ulrich, A. Simon, Appl. Opt., **18**, 2241 (1979)
10. C. D. Poole, J. H. Winters, J. A. Nagel, Opt. Lett., **16**, 372 (1991)
11. J. P. Gordon, H. Kogelnik, Proc. Natl. Acad. Sci. USA, **97**, 4541 (2000)
12. N. J. Frigo, IEEE J. Quantum Electron., **22**, 2131 (1986)

13. A. Galtarossa, L. Palmieri, M. Schiano, T. Tambosso, Opt. Lett., **26**, 962 (2001)
14. J. P. Gordon, J. Opt. Fiber Commun. Rep., **1**, 210–217 (2004) DOI: 10.1007/s10297-004-0003-3
15. F. Curti, B. Daino, G. De Marchis, F. Matera, J. Lightwave Technol., **9**, 1162 (1990)
16. M. Shtaif, A. Mecozzi, Opt. Lett., **25**, 707 (2000)
17. C. W. Gardiner, *Handbook of Stochastic Methods*, Springer Series in Synergetics 13 (Springer, Berlin/Heidelberg/New York, 1985)
18. M. Karlsson, J. Brentel, Opt. Lett., **24**, 939 (1999)
19. M. Shtaif, A. Mecozzi, J. Nagel, IEEE Photon. Technol. Lett., **12**, 53 (2000)
20. A. Mecozzi, M. Shtaif, IEEE Photon. Technol. Lett., **15**, December 2003, available at http://ieeexplore.ieee.org.
21. N. Gisin, B. Gisin, J. P. Won der Weid, R. Passy, IEEE Photon. Technol. Lett., **8**, 1671 (1996)
22. M. Boroditsky, M. Brodsky, P. Magill, N.J. Frigo, M. Shtaif, "Second order PMD statistics analyses improves the accuracy of mean DGD estimates," to be published in IEEE Photon. Technol. Lett. (2004)

DOI: 10.1007/s10297-004-0003-3
Originally published in J. Opt. Fiber. Commun. Rep. 1, 210–217 (2004)

Statistical properties of polarization mode dispersion

James P. Gordon

Consultant, Bell Labs, Lucent Technologies, Holmdel, NJ 07733, USA
Email: jamespgordon@att.net

Abstract. Polarization mode dispersion (PMD) is one of the effects which limit bit rates in optical communications systems. The statistics of PMD is complicated because it is multi-dimensional, however, many useful results have been obtained. Much work has been done, but the subject is not yet closed, as witnessed by the existence of this conference. In this talk I will review the way some statistical equations are treated. I will briefly discuss the methods of Ito and Lax, and their application to the PMD problem.

1. The Basic Equation

In this work I will mostly follow the nomenclature used in the review paper of Gordon and Kogelnik [1]. In Stokes space, the PMD vector is labeled $\boldsymbol{\tau}$, the birefringence vector is labeled $\boldsymbol{\beta}$, and a single letter subscript such as ω indicates differentiation. The length of $\boldsymbol{\tau}$, here simply τ, is the differential group delay (DGD). It is usually labeled $\Delta\tau$ in the literature. Unit vectors are distinguished by a hat.

The basic equation for the propagation of the PMD vector is [1]

$$\frac{\partial \boldsymbol{\tau}}{\partial z} = \boldsymbol{\beta}_\omega + \boldsymbol{\beta} \times \boldsymbol{\tau}. \tag{1}$$

In Eq. (1), z is the distance along the fiber, $\boldsymbol{\beta}_\omega$ is the local change of the PMD vector, and the cross product $\boldsymbol{\beta} \times \boldsymbol{\tau}$ describes the rotation of the PMD vector due to the birefringence of the fiber. The parameters $\boldsymbol{\beta}$ and $\boldsymbol{\beta}_\omega$ vary with both z and frequency. The frequency range of interest is usually small enough that one may consider $\boldsymbol{\beta}_\omega$ to be independent of frequency.

Many treatments of PMD use Eq. (1) as it stands. However, it is useful to simplify it by eliminating the birefringence rotation R_0 occuring at the central frequency ω_0.

To do this, we transform all the vectors of Eq. (1) using $\boldsymbol{v}(z) = R_0(z, 0)\boldsymbol{v}'(z)$, where $R_0(z, 0)$ is the 3 by 3 rotation matrix satisfying the equation

$$\frac{\partial R_0(z,0)}{\partial z} = \boldsymbol{\beta}_0(z) \times R_0(z,0), \tag{2}$$

where $\boldsymbol{\beta}_0$ is the birefringence vector at frequency ω_0. This transformation yields the modified equation

$$\frac{\partial \boldsymbol{\tau}'}{\partial z} = \boldsymbol{\beta}'_\omega + (\boldsymbol{\beta}' - \boldsymbol{\beta}'_0) \times \boldsymbol{\tau}'. \tag{3}$$

The final step is to use the first order approximation $\boldsymbol{\beta}' - \boldsymbol{\beta}'_0 = (\omega - \omega_0)\boldsymbol{\beta}'_\omega$ to get

$$\frac{\partial \boldsymbol{\tau}}{\partial z} = \boldsymbol{\beta}_\omega + \Delta\omega\, \boldsymbol{\beta}_\omega \times \boldsymbol{\tau}, \tag{4}$$

where $\Delta\omega = \omega - \omega_0$, and where we have suppressed the primes. This is the stochastic equation we will be concerned with. In the fibers used for long distance communications, $\boldsymbol{\beta}_\omega$ is some random function of distance, dependent on fiber imperfections and stresses.

To get ahead, we need a working assumption about the statistics of $\boldsymbol{\beta}_\omega$. For the purposes of this paper, we shall assume that the cartesian components of $\boldsymbol{\beta}_\omega$ satisfy the relations $\langle \beta_{\omega i}(z) \rangle = 0$ and

$$\langle \beta_{\omega i}(z)\, \beta_{\omega j}(z') \rangle = \frac{1}{3} D\, \delta(i,j)\, g(z'-z), \tag{5}$$

where the angular brackets signify the statistical mean (or expectation) value, and where D is a diffusion constant whose meaning will appear below. Also $g(z'-z)$ is a symmetric shape function normalized so that

$$\int_{-\infty}^{\infty} dz'\, g(z'-z) = 1.$$

The coherence length of $\boldsymbol{\beta}_\omega$ may then be defined as

$$L_c = \int_{-\infty}^{\infty} dz'\, |z'-z|\, g(z'-z). \tag{6}$$

Note that D and L_c may be slowly or piecewise dependent on z, on a length scale much larger than L_c. These assumptions lead to the relations

$$\langle \boldsymbol{\beta}_\omega(z) \cdot \boldsymbol{\beta}_\omega(z') \rangle = D\, g(z'-z) \tag{7}$$

and

$$\langle \boldsymbol{\beta}_\omega(z) \times \boldsymbol{\beta}_\omega(z') \rangle = 0. \tag{8}$$

For fibers long compared to L_c it is reasonable to treat $g(z'-z)$ as a Dirac delta function, that is, $g(z'-z) \rightarrow \delta(z'-z)$. The assumption about the statisical properties of $\boldsymbol{\beta}_\omega$ embodied in Eq. (5) is not locally correct, particularly in view of the transformation leading to Eq. (4). However, it is analytically simple and leads to many accepted results. The implication is that only the diffusion constant and the correlation length are of much significance.

2. Statistical Equations

At this point I would like to pause for a short discussion of a simple statistical equation that may be helpful in understanding the ways such equations are treated. A one dimensional analogy to Eq. (4) is

$$\frac{\partial X}{\partial t} = F(X)\, n(t). \tag{9}$$

The exercise is to integrate this equation over a small interval Δt, which is what one might want to do if $n(t)$ is a random function having a short auto-correlation time. Formally, the result is

$$\Delta X = \int_t^{t+\Delta t} dt'\; F(X(t'))\, n(t'). \tag{10}$$

The Lax procedure [2] is to express $F(X(t'))$ in Eq. (10) as $F(X(t))$ plus its integral from t to t', to iterate that procedure on the ensuing relation, and so on. The eventual result takes the form

$$\Delta X = F\;\Delta W + \frac{1}{2} F_X F\,(\Delta W)^2 + \frac{1}{3!}(F_X F)_X F\,(\Delta W)^3 + \cdots, \tag{11}$$

where F and its X derivatives are evaluated at the the beginning of the interval, and

$$\Delta W = \int_t^{t+\Delta t} dt'\; n(t').$$

Two points are relevant here. First, this expansion is exact, and independent of the particular functions $F(X)$ and $n(t)$. One can test it using $n(t) = 1$ or $F(X) = X$, for example. Second, since F and its X derivatives are evaluated at the beginning of the interval, they are uncorrelated with ΔW, which involves later values of $n(t)$. If $n(t)$ is a random function whose mean is zero, and whose correlation time is short compared to Δt, with diffusion constant D, we can take $\langle n(t)\, n(t')\rangle = D\,\delta(t - t')$. In this case $\langle \Delta W\rangle = 0$ and $\langle (\Delta W)^2\rangle = D\;\Delta t$, and so we get from Eq. (11) the relation

$$\Delta\langle X\rangle = \frac{1}{2} D\;\langle F_X F\rangle\;\Delta t + \cdots. \tag{12}$$

If the interval Δt is short enough, this can be treated as a differential equation.

The Ito procedure [3] is to approximate Eq. (11) by the iterative expression

$$\Delta X = F(X + \frac{1}{2}\Delta X)\;\Delta W. \tag{13}$$

Note that if X is taken at the beginning of the interval $(t, t + \Delta t)$, then $X + (1/2)\Delta X$ is near its center. Assuming that ΔX is small, Eq. (13) expands into

$$\Delta X = F\;\Delta W + \frac{1}{2} F_X F\;(\Delta W)^2 + \cdots. \tag{14}$$

Thus, to second order in ΔW, it mimics Eq. (11). The higher order terms differ.

If $F(X) = 0$ at $t = 0$, then in its first iteration, the Lax method [2] effectively replaces $F(X)$ in Eq. (9) by the integral of its motion from 0 to t. This gives

$$\frac{\partial X}{\partial t} = \int_0^t dt' \, (F_X F)_{t=t'} \, n(t') n(t). \tag{15}$$

So far this is exact. Suppose that the autocorrelation of $n(t)$ approximates a delta function. Taking mean values, and noting that the quantity $(F_X F)_{t=t'}$ varies slowly and has no correlation with $n(t)$ since it comes at an earlier time, we can replace $\langle (F_X F)_{t=t'} \, n(t') n(t) \rangle$ in first order by $\langle F_X F \rangle \langle n(t) n(t') \rangle$. Since the integral in Eq. (15) takes in only half of the delta function, the resulting equation is

$$\frac{\partial}{\partial t} \langle X \rangle = \frac{1}{2} D \langle F_X F \rangle, \tag{16}$$

which is the differential form of Eq. (12). I personally prefer the Lax method, since it is more straightforward, and is more versatile.

3. Statistics of $\boldsymbol{\tau}$

First we look at the statistics of $\boldsymbol{\tau}$. Since only one frequency is involved here, we can set $\Delta\omega = 0$. Equation (4) reduces simply to

$$\frac{\partial \boldsymbol{\tau}}{\partial z} = \boldsymbol{\beta}_\omega. \tag{17}$$

If $\boldsymbol{\beta}_\omega$ is a random function such as described above, this equation describes a random walk process in three dimensions. There is no rotation term, as we have eliminated the rotation at ω_0. A recurring parameter in what follows is the expectation value of $\langle \tau^2 \rangle \equiv \langle \boldsymbol{\tau} \cdot \boldsymbol{\tau} \rangle$. To find it, the Lax procedure is to take the z derivative, take the mean value, and then do the z integral. Thus we start with

$$\frac{\partial}{\partial z} \tau^2 = 2 \boldsymbol{\tau} \cdot \boldsymbol{\beta}_\omega. \tag{18}$$

Next, we replace $\boldsymbol{\tau}$ on the right-hand side by the integral of its motion using Eq. (17) and take the mean value using Eq. (7). The result is

$$\frac{\partial}{\partial z} \langle \tau^2 \rangle = 2D \int_0^z dz' \, g(z' - z). \tag{19}$$

If $g(z' - z)$ has the exponential form $(1/(2L_c)) \exp(-|z' - z|/L_c)$ we get

$$\frac{\partial}{\partial z} \langle \tau^2 \rangle = D \left[1 - \exp(-z/L_c) \right] \tag{20}$$

Integration of this result gives the standard formula [4, 5]

$$\langle \tau^2(z) \rangle = D \left[z - L_c (1 - \exp(-z/L_c)) \right]. \tag{21}$$

For $z >> L_c$ this simply reduces to $\langle \tau^2(z) \rangle = Dz$, which is the result we would have obtained by using the delta function auto-correlation function for $\boldsymbol{\beta}_\omega$. This result really defines the diffusion constant D. In what follows, we will use the delta function correlation in the interests of simplicity. The premise is that the fibers must be much longer than L_c.

Characteristic functions (CF's) are very important to the discussion of statistics. They are moment generating functions, and can be transformed to get probability distribution functions (PDF's). In the case of $\boldsymbol{\tau}$, we may expect that the PDF will be a three dimensional Gaussian, as a result of its random walk. The CF in this case is

$$C(\boldsymbol{\lambda}) = \langle \exp(i\boldsymbol{\lambda} \cdot \boldsymbol{\tau}) \rangle = \langle Q \rangle, \tag{22}$$

where $\boldsymbol{\lambda}$ is an expansion parameter and $Q \equiv \exp(i\boldsymbol{\lambda} \cdot \boldsymbol{\tau})$. Note that one can get, from the CF, moments of the various components of $\boldsymbol{\tau}$ by differentiation with respect to the corresponding components of $\boldsymbol{\lambda}$, followed by setting $\boldsymbol{\lambda}$ to zero. The easiest way to do this is often to expand the CF as a power series in the components of $\boldsymbol{\lambda}$.

To evaluate this CF, we go through the same procedure as before, i.e, take the z derivative, take the mean value, and integrate. The z derivative of Q is $i\lambda \cdot \boldsymbol{\beta}_\omega Q$, so we get

$$\frac{\partial}{\partial z} C(\boldsymbol{\lambda}) = i \, \langle \boldsymbol{\lambda} \cdot \boldsymbol{\beta}_\omega Q \rangle. \tag{23}$$

As before, we express Q as the integral of it's motion, arriving at

$$\frac{\partial}{\partial z} C(\boldsymbol{\lambda}) = - \int_0^z dz' \, \langle (\boldsymbol{\lambda} \cdot \boldsymbol{\beta}_\omega(z))(\boldsymbol{\lambda} \cdot \boldsymbol{\beta}_\omega(z'))Q(z') \rangle. \tag{24}$$

Using the correlation Eq. (5) with the delta function approximation gives

$$\frac{\partial}{\partial z} C(\boldsymbol{\lambda}) = -\frac{1}{6} D \, \lambda^2 \, C(\boldsymbol{\lambda}), \tag{25}$$

which integrates easily to give

$$C(\boldsymbol{\lambda}) = \exp(-\frac{1}{6}\lambda^2 \langle \tau^2 \rangle), \tag{26}$$

where we have used $Dz = \langle \tau^2 \rangle$, and $\lambda^2 \equiv \boldsymbol{\lambda} \cdot \boldsymbol{\lambda}$. This indeed is the CF representing a three-dimensional (3-D) Gaussian PDF. To see this, rewrite Eq. (22) in the form

$$C(\boldsymbol{\lambda}) = \int^{(3)} d^{(3)}\boldsymbol{\tau} \; P(\boldsymbol{\tau}) \exp(i\boldsymbol{\lambda} \cdot \boldsymbol{\tau}), \tag{27}$$

where the 3-D integral is complete, and $P(\boldsymbol{\tau})$ is the 3-D PDF of $\boldsymbol{\tau}$. Equation (27) may be inverted by Fourier transform, using Eq. (26), to yield

$$P(\boldsymbol{\tau}) = \left(\frac{3}{2\pi \langle \tau^2 \rangle} \right)^{3/2} \exp\left(-\frac{3\tau^2}{2\langle \tau^2 \rangle} \right), \tag{28}$$

representing the 3-D Gaussian distribution of the vector $\boldsymbol{\tau}$. To get the PDF of the magnitude of $\boldsymbol{\tau}$, one simply multiplies the PDF of Eq. (28) by $4\pi\tau^2$, yielding the well-known Maxwellian distribution. The mean value of τ is given by $\langle \tau \rangle = \sqrt{8\langle \tau^2 \rangle / 3\pi}$, and is conventionally labeled τ_{DGD}.

4. Statistics of $\boldsymbol{\tau}(\omega_0)$ and $\boldsymbol{\tau}(\omega)$

This section addresses the question of the combined statistics of $\boldsymbol{\tau}$ at two nearby frequencies. Given a measured value of $\boldsymbol{\tau}(\omega_0)$, one might want to know the conditional statistics of $\boldsymbol{\tau}(\omega)$, at a nearby frequency. This question has not been resolved; although one can get an equation for the propagation of the appropriate 6-D characteristic function, it has not been integrated. What has been done is to solve the problem of $\boldsymbol{\tau}$ and $\boldsymbol{\tau}_\omega$, both at the same frequency.

For the purposes of this paragraph, let $\boldsymbol{\tau}_0 \equiv \boldsymbol{\tau}(\omega_0)$, and $\boldsymbol{\tau} \equiv \boldsymbol{\tau}(\omega_0 + \Delta\omega)$. One important result is the correlation of $\boldsymbol{\tau}$ with $\boldsymbol{\tau}_0$ [6, 7]. Using our standard scheme, we have

$$\frac{\partial}{\partial z}\langle \boldsymbol{\tau}\cdot\boldsymbol{\tau}_0\rangle = \langle (\boldsymbol{\beta}_\omega + \Delta\omega\, \boldsymbol{\beta}_\omega \times \boldsymbol{\tau})\cdot\boldsymbol{\tau}_0 + \boldsymbol{\tau}\cdot\boldsymbol{\beta}_\omega\rangle = \langle (\boldsymbol{\tau}+\boldsymbol{\tau}_0+\Delta\omega\, \boldsymbol{\tau}\times\boldsymbol{\tau}_0)\cdot\boldsymbol{\beta}_\omega\rangle. \tag{29}$$

Taking the mean value here is a little more complicated than before, however the method is the same. Using Eq. (4), we find that $\langle \boldsymbol{\tau}\cdot\boldsymbol{\beta}_\omega\rangle = \langle \boldsymbol{\tau}_0\cdot\boldsymbol{\beta}_\omega\rangle = D/2$. For $\langle \boldsymbol{\tau}\times\boldsymbol{\tau}_0\cdot\boldsymbol{\beta}_\omega\rangle$, we have

$$\langle \boldsymbol{\tau}\times\boldsymbol{\tau}_0\cdot\boldsymbol{\beta}_\omega\rangle = \int_0^z dz'\langle \left(\frac{\partial\boldsymbol{\tau}}{\partial z}\times\boldsymbol{\tau}_0 + \boldsymbol{\tau}\times\frac{\partial\boldsymbol{\tau}_0}{\partial z}\right)_{z=z'}\cdot\boldsymbol{\beta}_\omega(z)\rangle. \tag{30}$$

From Eq. (30), it can be seen that in effect one takes the mean of $\boldsymbol{\beta}_\omega$ separately with each other z-dependent term. The result, again using Eq. (4), is $\langle \boldsymbol{\tau}\times\boldsymbol{\tau}_0\cdot\boldsymbol{\beta}_\omega\rangle = -\Delta\omega\,(D/3)\,\langle \boldsymbol{\tau}\cdot\boldsymbol{\tau}_0\rangle$. Thus we get the relation

$$\frac{\partial}{\partial z}\langle \boldsymbol{\tau}\cdot\boldsymbol{\tau}_0\rangle = D\left(1 - \frac{(\Delta\omega)^2}{3}\right)\langle \boldsymbol{\tau}\cdot\boldsymbol{\tau}_0\rangle). \tag{31}$$

Integration of this equation yields

$$\langle \boldsymbol{\tau}\cdot\boldsymbol{\tau}_0\rangle = \frac{3}{(\Delta\omega)^2}\left(1 - \exp\left(-\frac{(\Delta\omega)^2}{3}\langle\tau^2\rangle\right)\right), \tag{32}$$

where again we have used $Dz = \langle\tau^2\rangle$. This relation describes the frequency separation needed to get independent values of $\boldsymbol{\tau}$.

Finally, we consider the 6-D characteristic function for $\boldsymbol{\tau}$ and $\boldsymbol{\tau}_\omega$ [4]. Since $\boldsymbol{\tau}$ and $\boldsymbol{\tau}_\omega$ are evaluated at the same frequency, we can set $\Delta\omega = 0$. In this case the CF is

$$C(\boldsymbol{\lambda},\boldsymbol{\eta}) = \langle \exp(i\boldsymbol{\lambda}\cdot\boldsymbol{\tau} + i\boldsymbol{\eta}\cdot\boldsymbol{\tau}_\omega)\rangle = \langle Q\rangle, \tag{33}$$

where now $Q = \exp(i\boldsymbol{\lambda}\cdot\boldsymbol{\tau} + i\boldsymbol{\eta}\cdot\boldsymbol{\tau}_\omega)$. From the equation of motion for $\boldsymbol{\tau}$, we have $\partial\boldsymbol{\tau}/\partial z = \boldsymbol{\beta}_\omega$, and $\partial\boldsymbol{\tau}_\omega/\partial z = \boldsymbol{\beta}_\omega\times\boldsymbol{\tau}$. Thus we get

$$\frac{\partial Q}{\partial z} = i(\boldsymbol{\lambda} - \boldsymbol{\eta}\times\boldsymbol{\tau})\cdot\boldsymbol{\beta}_\omega Q. \tag{34}$$

Using the same scheme as in section 3, we arrive at

$$\frac{\partial}{\partial z}C(\boldsymbol{\lambda},\boldsymbol{\eta}) = -\frac{1}{6}D\,\langle (\boldsymbol{\lambda} - \boldsymbol{\eta}\times\boldsymbol{\tau})\cdot(\boldsymbol{\lambda} - \boldsymbol{\eta}\times\boldsymbol{\tau})Q\rangle. \tag{35}$$

The last step is to use $\boldsymbol{\tau}Q = -i\boldsymbol{\nabla}_\lambda Q$ to get

$$\frac{\partial}{\partial z}C(\boldsymbol{\lambda},\boldsymbol{\eta}) = -\frac{1}{6}D\left(\lambda^2 + i2\boldsymbol{\lambda}\times\boldsymbol{\eta}\cdot\boldsymbol{\nabla}_\lambda - (\boldsymbol{\eta}\times\boldsymbol{\nabla}_\lambda)^2\right)C(\boldsymbol{\lambda},\boldsymbol{\eta}). \tag{36}$$

This equation has a solution of the form $C = \exp(-A\lambda^2 - B(\boldsymbol{\lambda}\cdot\hat{\eta})^2 - G)$. The result is [4]

$$C(\boldsymbol{\lambda}, \boldsymbol{\eta}) = \operatorname{sech}\gamma \, \exp\left(-\frac{1}{2\eta}\left(\lambda^2 \tanh\gamma + (\boldsymbol{\lambda}\cdot\hat{\eta})^2 (\gamma - \tanh\gamma)\right)\right), \tag{37}$$

where $\gamma = \eta\,\langle\tau^2\rangle/3$, where η is the magnitude of $\boldsymbol{\eta}$, and where $\hat{\eta}$ is the unit vector $\boldsymbol{\eta}/\eta$. This 6-D CF can be manipulated in a variety of ways. If we take $\boldsymbol{\eta}$ to zero, it reverts to the CF corresponding to $\boldsymbol{\tau}$, Eq. (26). If we take $\boldsymbol{\lambda}$ to zero, we obtain the CF corresponding to $\boldsymbol{\tau}_\omega$ averaged over the PDF of $\boldsymbol{\tau}$, namely $C(\boldsymbol{\eta}) = \langle\exp(i\boldsymbol{\eta}\cdot\boldsymbol{\tau}_\omega)\rangle = \operatorname{sech}\gamma$. Since the Fourier transform of a hyperbolic secant is a hyperbolic secant, this result shows that the cartesian components of $\boldsymbol{\tau}_\omega$ have PDF's of hyperbolic secant shape. The corresponding 3-D PDF is

$$P(\boldsymbol{\tau}_\omega) = \frac{1}{8a^2\tau_\omega}\tanh\left(\frac{\pi}{2a}\tau_\omega\right)\operatorname{sech}\left(\frac{\pi}{2a}\tau_\omega\right), \tag{38}$$

where $a = \langle\tau^2\rangle/3$. Of more significance, perhaps, is that since the exponent in $C(\boldsymbol{\lambda}, \boldsymbol{\eta})$ is quadratic in the components of $\boldsymbol{\lambda}$, it can be inverted with respect to $\boldsymbol{\lambda}$, leading to the CF of $\boldsymbol{\eta}$ conditional on the value of $\boldsymbol{\tau}$. The result of this exercise is [4]

$$C(\boldsymbol{\eta}/\boldsymbol{\tau}) = \frac{a\eta}{\sinh(a\eta)}\exp\left(-\frac{1}{2a}(\hat{\eta}\times\boldsymbol{\tau})^2\left(\frac{a\eta}{\tanh(a\eta)} - 1\right)\right). \tag{39}$$

This CF can be considered as the product of two CF's, corresponding to two statistically independent elements of $\boldsymbol{\tau}_\omega$. If we take $\boldsymbol{\eta}\cdot\boldsymbol{\tau}_\omega = \boldsymbol{\eta}\cdot(\boldsymbol{\tau}_{\omega1} + \boldsymbol{\tau}_{\omega2})$, it can be seen that the CF becomes the product of two CF's if $\boldsymbol{\tau}_{\omega1}$ and $\boldsymbol{\tau}_{\omega2}$ are statistically independent. In Eq. (39), we can take $C_1(\boldsymbol{\eta}/\boldsymbol{\tau}) = a\eta/\sinh(a\eta)$, and $C_2(\boldsymbol{\eta}/\boldsymbol{\tau})$ as the exponential term. Here C_1 is the CF corresponding to $\boldsymbol{\tau}_{\omega1}$, and one can see that $\boldsymbol{\tau}_{\omega1}$ is independent of the value of $\boldsymbol{\tau}$. It is dependent only on $\langle\tau^2\rangle$. Also, C_2 is the CF corresponding to $\boldsymbol{\tau}_{\omega2}$ and shows that $\boldsymbol{\tau}_{\omega2}$ is perpendicular to $\boldsymbol{\tau}$ and proportional to τ. It has been reasonably argued [8] that $\boldsymbol{\tau}_{\omega1}$ comes mainly from the early part of the fiber and becomes independent of $\boldsymbol{\tau}$ after sufficient propagation, while $\boldsymbol{\tau}_{\omega2}$ comes mainly from the later part of the fiber, since $\boldsymbol{\tau}_\omega$ is created perpendicular to $\boldsymbol{\tau}$. The moments of these distributions are easily found by expanding their CF's in power series of their arguments. In particular, if we take $\boldsymbol{\eta} = \eta_1\hat{e}_1 + \eta_2\hat{e}_2 + \eta_3\hat{e}_3$, where the $\hat{e}_i$ form an orthogonal set of unit vectors, with $\hat{e}_3 = \hat{\tau}$, then the second moments of the distributions of the components of $\boldsymbol{\tau}_\omega$ are

$$\begin{aligned}\langle\tau_{\omega11}^2/\boldsymbol{\tau}\rangle = \langle\tau_{\omega12}^2/\boldsymbol{\tau}\rangle = \langle\tau_{\omega13}^2/\boldsymbol{\tau}\rangle &= \frac{1}{27}\langle\tau^2\rangle^2 \\ \langle\tau_{\omega21}^2/\boldsymbol{\tau}\rangle = \langle\tau_{\omega22}^2/\boldsymbol{\tau}\rangle &= \frac{1}{9}\tau^2\langle\tau^2\rangle, \quad \langle\tau_{\omega23}^2/\boldsymbol{\tau}\rangle = 0.\end{aligned} \tag{40}$$

PDFs for the components of $\boldsymbol{\tau}_\omega$ parallel and perpendicular to $\boldsymbol{\tau}$ have been obtained from the CF of Eq. (39) [4]. The parallel component is the parallel component of $\boldsymbol{\tau}_{\omega1}$, while the perpendicular component is the sum of the perpendicular components of $\boldsymbol{\tau}_{\omega1}$ and $\boldsymbol{\tau}_{\omega2}$.

In summary, I have attempted to demonstrate how the statistical properties of PMD are derived in a straightforward way. The main thing one might object to is the assumption about the correlation properties of $\boldsymbol{\beta}_\omega$, which is obviously inaccurate. More realistic assumptions change the local statistics, but for the long fibers of importance to communications little difference is to be expected.

The author is pleased to acknowledge good advice and stimulation from Herwig Kogelnik and Mark Shtaif.

References

1. J. P. Gordon and H. Kogelnik, PMD Fundamentals: Polarization Mode Dispersion in Optical Fibers, Proc. Natl. Acad. Sci. USA, **97**, 4541, (2000)
2. M. Lax, Classical Noise IV: Langevin Methods, Rev. Mod. Phys., **38**, 541, (1966)
3. C. W. Gardiner, *Handbook of Stochastic Methods*, 2nd ed., Vol. 13 of Springer Series in Synergetics, Springer-Verlag, Berlin (1985)
4. G. J. Foschini et.al., Statistics of Second Order PMD Polarization, J. Lightwave Techn. **19**, 1882 (2001)
5. P. K. A. Wai and C. R. Menyuk, Polarization Mode Dispersion, Decorrelation, and Diffusion in Optical Fibers with Randomly Varying Birefringence. IEEE J. Lightwave Techn. **14**, 148 (1996)
6. M. Karlsson and J. Brentel, Autocorrelation Function of the Polarization-mode Dispersion Vector. Opt. Lett. **24** 939 (1999)
7. M. Shtaif, A. Mecozzi and J. Nagel, Mean-square Magnitude of All Orders of PMD and the Relation with the Bandwidth of the Principal States, IEEE Photon. Lett. **12**, 53 (2000)

DOI: 10.1007/s10297-004-0004-2
Originally published in J. Opt. Fiber. Commun. Rep. 1, 345–354 (2004)

Three Representations of Polarization Mode Dispersion

H. A. Haus and P. B. Phua

Department of Electrical Engineering and Computer Science
and Research Laboratory of Electronics
Massachusetts Institute of Technology
Cambridge, Massachusetts 02139 USA
Email: haus@mit.edu

Abstract. Polarization Mode Dispersion (PMD) can be described in Jones space, Stokes space, or energy space. The three representations are compared and it is shown that equivalent information required for the prediction of the transmission eye-diagram is contained in the Jones space and energy space descriptions. Isotropic dispersion is suppressed in the Stokes space formulation.

1. Introduction

The Polarization Mode Dispersion (PMD) of a fiber is usually described in Stokes space[1]. The intersection with the Poincaré sphere of a straight line along the PMD vector $\mathbf{\Omega}$ defines the two orthogonal principal polarizations. The magnitude of the vector defines the difference of the delays experienced by the two principal polarizations. Nth order PMD is characterized by the Nth derivative of the PMD vector with respect to frequency.

Another way of describing PMD is in Jones space. The Stokes vectors are constructed from products of the Jones vectors (column matrices). The relation between input and output is written in terms of the two-by-two transfer matrix $\mathbb{T}$. The vector $\mathbf{\Omega}$ is related to the product of $d\mathbb{T}/d\omega$ and $\mathbb{T}^\dagger$. Nth-order dispersion is described by derivatives up to $N+1$ order of $\mathbb{T}$. The eye-diagram at the output is determined from the transfer function $\mathbb{T}^\dagger(\omega)\mathbb{T}(\omega')$.

The physical origin of PMD is the difference of group velocities of the modes in the two principal polarizations. Hence, the concept of group velocity is central to the theory of PMD. The group velocity is the velocity of energy propagation. Hence one must expect a close connection between the characterization of energy, and the PMD vector and its derivatives. Energy is a quadratic form of the field vectors. The characterization of PMD in terms of group velocity and energy may be dubbed a description of PMD in "Poynting space."

This paper reviews briefly the various definitions in the three formulations: Stokes space, Jones space, and Poynting space. We establish the correspondences among these three formulations. As is well known, the Stokes space description misses isotropic dispersive effects. Full information is contained in all three formulations for the construction of a PMD compensator. The evaluation of the eye-diagram in the presence of higher order dispersion requires knowledge of the Jones transformation matrix.

In their elegant paper, Gordon and Kogelnik[2] used kets to denote column matrices in Jones space, such as $|a\rangle$. The operations employed are standard matrix multiplications in 2-space and vector dot-products in 3-space. We are compelled to use column matrices in Jones space, which we denote by double-strike lower case letters (e.g. $\mathbb{a}$). 2×2 matrices will be denoted by double-strike capital letters (e.g. $\mathbb{T}$), vectors by boldface letters, and tensors by two over-bars $^{=}$.

2. Stokes Space and Jones Space

In Stokes space, an input polarization vector $\mathbf{s}_a$ produces an output polarization vector $\mathbf{s}_b$. When the frequency is raised to $\omega + d\omega$, and the input polarization vector is assumed frequency independent, the change of the output polarization vector is

$$\frac{d}{d\omega}\mathbf{s}_b = \mathbf{\Omega} \times \mathbf{s}_b. \tag{2.1}$$

The Stokes PMD vector $\mathbf{\Omega}$ is a function of frequency. N-th order dispersion is characterized by the N-th derivative of the vector with respect to frequency.

Another description of PMD is in Jones space. Denote the column matrix describing the output by $\mathbb{b}$, and the column matrix of the input by $\mathbb{a}$. The Jones matrix $\mathbb{T}$ relates input and output:

$$\mathbb{b} = \mathbb{T}\mathbb{a}. \tag{2.2}$$

It is customary to normalize the input to unit power. We consider only the loss-free case in which power is conserved. One must have

$$\mathbb{T}^\dagger\mathbb{T} = \mathbb{T}\mathbb{T}^\dagger = \mathbb{1}. \tag{2.3}$$

$\mathbb{T}$ is unitary. The unitarity of $\mathbb{T}$, Eq. (2.3), provides four real scalar equations as constraints for the eight real scalars required to specify a complex two-by-two matrix. Thus, there are four real parameters left. The most general form of the matrix is

$$\mathbb{T} = \begin{bmatrix} re^{-j\theta} & jte^{j\psi} \\ -jte^{-i\psi} & re^{j\theta} \end{bmatrix} \exp(-j\phi), \tag{2.4}$$

with $t = \sqrt{1 - r^2}$. The matrix has four free parameters r, θ, ψ, and ϕ. The common phase factor $\exp(-j\phi)$ is suppressed in the characterization of PMD in Stokes space in which only relative delays are of interest. Note that energy conservation gives

$$\mathbb{T}^\dagger \frac{d\mathbb{T}}{d\omega} = -\frac{d\mathbb{T}^\dagger}{d\omega}\mathbb{T}. \tag{2.5}$$

The matrix $\mathbb{T}^\dagger \frac{d\mathbb{T}}{d\omega}$ is anti-Hermitian. A Hermitian or anti-Hermitian two-by-two matrix is also represented by four real scalars since the Hermitian matrix $\mathbb{H}$ and anti-Hermite matrix $\mathbb{A}$ obey the relations

$$\mathbb{H}^\dagger = \mathbb{H}; \quad \mathbb{A}^\dagger = -\mathbb{A}. \tag{2.6}$$

The relation between Stokes space vectors and Jones space matrices is accomplished through the use of the "vector matrix" $\vec{\mathbf{Q}}$.[2] We denote the vector matrix by boldface capital letters with a $\vec{\ }$. $\vec{\mathbf{Q}}$ is defined in terms of the Pauli matrices

$$\vec{\mathbf{Q}} = \mathbf{i}_x \begin{bmatrix} 1 & 0 \\ 0 & -1 \end{bmatrix} + \mathbf{i}_y \begin{bmatrix} 0 & 1 \\ 1 & 0 \end{bmatrix} + \mathbf{i}_z \begin{bmatrix} 0 & -j \\ j & 0 \end{bmatrix} = \mathbf{i}_x\sigma_x + \mathbf{i}_y\sigma_y + \mathbf{i}_z\sigma_z. \tag{2.7}$$

The Pauli matrices obey the commutation relations:

$$[\sigma_x, \sigma_y] = 2j\sigma_z \tag{2.8a}$$

$$[\sigma_y, \sigma_z] = 2j\sigma_x \tag{2.8b}$$

$$[\sigma_z, \sigma_x] = 2j\sigma_y. \tag{2.8c}$$

Now, any Hermitean matrix $\mathbb{H}$ in Jones space can be expressed in terms of four real scalars X, Y, Z, and U as follows:

$$\vec{\mathbf{H}} = X\mathbf{i}_x\sigma_x + Y\mathbf{i}_y\sigma_y + Z\mathbf{i}_z\sigma_z + U\mathbb{1}, \tag{2.9}$$

where

$$X = \frac{1}{2}(H_{11} - H_{22}) \tag{2.10a}$$

$$Y = \mathrm{Re}(H_{12}) = \mathrm{Re}(H_{21}) \tag{2.10b}$$

$$Z = \mathrm{Im}(H_{12}) \tag{2.20c}$$

$$U = \frac{1}{2}(H_{11} + H_{22}). \tag{2.10d}$$

Since an anti-Hermitian matrix is equal to a Hermitian matrix times j, an analogous decomposition of an anti-Hermitean matrix is equally possible. We use the letter X, Y, and Z in anticipation of the fact that they define three components of a vector. The transformation from Jones space to Stokes space is accomplished via the vector-matrix $\vec{\mathbf{Q}}$:

$$\mathbf{s}_b = \mathbb{b}^\dagger \vec{\mathbf{Q}} \mathbb{b}. \tag{2.11}$$

Equation (2.1) is obtained from the derivative of Eq. (2.11).

$$\begin{aligned} \frac{d\mathbf{s}_b}{d\omega} &= \frac{d\mathbb{b}^\dagger}{d\omega}\vec{\mathbf{Q}}\mathbb{b} + \mathbb{b}^\dagger\vec{\mathbf{Q}}\frac{d\mathbb{b}}{d\omega} = \mathbb{a}^\dagger\frac{d\mathbb{T}^\dagger}{d\omega}\vec{\mathbf{Q}}\mathbb{T}\mathbb{a} + \mathbb{a}^\dagger\mathbb{T}^\dagger\vec{\mathbf{Q}}\frac{d\mathbb{T}}{d\omega}\mathbb{a} \\ &= \mathbb{b}^\dagger\mathbb{T}\frac{d\mathbb{T}^\dagger}{d\omega}\vec{\mathbf{Q}}\mathbb{b} + \mathbb{b}^\dagger\vec{\mathbf{Q}}\frac{d\mathbb{T}}{d\omega}\mathbb{T}^\dagger\mathbb{b} = \mathbb{b}^\dagger\left[\mathbb{T}\frac{d\mathbb{T}^\dagger}{d\omega}, \vec{\mathbf{Q}}\right]\mathbb{b}. \end{aligned} \tag{2.12}$$

The symbol $[,]$ denotes a commutator. Here we used the fact that $d\mathbb{a}/d\omega = 0$. Let us write the Hermitian matrix $-j\mathbb{T}\frac{d\mathbb{T}^\dagger}{d\omega}$ in the form

$$-j\mathbb{T}\frac{d\mathbb{T}^\dagger}{d\omega} = j\frac{d\mathbb{T}}{d\omega}\mathbb{T}^\dagger = \frac{1}{2}[\Omega_x\sigma_x + \Omega_y\sigma_y + \Omega_z\sigma_z] + U\mathbb{1}. \quad (2.13)$$

where Ω_x, Ω_y, and Ω_z are the components of an as yet undetermined vector and determine the x-component of expression (2.13). We find

$$\begin{aligned}\mathbf{i}_x \cdot \mathbb{b}^\dagger \left[\mathbb{T}\frac{d\mathbb{T}^\dagger}{d\omega}, \vec{\mathbf{Q}}\right]\mathbb{b} = \mathbb{b}^\dagger \left[\mathbb{T}\frac{d\mathbb{T}^\dagger}{d\omega}, \sigma_x\right]\mathbb{b} &= \frac{1}{2}j\mathbb{b}^\dagger[(\Omega_y\sigma_y + \Omega_z\sigma_z), \sigma_x]\mathbb{b} \\ &= -\mathbb{b}^\dagger(\Omega_z\sigma_y - \Omega_y\sigma_z)\mathbb{b} = \Omega_y s_z^{(b)} - \Omega_z s_y^{(b)} \\ &= \mathbf{i}_x \cdot (\mathbf{\Omega} \times \mathbf{s}_b).\end{aligned} \quad (2.14)$$

Thus, the x-component of the vector (2.12) is the triple scalar product of $\mathbf{i}_x$, $\mathbf{\Omega}$, and $\mathbf{s}_b$. The product is one of the components of Eq. (2.1) and hence we have proven that the three coefficients of the Pauli matrices composing the matrix $-j\mathbb{T}\frac{d\mathbb{T}^\dagger}{d\omega}$ are the components of the vector $\mathbf{\Omega}$. It is convenient to write this matrix in the Jones representation in the form

$$-j\mathbb{T}\frac{d\mathbb{T}^\dagger}{d\omega} = \frac{1}{2}\begin{bmatrix}\Omega_x & \Omega_y - j\Omega_z \\ \Omega_y + j\Omega_z & -\Omega_x\end{bmatrix} + U\mathbb{1} = \frac{1}{2}\mathbf{\Omega} + U\mathbb{1}. \quad (2.15)$$

This defines a matrix $\mathbf{\Omega}$ in Jones space. Note that the term multiplying the identity matrix in Eq. (2.15) has disappeared in the triple scalar product. This is a component describing the isotropic part of transmission. We shall return to its discussion later.

3. Poynting Space

The matrix $\mathbb{T}\frac{d\mathbb{T}^\dagger}{d\omega}$ plays a role in the propagation of wave-packets formed of two sinusoids of equal magnitude and slightly different frequencies. Such wave-packets do not exchange energy with their neighbors and the energy contained in any one of the wave-packets travels with the speed of the envelope, the group velocity. If two degenerate modes of different polarization are superimposed, wave-packets cannot be formed unless they are of a specific "principal" polarization. We review briefly the description of PMD in Poynting space. The energy theorem[4] follows from Maxwell's equations. If $\mathbf{E}$ and $\mathbf{H}$ are the electric and magnetic field, the energy theorem states that

$$\begin{aligned}\frac{1}{4}\oint d\mathbf{S} \cdot \left\{\frac{\partial \mathbf{E}}{\partial\omega} \times \mathbf{H}^* + \mathbf{E}^*\frac{\partial \mathbf{H}}{\partial\omega}\right\} \\ = -\frac{j}{4}\int dv\left[\mathbf{H}^* \cdot \frac{\partial}{\partial\omega}(\omega\overline{\overline{\mu}}) \cdot \mathbf{H} + \mathbf{E}^* \cdot \frac{\partial}{\partial\omega}(\omega\overline{\overline{\epsilon}}) \cdot \mathbf{E}\right] \\ = -jw.\end{aligned} \quad (3.1)$$

The right-hand side is equal to the energy w in the volume enclosed by the surface. The surface integral is taken over the closed surface containing the input and output cross-sections and the volume integral is over the volume enclosed by the surface. The volume integral is $-j$ times the energy w. Tensor dielectric constant and magnetic permeability are assumed. In the case of an optical fiber the magnetic permeability is that of free space. The dielectric tensor causes polarization transformation, as the waves propagate. The $\bar{\bar{\epsilon}}$-tensor is Hermitian.

We specialize to a wave-guide or fiber supporting two modes of orthogonal polarizations. Consider propagation in Jones space described by Eq. (2.2). Assume an excitation $a_i(\omega)$ that forms a column matrix in Jones space. The field patterns of the two modes are $\mathbf{e}_i(\mathbf{r}_T)$ and $\mathbf{h}_i(\mathbf{r}_T)$, where $\mathbf{r}_T$ is the coordinate vector in the wave-guide cross-section. They have the same pattern, but their transverse components are rotated by 90° with respect to each other. It is convenient to describe simultaneous excitations in two polarizations in terms of vector-Jones matrices. These then also relate the pure field vector space of Maxwell's equations to Jones space. Define the vector column matrices

$$\vec{\mathbf{E}}_a \equiv \begin{bmatrix} a_1\mathbf{e}_1 \\ a_2\mathbf{e}_2 \end{bmatrix} \tag{3.2a}$$

and

$$\vec{\mathbf{H}}_a = \begin{bmatrix} a_1\mathbf{h}_1 \\ a_2\mathbf{h}_2 \end{bmatrix}. \tag{3.2b}$$

The field patterns are normalized so that the integral of their Poynting vector over a cross-section gives unity

$$\int d\mathbf{S}\cdot\mathbf{e}_i\times\mathbf{h}_i^* = \int d\mathbf{S}\cdot\mathbf{e}_i\times\mathbf{h}_i = 1. \tag{3.3}$$

The complex conjugation can be dropped because the transverse field patterns have planar wave-fronts in an isotropic guide and are in phase. The phase assignment is taken up by the amplitudes a_j. The fact that the fiber is weakly anisotropic, so that PMD is produced need not be incorporated in the field pattern, since the anisotropy only manifests itself over many millions of wavelengths. The power flowing through a guide cross-section is given by

$$-\frac{1}{2}\int d\mathbf{S}\cdot\vec{\mathbf{E}}_a^\dagger\times\vec{\mathbf{H}}_a = \frac{1}{2}\int d\mathbf{S}\cdot\vec{\mathbf{H}}_a^\dagger\times\vec{\mathbf{E}}_a = \mathbb{a}^\dagger\mathbb{a}. \tag{3.4}$$

The dagger denotes a complex conjugate transpose. With the aid of this compact notation we may evaluate Eq. (3.1). Consider propagation described in Jones space by Eq. (2.2). Assume excitations $a_i(\omega)$ and $a_i(\omega+d\omega) = a_i(\omega)$; $i = 1, 2$. We have

$$\frac{d}{d\omega}\mathbb{a} = 0. \tag{3.5}$$

This does not imply that $\frac{\partial\mathbf{e}_i}{\partial\omega} = 0$ or $\frac{\partial\mathbf{h}_i}{\partial\omega} = 0$, since the field patterns can be functions of frequency. However, the transverse field patterns $\mathbf{e}_i(x,y)$ and $\mathbf{h}_i(x,y)$ have been made real (the phase fronts are planar). The normalization condition (3.3) assures, that even in the case of frequency dependent mode patterns, the contribution to the surface integral of the excitation on side (a) vanishes. We find that

$$\frac{1}{4}\int d\mathbf{S}\cdot\left\{\frac{\partial\mathbf{E}}{\partial\omega}\times\mathbf{H}^* + \mathbf{E}^*\times\frac{\partial\mathbf{H}}{\partial\omega}\right\} = \mathbb{b}^\dagger\frac{d}{d\omega}\mathbb{b}.$$

Further,

$$\frac{d\mathbb{b}}{d\omega} = \frac{d\mathbb{T}}{d\omega}\mathbb{a}. \tag{3.6}$$

When these relations are introduced into the energy theorem, we find

$$\mathbb{a}^\dagger j\mathbb{T}^\dagger \frac{d\mathbb{T}}{d\omega}\mathbb{a} = w \tag{3.7}$$

where w is the energy in the segment due to the excitations a_i. The matrix that emerged from the PMD description in Jones space is $j\mathbb{T}\frac{d\mathbb{T}^\dagger}{d\omega}$. Referring (3.7) to the output excitation, we may introduce this very same matrix into the energy theorem:

$$\mathbb{a}^\dagger j\mathbb{T}^\dagger \frac{d\mathbb{T}}{d\omega}\mathbb{a} = -\mathbb{a}^\dagger j\frac{d\mathbb{T}^\dagger}{d\omega}\mathbb{T}\mathbb{a} = -j\mathbb{a}\mathbb{T}^\dagger\mathbb{T}\frac{d\mathbb{T}^\dagger}{d\omega}\mathbb{T}\mathbb{a} = -j\mathbb{b}^\dagger\mathbb{T}\frac{d\mathbb{T}^\dagger}{d\omega}\mathbb{b} = w. \tag{3.8}$$

Hence, the matrix $-j\mathbb{T}\frac{d\mathbb{T}^\dagger}{d\omega}$ expresses the energy w in the segment in terms of the output excitation. We may introduce the matrices $\mathbb{W}_a$ and $\mathbb{W}_b$ such that the energy w is written as a quadratic form of the excitations $\mathbb{a}$ or $\mathbb{b}$:

$$w = \mathbb{b}^\dagger\mathbb{W}_b\mathbb{b} = \mathbb{a}^\dagger\mathbb{W}_a\mathbb{a}. \tag{3.9}$$

The energy matrix $\mathbb{W}_a$ is

$$\mathbb{W}_a = -j\frac{d\mathbb{T}^\dagger}{d\omega}\mathbb{T}. \tag{3.10a}$$

The energy matrix $\mathbb{W}_b$ in Poynting space is equal to the matrix $-j\mathbb{T}\frac{d\mathbb{T}^\dagger}{d\omega}$ defined by Eq. (2.15) in Jones space

$$-j\mathbb{T}\frac{d\mathbb{T}^\dagger}{d\omega} = \mathbb{W}_b. \tag{3.10b}$$

The elements of $\mathbb{W}_b$ are in units of time.

The energy is an integral over the volume between the two reference planes. It involves the parameters of the medium and the field patterns. We have mentioned earlier that the field patterns may be frequency dependent. They are so in fibers, and more so in dispersion shifted fibers. This frequency dependence gives rise to isotropic geometric (as opposed to material) group velocity dispersion (GVD). Twisting of the fiber and anisotropic perturbations of the fiber cause coupling between the two polarizations. Transformations of the polarization of the wave are angular momentum changes of the wave. These angular momentum changes do result in changes of the energy matrix, because waves of different polarizations have different energy densities in a fiber of deformed cross-section and/or containing an anisotropic dielectric. A frequency dependence is introduced when the polarizations $i = 1$, and $i = 2$ rotate with a rotation pattern that is frequency dependent.

The energy matrix $\mathbb{W}_a$ can be written concisely in terms of the electric and magnetic field mode patterns by introducing the vector column matrices

$$\vec{\mathbf{e}} = \begin{bmatrix} \mathbf{e}_1(x,y) \\ \mathbf{e}_2(x,y) \end{bmatrix} \tag{3.11a}$$

and

$$\vec{\mathbf{h}} = \begin{bmatrix} \mathbf{h}_1(x,y) \\ \mathbf{h}_2(x,y) \end{bmatrix}. \tag{3.11b}$$

The Jones vector $\mathbb{a}$ transforms as the mode travels along the system, $\mathbb{a}(z) = \mathbb{T}(z)\mathbb{a}(0)$. With aid of these matrices, we have

$$\mathbb{W}_a = \frac{1}{4}\int dv \mathbb{T}^\dagger(z)\left[\vec{\mathbf{h}}^\dagger \cdot \frac{\partial}{\partial\omega}(\omega\overline{\overline{\mu}}) \cdot \vec{\mathbf{h}} + \vec{\mathbf{e}}^\dagger \cdot \frac{\partial}{\partial\omega}(\omega\overline{\overline{\epsilon}}) \cdot \vec{\mathbf{e}}\right]\mathbb{T}(z). \tag{3.12}$$

Energy theorems put constraints on the solutions of Maxwell's equations. They do not give full information on the field solutions. The input excitation characterized by the Jones column matrix $\mathbb{a}$ evolves as it travels along the fiber. To find this evolution one would need to solve Maxwell's equations. The Jones matrix $\mathbb{T}(z)$ gives part of this information. Further, the field patterns $\vec{\mathbf{e}}$ and $\vec{\mathbf{h}}$ experience changes of their propagation constant because of the changing geometric dispersion of the fiber. As they propagate along an anisotropic medium they contribute anisotropic terms to the energy matrix. The $\mathbb{T}$-matrix is frequency dependent, and so is the propagation constant. This leads to a frequency dependence of the energy matrix.

Summarizing the derivations thus far, we note that the three descriptions each contain information about PMD. In Stokes space it is the $\mathbf{\Omega}$ vector. In Jones space it is the matrix $-j\mathbb{T}\frac{d\mathbb{T}^\dagger}{d\omega}$. This matrix is uniquely related to the energy matrix determined from the energy theorem in Poynting space, the matrix $\mathbb{W}_b$. Note that Stokes space misses one parameter U, which we are now able to identify as the isotropic contribution to the energy, $(1/2)(W_{11}^{(b)} + W_{22}^{(b)})$.

We now show that $j\mathbb{T}\frac{d\mathbb{T}^\dagger}{d\omega}$ forms the basis for the definition of energy velocity in an anisotropic waveguide. To derive an energy velocity of a mode with one polarization one forms a wave-packet with perfect nodes using excitations of equal amplitude at two close frequencies, $\mathbb{a}(\omega) = \mathbb{a}(\omega + d\omega)$. The energy is trapped between the nodes and travels at the group velocity. One can show that even in the case when the mode patterns are frequency dependent, no power crosses the nodes. This argument can be extended to modes of two polarizations. Here we encounter the difficulty that a pair of nodes at cross-section (a) does not necessarily assure a pair of nodes at cross-section (b). Nodes are preserved only when $\mathbb{b}(\omega + d\omega) = \mathbb{b}(\omega) - j\lambda d\omega \mathbb{b}(\omega)$ for $\mathbb{a}(\omega + d\omega) = \mathbb{a}(\omega)$, with λ real; i.e. the output is only shifted in phase, not changed in amplitude. This condition requires

$$\frac{d\mathbb{T}}{d\omega}\mathbb{T}^\dagger\mathbb{b} = -j\lambda\mathbb{b} \tag{3.13a}$$

or, using Eq. (2.5),

$$j\mathbb{T}\frac{d\mathbb{T}^\dagger}{d\omega}\mathbb{b} = \mathbb{W}_b\mathbb{b} = \lambda\mathbb{b}. \tag{3.13b}$$

The eigenvectors of this matrix relation are the two polarizations for which a group velocity can be defined. Note that the same matrix enters into the wave-packet argument as into the energy theorem. This means that the polarizations for which a group velocity can be defined are also polarizations that diagonalize the energy matrix. Energy flow can be defined only for these two "principal" polarizations.

4. The Jones Matrix of Segment with First Order PMD

The energy matrix $\mathbb{W}_b$ is in one to one correspondence with the PMD vector. If the system possesses only first order PMD, then both the PMD vector and the energy matrix are frequency independent. We now show that the Jones matrix of a system with first order PMD, and no higher order PMD, can be expressed as follows

$$\mathbb{T}(\omega_o + \Delta\omega) = \mathbb{T}(\omega_o)\exp(-j\Delta\omega\mathbb{W}_a), \tag{4.1}$$

where $d\mathbb{W}_a/d\omega = 0$. Indeed when we construct the matrix $-jd\mathbb{T}^\dagger/d\Delta\omega_{|\Delta\omega\to 0}\mathbb{T}$, we find

$$-j\frac{d\mathbb{T}^\dagger}{d\omega}\bigg|_{\Delta\omega\to 0}\mathbb{T} = -j\exp(j\Delta\omega\mathbb{W}_a)\exp(-j\Delta\omega\mathbb{W}_a)j\mathbb{W}_a\mathbb{T}^\dagger(\omega_o)\mathbb{T}(\omega_o) = \mathbb{W}_a \tag{4.2}$$

Here we have taken advantage of the fact that any power of the matrix $\mathbb{W}_a$ commutes with any other power. The result is equal to the energy matrix evaluated in terms of the input variables. If the energy matrix is frequency independent, we have to get the same result at all frequencies. This proves that Eq. (4.1) is the correct representation of the Jones matrix for all frequencies for a system with only first order PMD.

Suppose we concatenate two segments with only first order PMD. The net Jones matrix is

$$\begin{aligned}\mathbb{T} &= \mathbb{T}_2(\omega_o)e^{-j\Delta\omega\mathbb{W}_a^{(2)}}\mathbb{T}_1(\omega_o)e^{-j\Delta\omega\mathbb{W}_a^{(1)}}\\ &= \mathbb{T}_2(\omega_o)\mathbb{T}_1(\omega_o)e^{-j\Delta\omega\mathbb{T}_1^\dagger\mathbb{W}_a^{(2)}\mathbb{T}_1}e^{-j\Delta\omega\mathbb{W}_a^{(1)}}.\end{aligned} \tag{4.3}$$

The Jones matrix of the concatenation would exhibit only first order PMD if it could be written in the form $\exp(-j\Delta\omega\mathbb{W})$. However, this is possible only in the case when the two matrices $\mathbb{T}_1^\dagger\mathbb{W}_a^{(2)}\mathbb{T}_1$ and $\mathbb{W}_a^{(1)}$ commute. In general, one cannot write the product of exponentials in this form and thus higher order PMD is automatically produced by the concatenation of two segments of first order PMD.

5. The Detector Current

If both polarizations propagating from (a) to (b) are subject to the transformation matrix $\mathbb{T}$, and are detected by a diode detector that produces a current proportional to the sum of the powers in the two polarizations, the detector current is given by

$$i(t) = q\eta 2\pi\int \mathbb{b}^\dagger(\omega')e^{-j\omega' t}d\omega'\int \mathbb{b}(\omega)e^{j\omega t}d\omega. \tag{5.1}$$

where q is the charge, η is the quantum efficiency of the detector, and power and energy are measured in terms of photon flux and photon number. It is easy to check that energy is conserved, i.e., that the current integrated over all time is equal to ηq times the incident photon flow:

$$\int_{-\infty}^{\infty} dt i(t) = 2\pi q\eta\int d\omega\mathbb{a}^\dagger(\omega)\mathbb{a}(\omega). \tag{5.2}$$

The current as a function of time is given by the double integral

$$i(t) = q\eta 2\pi\int d\omega'\int d\omega e^{j(\omega-\omega')t}\mathbb{a}^\dagger(\omega')\mathbb{T}^\dagger(\omega')\mathbb{T}(\omega)\mathbb{a}(\omega). \tag{5.3}$$

When the factor representing the carrier, $\exp(j\omega_o t)$, is removed, the integral can be written in terms of the frequency deviation $\Delta\omega = \omega - \omega_o$:

$$i(t) = q\eta 2\pi \int d\Delta\omega' \int d\Delta\omega e^{i(\Delta\omega - \Delta\omega')t} \mathbb{a}^\dagger(\Delta\omega')\mathbb{T}^\dagger(\Delta\omega')\mathbb{T}(\Delta\omega)\mathbb{a}(\Delta\omega). \quad (5.4)$$

Consider first the simple case of a segment with only first order PMD. The Jones matrix is then given by (4.1). In this case we obtain for the detector current:

$$\begin{aligned} i(t) = {} & q\eta^2 2\pi \int d\Delta\omega \int d\Delta\omega' \\ & \times \exp[j(\Delta\omega - \Delta\omega')t]\mathbb{a}^\dagger(\Delta\omega') \exp[-j(\Delta\omega - \Delta\omega')\mathbb{W}_a]\mathbb{a}(\Delta\omega). \quad (5.5) \end{aligned}$$

Suppose $\mathbb{R}$ is the rotation matrix that puts $\mathbb{W}_a$ into diagonal form.

$$\mathbb{R}\mathbb{W}_a\mathbb{R}^\dagger = \begin{bmatrix} \tau_1 & 0 \\ 0 & \tau_2 \end{bmatrix}. \quad (5.6)$$

We may then rewrite Eq. (5.5) in a more explicit form that shows the delays in the two principal polarizations:

$$\begin{aligned} i(t) = {} & q\eta 2\pi \int d\Delta\omega \int d\Delta\omega' \mathbb{a}^\dagger(\Delta\omega')\mathbb{R}^\dagger \\ & \times \begin{bmatrix} \exp[j(\Delta\omega - \Delta\omega')(t - \tau_1) & 0 \\ 0 & \exp[j(\Delta\omega - \Delta\omega')(t - \tau_2) \end{bmatrix} \mathbb{R}\mathbb{a}(\Delta\omega). \end{aligned} \quad (5.7)$$

When the PMD is of all orders, the current cannot be written in such a simple form. Instead, one may start with Eq. (5.4) and Taylor expand the Jones matrices $\mathbb{T}$ and $\mathbb{T}^\dagger$:

$$i(t) = q\eta 2\pi \int d\Delta\omega \int d\Delta\omega' \mathbb{a}^\dagger(\Delta\omega') \sum_{m,n} \frac{\Delta\omega'^m \Delta\omega^n}{m!n!} \frac{d^m\mathbb{T}^\dagger}{d\omega^m} \frac{d\mathbb{T}^n}{d\omega^n} \mathbb{a}(\Delta\omega). \quad (5.8)$$

We now show that all coefficients in the expansion can be derived from the $\mathbb{W}_a$ matrix alone. The product $-j(d\mathbb{T}^\dagger/d\omega)\mathbb{T}$ is, of course, equal to the energy matrix:

$$\frac{d\mathbb{T}^\dagger}{d\omega}\mathbb{T} = j\mathbb{W}_a. \quad (5.9)$$

The derivative of this product gives

$$\frac{d^2\mathbb{T}^\dagger}{d\omega^2}\mathbb{T} + \frac{d\mathbb{T}^\dagger}{d\omega}\frac{d\mathbb{T}}{d\omega} = j\frac{d\mathbb{W}_a}{d\omega}. \quad (5.10)$$

The second term on the left-hand side, $(d\mathbb{T}^\dagger/d\omega)(d\mathbb{T}/d\omega)$, is equal to the energy matrix squared. Thus we find for $d^2\mathbb{T}^\dagger/d\omega^2\mathbb{T}$:

$$\frac{d^2\mathbb{T}^\dagger}{d\omega^2}\mathbb{T} = j\left(\frac{d\mathbb{W}_a}{d\omega} + j\mathbb{W}_a^2\right). \quad (5.11)$$

The third derivative is obtained by differentiating Eq. (5.11). One obtains

$$\frac{d^3\mathbb{T}^\dagger}{d\omega^3}\mathbb{T} + \frac{d^2\mathbb{T}^\dagger}{d\omega^2}\frac{d\mathbb{T}}{d\omega} = j\frac{d^2\mathbb{W}_a}{d\omega^2} - \left(\frac{d\mathbb{W}_a}{d\omega}\mathbb{W}_a + \mathbb{W}_a\frac{d\mathbb{W}_a}{d\omega}\right). \tag{5.12}$$

The second term on the left-hand side may be obtained from Eq. (5.11) in the same way as the second term in Eq. (5.10) was obtained from Eq. (5.9). In this way we obtain

$$\frac{d^3\mathbb{T}^\dagger}{d\omega^3}\mathbb{T} = j\frac{d^2\mathbb{W}_a}{d\omega^2} - 2\frac{d\mathbb{W}_a}{d\omega}\mathbb{W}_a - \mathbb{W}_a\frac{d\mathbb{W}_a}{d\omega} - j\mathbb{W}_a^3. \tag{5.13}$$

In this way one may construct $d^n\mathbb{T}^\dagger/d\omega^n T$ to any order n. The terms in the kernel (5.8) are obtained from these and their Hermitian conjugates.

6. Summary and Conclusions

We have related the three formalisms of PMD in Stokes space, Jones space and Poynting space. The formulation in Stokes space describes only PMD in ascending order, but ignores isotropic dispersion. The information is incomplete. The formulation in Jones space contains full information in the Jones Matrix $\mathbb{T}$ as a function of frequency. The Stokes vector as a function of frequency is contained in the matrix $\mathbb{T}d\mathbb{T}^\dagger/d\omega$. Poynting space provides full information about the PMD vector in the energy matrix $\mathbb{W}_b$ evaluated in terms of the output polarizations. The matrix $\mathbb{W}_b$ contains the information of the PMD vector to all orders. In addition, it has information on the isotropic dispersion. On the basis of this information alone one may construct a compensation structure that eliminates PMD and dispersion to all orders. Knowledge of the PMD vector as a function frequency does not include information on the isotropic dispersion.

Construction of the eye-diagram requires knowledge of $\mathbb{W}_a$. This information is available in Jones space in terms of the matrix $\mathbb{T}(\omega)$, but is not contained in the PMD vector.

References

[1] C. D. Poole, N. S. Bergano, R. E. Wagner, and H. J. Schulte, "Polarization dispersion and principle states in a 147 km undersea lightwave cable," J. Lightwave Technol. **6**, 1185-1190 (1988).

[2] J. P. Gordon and H. Kogelnik, "PMD fundamentals: polarization mode dispersion in optical fibers," Proc. Natl. Acad. Sci. **97**, 4541-4550 (2000).

[3] H. A. Haus, "Group velocity, energy and polarization mode dispersion," J. Opt. Soc. Am. B **16**, 1863-1867 (1999).

[4] L. D. Landau and E. M. Lifshitz, *Electrodynamics of Continus Media* (Pergamon, Oxford; Addison Wesley, Reading, MA, 1960).

DOI: 10.1007/s10297-004-0005-1
Originally published in J. Opt. Fiber. Commun. Rep. 1, 107–122 (2004)

The inverse PMD problem

H. Kogelnik[1], L. E. Nelson[2], and J. P. Gordon[1]

[1]Bell Labs, Lucent Technologies
Holmdel, NJ 07733 USA
[2]OFS Laboratories, Somerset, NJ 08873 USA

Abstract. When a fiber is characterized by measured polarization mode dispersion (PMD) vector data, inversion of these data is required to determine the frequency dependence of the fiber's Jones matrix and, thereby, its pulse response. We briefly review approaches to PMD inversion and discuss three second-order models used for this purpose. We report extension of inversion to fourth-order PMD using higher-order concatenation rules, rotations of higher power designating higher rates of acceleration with frequency, and representation of these rotations by Stokes vectors.

1. Introduction

In the past 15 years the subject of polarization mode dispersion (PMD) in optical fibers has seen the development of a considerable knowledge base numbering over 400 publications [1,2]. We have clear definitions, tested measurement and simulation methods, a well developed conceptual and theoretical framework based on the principal state of polarization (PSP) allowing simple physical interpretation and visualization on the Poincaré sphere, good understanding of the statistics, simple concatenation rules for combinations of fibers and components as well as for the depiction of compensator concepts, etc. Yet there seemingly exists some dissatisfaction with the current state of affairs as evidenced by a continuing stream of proposals for new PMD representations and formalisms [3–7]. The cause for much of this dissatisfaction appears to be the "Inverse PMD Problem": the lack of a simple and transparent method to derive or construct a Jones matrix representing given higher-order PMD, and the associated lack of a simple physical interpretation of higher-order PMD in the time domain. It is the goal of this communication to propose a solution for this problem. For our discussion, we will conform to the notation of [2,8] unless specifically noted.[1]

As an illustration of the problem, consider Fig. 1 showing the measured output PMD vector $\boldsymbol{\tau}(\omega)$ of a fiber as a function of optical frequency ω (or wavelength). These data represent the fiber's PMD. The PMD vector

$$\boldsymbol{\tau} = \Delta\tau\hat{p} \tag{1}$$

defines the polarization of the slow PSP via the unit Stokes vector $\hat{p}$, and the differential group delay (DGD), $\Delta\tau$, of pulses launched with the two PSP polarizations. The DGD is also shown in the figure. Higher-order PMD vectors, such as the second-order vector $\boldsymbol{\tau}_\omega$, are obtained by differentiation of $\boldsymbol{\tau}(\omega)$ as indicated by the subscript ω.

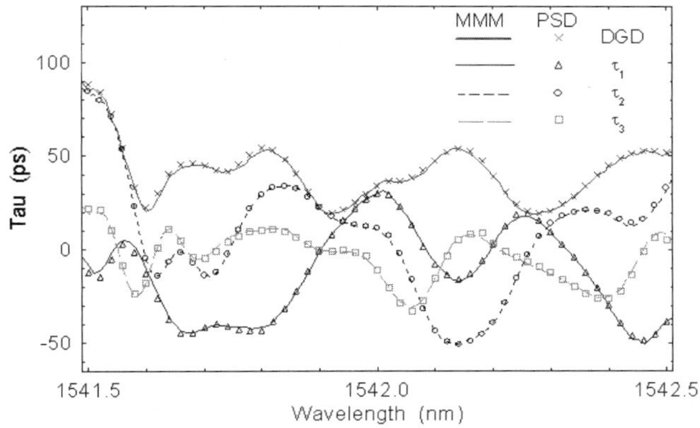

Fig. 1. Output PMD vector $\boldsymbol{\tau}(\omega)$ of a fiber with a mean DGD of 35 ps as a function of wavelength [11]. Measurements were performed with the PSD and the MMM methods. The figure shows the DGD and the three vector components τ_i.

We note that some PMD measurement techniques, such as the Jones Matrix Eigenanalysis (JME) [9] and the Müller Matrix Method (MMM) [10], determine the fiber's Jones (or the corresponding Müller) matrices at selected wavelengths as an intermediate step, while others, for example, the Polarization Dependent Signal Delay (PSD) method [11], do not. The task of "inverse PMD" is to start with $\boldsymbol{\tau}(\omega)$ data such as these and, near a selected carrier frequency ω_0, to deduce or construct a Jones matrix of sufficient accuracy to represent the fiber's pulse response for any desired order of PMD.

The Jones matrix is contained in the fiber's transmission matrix T relating the output Jones vector $\mid t\rangle$ to the input vector $\mid s\rangle$ via

$$\mid t\rangle = T \mid s\rangle. \tag{2}$$

To keep our notation simple, we focus on the frequency dependent part $U(\omega)$ of the transmission matrix $T(\omega)$,

$$T = e^{-j\phi}UU_0, \tag{3}$$

[1] A notable exception being the notation for Stokes vectors that are shown in boldface and not with the usual arrow superscripts.

where $\phi(\omega)$ is the common phase, $U(\omega_0) = I$, and U_0 is a frequency-independent term representing the polarization rotation through the fiber at the fixed carrier frequency ω_0. $U(\omega)$ describes the difference rotation of the polarization at the fiber output as the frequency $\omega = \omega_0 + \Delta\omega$ is changed from ω_0 to $\omega_0 + \Delta\omega$. This difference rotation is employed in the JME [9] and MMM [10] measurement techniques and in the second-order models by Bruyere [12–14] and Eyal et al. [4]. U is unitary with $UU^\dagger = I$ and $\det U = 1$.

2. Eigenvectors and PMD Vectors of *U*

The distinction and relationship between the eigenvectors and the PMD vectors of U play a critical role in our discussion of PMD inversion. The Jones matrix U can be written in various physically equivalent forms that emphasize different characteristics. These forms and their connections are discussed in the review of [8] and listed for convenience in Appendix A. The eigenvectors of U are, of course, the same for all these forms. Two forms exhibiting the eigenvectors and eigenvalues of U explicitly are

$$U = e^{-j(\varphi/2)\hat{r}\cdot\boldsymbol{\sigma}} = I\cos\varphi/2 - j\hat{r}\cdot\boldsymbol{\sigma}\sin\varphi/2, \tag{4}$$

the compact exponential form and the rotational form. Here $\hat{r} = \langle r \mid \boldsymbol{\sigma} \mid r\rangle$ is the rotation axis of the polarization rotation due to U on the Poincaré sphere, φ is the rotation angle, and $\mid r\rangle$ is an eigenvector of U. The quantities $\varphi(\omega)$ and $\hat{r}(\omega)$ provide a full description of U and will be at the center of our discussion. The matrix $\hat{r}\cdot\boldsymbol{\sigma}$ appearing in these expressions is explained in [8] and Appendix A and can, for the present purposes, be read as a symbol for a matrix containing the components r_i of $\hat{r}$ in the form

$$\hat{r}\cdot\boldsymbol{\sigma} = \begin{vmatrix} r_1 & r_2 - jr_3 \\ r_2 + jr_3 & -r_1 \end{vmatrix}. \tag{5}$$

Note that $\varphi(\omega_0) = 0$ because $U(\omega_0) = I$. In the theory of the PSP [8], the PMD vector $\boldsymbol{\tau}$ is obtained via

$$jU_\omega U^\dagger = \frac{1}{2}\boldsymbol{\tau}\cdot\boldsymbol{\sigma} = \frac{1}{2}\begin{vmatrix} \tau_1 & \tau_2 - j\tau_3 \\ \tau_2 + j\tau_3 & -\tau_1 \end{vmatrix}, \tag{6}$$

when $U(\omega)$ is given. When $\varphi(\omega)$ and $\hat{r}(\omega)$ are given, $\boldsymbol{\tau}$ follows from the relation [8]

$$\boldsymbol{\tau} = \varphi_\omega\hat{r} + \sin\varphi\hat{r}_\omega - (1-\cos\varphi)\hat{r}_\omega \times \hat{r}. \tag{7}$$

This is reasonably straightforward. However when $\boldsymbol{\tau}(\omega)$ is given, as in our problem, we must find a way to invert the process. This is easy for first-order PMD, but becomes more intricate once higher orders are involved.

3. The First-order PMD Picture

Our standard physical picture of first-order PMD considers an input pulse split into the two PSPs according to the orientation of the input polarization relative to the polarization $\hat{p}$ of the PSP. On propagation through the fiber the two pulses are differentially delayed by the DGD $\Delta\tau$. Based on the first-order picture one can simply write down an expression for the corresponding U matrix. However to prepare for later, we first link to the eigenvalues and eigenvectors of U via Eq. (7). In Eq. (1) the PMD vector at ω_0 is defined by $\Delta\tau\hat{p}$, while Eq. (7) gives

$$\boldsymbol{\tau}(\omega_0) = \varphi_\omega(\omega_0)\hat{r}(\omega_0), \tag{8}$$

as $\varphi(\omega_0) = 0$. Equation (8) is the key part of the MMM algorithm [10]. In first order, $\Delta\tau$ and $\hat{p}$ are independent of frequency, PMD inversion is trivial, and we identify

$$\varphi = \Delta\tau\Delta\omega \text{ and } \hat{r} = \hat{p}, \tag{9}$$

when $\Delta\tau$ and $\hat{p}$ are given. To determine the U matrix, the identified φ and $\hat{r}$ are inserted into any of its forms. Usually, the diagonal form or the Caley-Klein form,

$$U = QDQ^\dagger = \begin{vmatrix} \alpha & \beta \\ -\beta^* & \alpha^* \end{vmatrix} \tag{10}$$

is chosen. The diagonal elements in the diagonal matrix D are the eigenvalues $e^{-j\Delta\tau\Delta\omega/2}$ and $e^{j\Delta\tau\Delta\omega/2}$ of U, and the column vectors of Q are the frequency independent eigenvectors of U. In the special case where $\hat{p}$ and $\hat{r}$ are aligned with $\hat{S}_1$, the parameters of the Caley-Klein matrix simplify to $\alpha = e^{-j\Delta\tau\Delta\omega/2}$ and $\beta = 0$. Considering the diagonal form of U together with U_0, one deduces the structure of the standard emulator of first-order PMD shown in Fig. 2, consisting of a polarization maintaining fiber (PMF) between two frequency independent polarization controllers.

Fig. 2. Emulation of a fiber with first-order PMD: a polarization maintaining fiber between two frequency independent polarization controllers, PC1, and PC2.

The above discussion already highlights two points of the wish list for PMD inversion at higher orders:

1. the mathematical elements of the resulting matrix should be realizable for implementation as higher-order PMD emulators.
2. Exponential terms in the frequency domain, such as the one obtained for the element α above, translate into simple pulse delays in the time domain. They should be retained during Taylor expansions, iterations, etc. as much as possible. We prefer, e.g., an expansion of the phase φ in the exponent to an expansion of the exponent itself. This is done in Eq. (7) and its derivatives.

4. Mathematical PMD Inversion

Multiplying Eq. (6) by U sets up a matrix differential equation

$$jU_\omega = \frac{1}{2}\boldsymbol{\tau}\cdot\boldsymbol{\sigma}U, \tag{11}$$

for the matrix $U(\omega)$ for given $\boldsymbol{\tau}(\omega)$. A problem of this kind is known as a system of differential equations and is treated extensively in the literature [15]. Various interesting methods for its solution, using expansions in powers of $\Delta\omega$ and iterations are described in [16,17]. It has also been pointed out [16] that Eq. (11) can be simplified to a vector differential equation

$$j \mid u\rangle_\omega = \frac{1}{2}\boldsymbol{\tau} \cdot \boldsymbol{\sigma} \mid u\rangle, \tag{12}$$

for the column vector $\mid u\rangle = (\alpha, -\beta^*)$, where α and β are the Caley-Klein parameters. It is useful to briefly consider yet another mathematical approach based on Eq. (7) and also using expansion in powers of $\Delta\omega$ as well as iteration. Differentiation of Eq. (7) and subsequent evaluation at ω_0 gives

$$\boldsymbol{\tau}_\omega(\omega_0) = \varphi_{\omega\omega}\hat{r} + 2\varphi_\omega \hat{r}_\omega, \tag{13}$$

$$\boldsymbol{\tau}_{\omega\omega}(\omega_0) = \varphi_{\omega\omega\omega}\hat{r} + 3\varphi_{\omega\omega}\hat{r}_\omega + 3\varphi_\omega\hat{r}_{\omega\omega} - \varphi_\omega^2 \hat{r}_\omega \times \hat{r}, \tag{14}$$

and so on, where all quantities are evaluated at ω_0 using $\varphi(\omega_0) = 0$. By combining Eq. (13) with the first-order PMD inversion results, second-order PMD inversion becomes an easy task. We write down the identity for the second-order PMD vector,

$$\boldsymbol{\tau}_\omega = \Delta\tau_\omega \hat{p} + \Delta\tau \hat{p}_\omega, \tag{15}$$

where $\Delta\tau_\omega$ is the term relating to polarization dependent dispersion (PCD) and $\hat{p}_\omega$ is the PSP depolarization. Using Eq. (9) and comparing Eqs. (13) and (15), we find that $\varphi_{\omega\omega} = \Delta\tau_\omega$ and $\hat{r}_\omega = \hat{p}_\omega/2$. With this we obtain the second-order inversion for the eigenvalues and eigenvectors,

$$\varphi(\omega_0 + \Delta\omega) = \Delta\tau\Delta\omega + \Delta\tau_\omega \Delta\omega^2/2 \tag{16}$$

and

$$\hat{r}(\omega_0 + \Delta\omega) = \hat{p} + \hat{p}_\omega \Delta\omega/2. \tag{17}$$

These can be inserted into any of the mentioned forms of $U(\omega)$ to get the corresponding second-order Jones matrix. Its Caley-Klein parameters, e.g., for the case where $\hat{p}$ and $\hat{p}_\omega$ are aligned with $\hat{S}_1$ and $\hat{S}_2$, are

$$\alpha(\omega) = e^{-j\varphi/2}, \quad \beta(\omega) = -(1/2)jp_\omega\Delta\omega \sin\varphi/2. \tag{18}$$

These results of the mathematical approach to PMD inversion are relatively satisfactory. They are of general validity as they do not depend on any specific model. The statistics of the appearing terms $\hat{p}$, $\hat{p}_\omega$, $\Delta\tau$ and $\Delta\tau_\omega$ are understood, and the exponential in the α-parameter is retained, satisfying point (2) of our wishlist. However point (1), physical realizability of U, is not met as $\hat{r} \cdot \hat{r}$, is not unity in $O(\Delta\omega^2)$ and $UU^\dagger$ is not unitary in $O(\Delta\omega^4)$. This point is satisfied by the model approaches to be discussed below. Extensions of the mathematical approach to higher than second PMD order appear to be possible in principle, but the results of those inversions seem to be lacking the desired simplicity.

5. Second-order Models

In the construction of second-order models one combines several realizable elements with unitary U matrix to match the prescribed first- and second-order PMD vectors of the fiber with those of the model. Once the parameters of the model are set to accomplish the match, there is no further control and the model's third- and higher-order PMD characteristics are automatically determined. There are at least three different second-order models, and all three have different third-order characteristics. To simplify the models, one usually chooses PMD vectors aligned with the coordinate system of the Poincaré sphere (Stokes space). Polarization controllers can be used to adapt the models to the specific situation in the laboratory.

5.1. *The Bruyere Model*

The Bruyere model consists of the three elements used for the diagonal form of U shown in Eq. (10). In distinction to the diagonal model for first-order PMD, the second-order model uses frequency dependent Q matrices [12–14]. This is expressed by the choice of a frequency-dependent eigenvector (rotation axis)

$$\hat{r}(\omega) = (\cos 2k\Delta\omega, \sin 2k\Delta\omega, 0), \tag{19}$$

rotating with $\Delta\omega$ along the $\hat{S}_1/\hat{S}_2$ equator. The model parameter k is chosen to match the fiber's depolarization by setting $p_\omega = 4k$. The rotation angle, appearing in the diagonal matrix D realized by a birefringent plate, is chosen to match the fiber's DGD and PCD as in Eq. (16), with the assumption that $\varphi_{\omega\omega\omega} = 0$. For these parameter choices, it is easy to check that Eq. (7) yields

$$\boldsymbol{\tau}(\omega_0) = \Delta\tau\hat{S}_1, \tag{20}$$

and that Eq. (13) leads to

$$\boldsymbol{\tau}_\omega(\omega_0) = \Delta\tau_\omega\hat{S}_1 + 4k\Delta\tau\hat{S}_2 \tag{21}$$

as prescribed. The third-order PMD vector due to these choices and the present model follows from Eq. (14) as

$$\boldsymbol{\tau}_{\omega\omega}(\omega_0) = 2k(-6k\Delta\tau, 3\Delta\tau_\omega, \Delta\tau^2). \tag{22}$$

Implementations of emulators based on the Bruyere model are described in [18].

5.2. *The Planar Sweep Model*

This model is a relatively simple emulator of fiber DGD and PSP depolarization constructed from two PM fibers or birefringent plates. The DGD values of the plates and the relative orientation of their birefringence axes are adjustable. The model emulates PSP depolarization producing a PMD vector that traces out a grand circle on the Poincaré sphere with $\Delta\omega$. This is called a planar sweep and it minimizes higher PMD orders. Consider a fiber with a DGD of $\Delta\tau$ and a depolarization of p_ω. The Jones matrices of the two plates are U_a and U_b, their PMD vectors are $\boldsymbol{\tau}_a$ and $\boldsymbol{\tau}_b$, and the orientation of their birefringence axes is set according to the PMD vector diagram shown in Fig. 3.

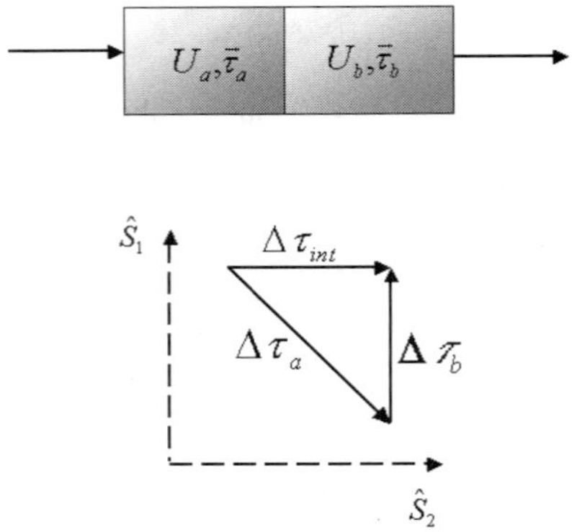

Fig. 3. Planar sweep emulator consisting of two birefingent sections. The birefringence axis of U_a is rotated according to the shown PMD vector diagram. The axis of U_b is aligned with $\hat{S}_1$. The diagram is in the $\hat{S}_1/\hat{S}_2$ plane and applies to the section interface.

The Jones matrix of the combination is

$$U = U_b U_a. \tag{23}$$

The axis of U_b is aligned with $\hat{S}_1$ so that $\hat{r}_b = \hat{S}_1$ and its DGD is set to match the fiber's sweep rate:

$$\Delta\tau_b = p_\omega = 4k. \tag{24}$$

The concatenation diagram applies to the interface between the two plates where the PMD vector $\boldsymbol{\tau}_{\text{int}}$ of the combination is adjusted to be aligned with $\hat{S}_2$ and set to the prescribed magnitude of the fiber's DGD, $\mid \boldsymbol{\tau}_{\text{int}} \mid = \Delta\tau$. This is accomplished by orienting $\boldsymbol{\tau}_a$ such that

$$\boldsymbol{\tau}_a = \boldsymbol{\tau}_{\text{int}} - \boldsymbol{\tau}_b, \tag{25}$$

and setting its magnitude to

$$\Delta\tau_a^2 = \Delta\tau^2 + p_\omega^2. \tag{26}$$

The PMD vector $\boldsymbol{\tau}_{\text{int}}$ at the interface does not change with ω. The concatenation rule says that the vector $\boldsymbol{\tau}_{\text{int}}$ will be transformed to the emulator output by U_b like any other Stokes vector. As it is perpendicular to the rotation axis $\hat{S}_1$, the output PMD vector $\boldsymbol{\tau}(\omega)$ will be swept with ω in the $\hat{S}_2/\hat{S}_3$ plane. It is described by

$$\boldsymbol{\tau}(\omega) = \Delta\tau(0, \cos 4k\Delta\omega, \sin 4k\Delta\omega). \tag{27}$$

From this we find the higher-order PMD vectors of the planar sweep model by differentiation. At the carrier frequency we have

$$\boldsymbol{\tau}(\omega_0) = \Delta\tau\hat{S}_2, \tag{28a}$$

$$\boldsymbol{\tau}_\omega(\omega_0) = \Delta\tau p_\omega \hat{S}_3, \tag{28b}$$

$$\boldsymbol{\tau}_{\omega\omega}(\omega_0) = \Delta\tau p_\omega^2 \hat{S}_2, \tag{28c}$$

The emulator's first- and second-order PMD vectors of Eqs. (28a) and (28b) match the prescribed PMD vectors of the fiber. Note that this model generates a third-order PMD vector that is aligned with the first-order vector and has a magnitude of $\Delta\tau p_\omega^2$.

A discussion of two-section emulators with general conical sweep can be found in [19], while the mathematical PMD inversion of the planar sweep is described in [13] and [17]. We should note that the planar sweep model does not emulate PCD as we have prescribed that $\Delta\tau_\omega = 0$. The latter can be considered a minor deficiency of these types of emulators as PSP depolarization is the statistically dominant second-order PMD component.

5.3. *The EMTY Model*

A third type of second-order PMD model has been proposed [4], which we name by the initials of the authors' surnames: EMTY. The model, with Jones matrix U, consists of two sections of different rotational power,

$$U = U_2 U_1, \tag{29}$$

with Jones matrices U_1 and U_2. The latter are represented by frequency independent rotation axes, $\hat{r}_1$ and $\hat{r}_2$, and rotation angles of the form

$$\varphi_1(\omega_0 + \Delta\omega) = k_1 \Delta\omega, \qquad \varphi_2(\omega_0 + \Delta\omega) = \frac{1}{2} k_2 \Delta\omega^2. \tag{30}$$

Here, the powers of $\Delta\omega$ indicate the power of rotation of each section. We designate this rotational power by the subscripts of U_i, $\hat{r}_i$, φ_i, etc. As the rotation axes are fixed, we have $\hat{r}_{i\omega} = 0$ and obtain the output PMD vectors of the individual sections from Eq. (7) in the simple form

$$\boldsymbol{\tau}_1(\omega) = k_1 \hat{r}_1, \qquad \boldsymbol{\tau}_2(\omega) = k_2 \hat{r}_2 \Delta\omega. \tag{31}$$

As $U_2(\omega_0) = U_1(\omega_0) = I$, the concatenation rules (more detail in Section 6) lead to a simple vector summation of the contributions of the two rotational sections to the PMD vector at the output of U. One gets

$$\boldsymbol{\tau}(\omega_0) = k_1 \hat{r}_1, \qquad \boldsymbol{\tau}_\omega(\omega_0) = k_2 \hat{r}_2. \tag{32}$$

These simple relations make it trivial to determine the model parameters when the fiber's $\boldsymbol{\tau}(\omega_0)$ and $\boldsymbol{\tau}_\omega(\omega_0)$ are known (or measured): $k_1 = \Delta\tau$, $\hat{r}_1 = \hat{p}$, etc. The third-order PMD vector generated by the model follows from the concatenation rules of Section 6 as

$$\boldsymbol{\tau}_{\omega\omega}(\omega_0) = k_1 k_2 \hat{r}_2 \times \hat{r}_1 = \boldsymbol{\tau}_\omega \times \boldsymbol{\tau}. \tag{33}$$

The right-hand side of Eq. (33) shows the known fiber parameters. It is obtained via the matching conditions of Eqs. (32). The emulator's $\boldsymbol{\tau}_{\omega\omega}(\omega_0)$ is perpendicular to the fiber's $\boldsymbol{\tau}(\omega_0)$ and $\boldsymbol{\tau}_\omega(\omega_0)$. Equations (1) and (15) give its magnitude as $\Delta\tau^2 p_\omega$, where p_ω is the magnitude of the fiber's depolarization vector $\hat{p}_\omega$.

While extensions of the earlier approaches to inversion of PMD vectors higher than second order appear complicated, we show in the following that there is a relatively simple path to higher-order inversion. The approach uses higher-order concatenation rules, concatenation of sections with higher rotational power and their EMTY matrices, as well as a Stokes vector description of these higher-power rotations.

6. Higher-order Concatenation

In the following discussion we use the model building approach to accomplish PMD inversion for higher PMD orders employing models with several concatenated sections. The motivation for understanding these higher orders is illustrated by the transmission systems experiment reported in [20]. Figure 4 shows the first- and second-order PMD components of the fiber used to determine PMD induced optical receiver penalties for transmission at 10 Gb/s. High penalties of several dB were measured near the depolarization spike in the vicinity of 1545.4 nm. The 0.05-nm width of this $\hat{p}_\omega$ spike is comparable to the bandwidth of the signal and clearly requires a PMD characterization of at least fourth order.

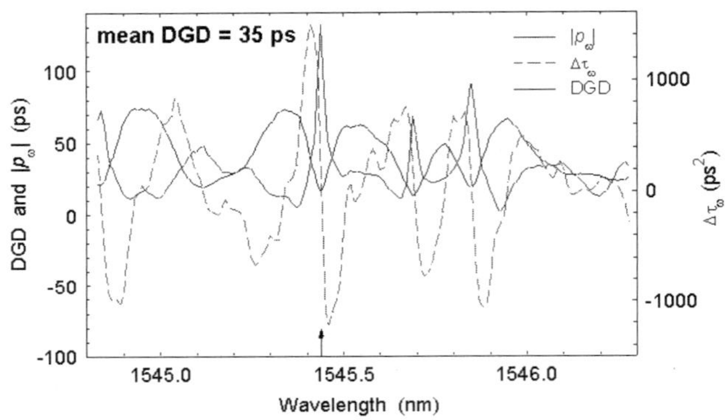

Fig. 4. Measured DGD, PSP depolarization, and PCD as a function of wavelength for a fiber with a mean DGD of 35 ps [20].

We demonstrate our approach for third-and fourth-order PMD. For this task, we will need the concatenation rules (see, e.g., [2,8]) and their extension to the higher orders. These rules say that PMD vectors transform like Stokes vectors, and the PMD vector of the combined system is equal to the sum of the PMD vectors of individual concatenated sections transformed to the output. The frequency-dependent transformation of the PMD vectors through a given section is governed by its Müller matrix R,

$$\mathbf{t}(\omega) = R(\omega)\mathbf{s}(\omega), \tag{34}$$

where R is isomorphically related to the Jones matrix U of the section [8]. Here $\mathbf{s}$ represents the sum of the PMD vectors of all preceding sections and $\mathbf{t}$ is the transformed PMD vector at the output of the section. To determine the higher-order PMD vectors at the carrier ω_0 we require higher-order transformation rules. These follow from the first-order relation of Eq. (34) by repeated differentiation. We also repeatedly substitute for the derivative R_ω using the identity [8]

$$R_\omega = \boldsymbol{\tau} \times R, \tag{35}$$

where $\boldsymbol{\tau}\times$ is a cross-product operator and $\boldsymbol{\tau}$ is the PMD vector of the transforming section. Using the simpler prime notation instead of the subscript ω, one obtains

$$t' = Rs' + \tau \times t, \tag{36}$$

$$t'' = Rs'' + \tau' \times t + 2\tau \times t' - \tau \times (\tau \times t), \tag{37}$$

$$\begin{aligned} t''' = {} & Rs''' + \tau'' \times t + 3\tau' \times t' + 3\tau \times t'' - \tau' \times (\tau \times t) \\ & -2\tau \times (\tau' \times t) - 3\tau \times (\tau \times t') + \tau \times (\tau \times (\tau \times t)), \end{aligned} \tag{38}$$

for the transformations of second, third, and fourth order through the section. Here we have omitted the boldface of the vectors to simplify the notation. These transformation rules will be used repeatedly in the following derivations.

7. Rotations of Higher Power

The properties of higher-power rotation sections have turned out to be very useful for higher-order PMD inversion. As in Section 5.3, these rotations are described by EMTY Jones matrices. The rotation of power (n) is defined by a frequency-independent rotation axis $\hat{r}_n$ and a frequency-dependent rotation angle

$$\varphi_n = k_n \Delta\omega^n / n!, \tag{39}$$

where the power of the frequency dependence equals the power of the rotation, and k_n is a constant. The subscript n in these quantities is used to designate the rotational power. When needed to avoid confusion we put this subscript in parentheses, as in $U_{(n)}$ and $\boldsymbol{\tau}_{(n)}$ to distinguish it from the subscript indicating placement of a section. The rotation of power (1) is, of course, equivalent to that of a birefringent plate. To simplify our expressions, we introduce the rotation vector

$$\boldsymbol{\rho}_n = k_n \hat{r}_n, \tag{40}$$

a Stokes vector characterizing the nth-power rotation. This is similar to the definition of a PMD vector, and indeed we find that there is a close relationship. As $r_{n\omega} = 0$, Eq. (7) simplifies and yields the PMD vector of a rotation section,

$$\boldsymbol{\tau}_{(n)}(\omega) = \Delta\omega^{n-1} \cdot \boldsymbol{\rho}_n / (n-1)!. \tag{41}$$

We can differentiate this simple expression with respect to frequency, evaluate the PMD vectors of all orders at ω_0, and find

$$\boldsymbol{\tau}_{(n)}^{(n-1)}(\omega_0) = \boldsymbol{\rho}_n, \tag{42}$$

where the superscript in parentheses indicates the order of differentiation. The nth-order PMD vector of a rotation of power (n) is equal to the rotation vector $\boldsymbol{\rho}_n$. At ω_0 all other PMD orders of this section are zero: An EMTY matrix of power (n) is empty of all PMD orders except for one, the nth-order PMD vector. It represents an element of pure high-order PMD. For the first- to third-power rotations we have, for example, $\boldsymbol{\tau}_{(1)} = \boldsymbol{\rho}_1$ for the PMD vector of a power (1) section, $\boldsymbol{\tau}_{(2)\omega} = \boldsymbol{\rho}_2$ for the second-order PMD vector of a power (2) rotation, and $\boldsymbol{\tau}_{(3)\omega\omega} = \boldsymbol{\rho}_3$ for the third-order vector of a power (3) rotation. All other PMD vectors are zero.

These properties of nth-power rotations allow a simple physical interpretation of pure high-order PMD. Refer to the diagonal form of the Jones matrix as in Eq. (10) and the corresponding Fig. 2. As the rotation axes $\hat{r}_n$ are frequency independent, the corresponding Q matrices are also frequency independent and can be represented by frequency independent polarization controllers. The eigenvalues of the diagonal matrix D are $e^{-j\varphi_n/2}$ and $e^{j\varphi_n/2}$. For pure first-order PMD we have $\varphi_1 = \Delta\tau\Delta\omega$, pictured as the differential group delay (DGD). Pure second-order PMD can thus be pictured as differential group velocity dispersion, and pure third-order PMD can be seen as differential dispersion slope. Here the "differentials" refer to differences of the mentioned dispersion characteristics between the polarizations identified by $\pm\hat{r}_n$.

There is another advantage of practical importance that derives from the PMD purity of the nth-power rotations. The transformation rules needed for PMD vector concatenation of several sections contain many terms with high-order PMD vectors of the transforming section. When nth-power rotations are used for those sections, most of these higher-order terms disappear as each section has only one PMD order, greatly simplifying model building and PMD inversion. We exploit this in Section 8.

8. Higher-order PMD Inversion

To accomplish PMD inversion for higher orders, we use model building and aim to match the first n orders of the fiber's PMD vector at ω_0. For the specific example of $n = 4$ we represent the fiber by a concatenation of the lowest four rotational powers as shown in Fig. 5, where U_0 represents a frequency-independent polarization controller as before. The U matrix of the model is

$$U = U_{(4)}U_{(3)}U_{(2)}U_{(1)}, \tag{43}$$

where the $U_{(n)}$ are EMTY matrices with the subscript indicating the power of rotation. The matrices $U_{(n)}$ are completely characterized by their rotation vector $\boldsymbol{\rho}_n$ defined in Eq. (40).

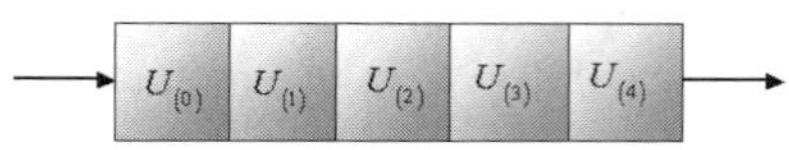

Fig. 5. Concatenation of the first four EMTY rotations, placed in increasing rotational power.

Note that the EMTY matrices do not commute, and that each specific ordering leads to a different model. We chose the ordering that we found to give the simplest inversion algorithm. The output PMD vectors of the model are found by the concatenation rules using the iteration

$$\boldsymbol{\tau}(n) = \boldsymbol{\tau}_n + R_n\boldsymbol{\tau}(n-1), \tag{44}$$

where $\boldsymbol{\tau}(n)$ is the ω-dependent PMD vector after the nth section, $\boldsymbol{\tau}_n(\omega)$ is the PMD vector of the nth section and $R(\omega)$ is the Müller matrix of that section. The vector $\boldsymbol{\tau}(n-1)$ is the output PMD vector of all sections preceding the nth, and corresponds to the input vector $\mathbf{s}(\omega)$ used in Section 6. The transformed output vector $R_n\boldsymbol{\tau}(n-1)$ corresponds to $\mathbf{t}(\omega)$. The transformation rules for the higher PMD orders listed in Eqs. (36) to (38) simplify for the present model because $R_{(n)}(\omega_0) = I$ and because of the PMD-purity of the nth-power rotations, as noted.

The iteration at ω_0 has to carry all required PMD orders, and here we choose four. We start at the output of U_1, where

$$\boldsymbol{\tau}(1) = \boldsymbol{\tau}_1 = \boldsymbol{\rho}_1, \tag{45}$$

and all higher PMD orders vanish at ω_0. Proceeding to two sections, $\boldsymbol{\tau}(1)$ becomes the input $\mathbf{s}$ for the transformation by R_2 and $\boldsymbol{\tau}_{(2)\omega} = \boldsymbol{\rho}_2$ is added. Using Eqs. (36)–(38) we find the first four orders of the output PMD vector of U_2U_1 [leaving out the (2) designation]:

$$\boldsymbol{\tau} = \boldsymbol{\rho}_1, \tag{46a}$$
$$\boldsymbol{\tau}_\omega = \boldsymbol{\rho}_2, \tag{46b}$$
$$\boldsymbol{\tau}_{\omega\omega} = \boldsymbol{\rho}_2 \times \boldsymbol{\rho}_1, \tag{46c}$$
$$\boldsymbol{\tau}_{\omega\omega\omega} = 0. \tag{46d}$$

The results (46a)–(46c) have already been used in the discussion of second-order models in Section 5.3. We should note that while we obtain a zero fourth-order PMD vector for the two-section model, the fifth and higher orders are nonzero vectors.

The output PMD vectors of $U_{(2)}U_{(1)}$ are used as the inputs $\mathbf{s}$ to the third section to proceed with the iteration, and so on. After four sections we obtain the output PMD vectors of $U_{(4)}U_{(3)}U_{(2)}U_{(1)}$:

$$\boldsymbol{\tau} = \boldsymbol{\rho}_1, \tag{47a}$$
$$\boldsymbol{\tau}_\omega = \boldsymbol{\rho}_2, \tag{47b}$$
$$\boldsymbol{\tau}_{\omega\omega} = \boldsymbol{\rho}_3 + \boldsymbol{\rho}_2 \times \boldsymbol{\rho}_1, \tag{47c}$$
$$\boldsymbol{\tau}_{\omega\omega\omega} = \boldsymbol{\rho}_4 + \boldsymbol{\rho}_3 \times \boldsymbol{\rho}_1. \tag{47d}$$

The PMD vectors for three sections $(U_{(3)}U_{(2)}U_{(1)})$ are obtained from these expressions by setting $\boldsymbol{\rho}_4 = 0$. The simplicity of the expressions is rooted in the PMD-purity of the nth-power rotations and in the chosen ordering. The expressions of Eqs. (47) have a particularly appealing structure as the addition of any nth-power rotation does not add terms to PMD vectors of lower order than n.

The above results give us the PMD vectors of the four-section model for given U, i.e., for given $\boldsymbol{\rho}_1$ through $\boldsymbol{\rho}_4$. However, the simplicity of these expressions makes PMD inversion an easy task. If the PMD vectors are given, e.g., as derived from a PMD measurement, the parameters of the $U_{(n)}$ matrices of the model matching these PMD vectors are determined by the inversion algorithm

$$\boldsymbol{\rho}_1 = \boldsymbol{\tau}, \tag{48a}$$
$$\boldsymbol{\rho}_2 = \boldsymbol{\tau}_\omega, \tag{48b}$$
$$\boldsymbol{\rho}_3 = \boldsymbol{\tau}_{\omega\omega} + \boldsymbol{\tau} \times \boldsymbol{\tau}_\omega, \tag{48c}$$
$$\boldsymbol{\rho}_4 = \boldsymbol{\tau}_{\omega\omega\omega} + \boldsymbol{\tau} \times \boldsymbol{\tau}_{\omega\omega} + \boldsymbol{\tau} \times (\boldsymbol{\tau} \times \boldsymbol{\tau}_\omega). \tag{48d}$$

For numerical pulse modeling based on these expressions, the resulting Jones matrix should, probably, be left in the product forms of Eqs. (3) and (43). Some programs may even be capable of handling the compact exponential form of Eq. (4) for $U_{(n)}$. As an illustration, consider the Jones matrix $U_{(4)}(\omega_0 + \Delta\omega)$ of the last element in our current model, the rotation section of power (4). Its exponential form is

$$U_{(4)} = e^{-j\Delta\omega^4 \boldsymbol{\rho}_4 \cdot \boldsymbol{\sigma}/24}. \tag{49}$$

Our fourth-order model will match the prescribed PMD vectors if we set the rotation vector of $U_{(4)}$ as required by Eq. (48d).

The algorithm for third- and second-order PMD inversion are contained in the above fourth-order algorithm. One simply eliminates $\boldsymbol{\rho}_4$ and $U_{(4)}$ for the third-order model, and $\boldsymbol{\rho}_3$ and $U_{(3)}$ for the second-order model. On the other hand, if a second-order model is required that has zero third- and fourth-order PMD, one can prescribe $\boldsymbol{\tau}_{\omega\omega} = \boldsymbol{\tau}_{\omega\omega\omega} = 0$. This latter assumption would be similar to the one made in some mathematical approaches [16].

9. Conclusions

When a fiber is characterized by measured PMD vector data, PMD inversion is required to determine the fiber's pulse response, i.e., its Jones matrix. Once this matrix is known, it can be used for such tasks as modeling system impairments and understanding the physical picture of higher-order PMD. We have briefly reviewed various approaches to PMD inversion including a discussion of three second-order models. While all three models can match the fiber's first-and second-order PMD vectors, they are distinguished by the specific characteristics of the Jones matrix eigenvectors and eigenvalues and by the third-order PMD vectors generated by each model. We show how PMD inversion can be extended to higher orders using higher-order concatenation rules and concatenation of higher-power rotations represented by EMTY matrices. These higher-power rotations are represented by rotation (Stokes) vectors. Relatively simple relationships are found between these rotation vectors and the higher-order PMD vectors of the model. This simplicity leads to compact inversion algorithms. We report results up to fourth-order PMD. Available statistics for first- and second-order PMD vectors can be applied to the elements of the Jones matrices of the first- and second-power rotations as their rotation vectors equal the PMD vectors. The higher-power rotations represent cases of pure high-order PMD, allowing their interpretation as a differential group delay for the first-order, differential group velocity dispersion for second order, differential dispersion slope for third order and so on for the higher orders.

For a more detailed discussion of these and related issues the reader is referred to [21].

Acknowledgments

The authors are grateful to Jerry Foschini, Bob Jopson, and Peter Winzer for valuable discussions and shared insight.

Appendix A: Formulations of the matrix U

We list here, for convenience, various equivalent forms of the Jones matrix U. Their connections are discussed in [8]. Special emphasis is placed on the relations between the matrix elements and the eigenvectors, $|\ r\rangle$ and $|\ r_-\rangle$, of U, the corresponding rotation axis represented by the unit Stokes vector $\hat{r} = \langle r \mid \boldsymbol{\sigma} \mid r \rangle$, and the rotation angle φ. The compact exponential form is

$$U = e^{-j(\varphi/2)\hat{r}\cdot\boldsymbol{\sigma}}, \tag{50}$$

where the matrix appearing in the exponent is defined as

$$\hat{r} \cdot \boldsymbol{\sigma} = r_1\sigma_1 + r_2\sigma_2 + r_3\sigma_3, \tag{51}$$

and the Pauli spin matrices are

$$\sigma_1 = \begin{vmatrix} 1 & 0 \\ 0 & -1 \end{vmatrix}, \qquad \sigma_2 = \begin{vmatrix} 0 & 1 \\ 1 & 0 \end{vmatrix}, \qquad \sigma_3 = \begin{vmatrix} 0 & -j \\ j & 0 \end{vmatrix}. \tag{52}$$

The closely related rotational form is

$$U = I\cos\varphi/2 - j\hat{r}\cdot\boldsymbol{\sigma}\sin\varphi/2. \tag{53}$$

Using the relation

$$\mid r\rangle\langle r \mid = \frac{1}{2}(I + \hat{r}\cdot\boldsymbol{\sigma}), \tag{54}$$

the rotational expression can be written in the dyadic form

$$U = e^{-j\varphi/2} \mid r\rangle\langle r \mid + e^{j\varphi/2} \mid r_-\rangle\langle r_- \mid. \tag{55}$$

The Caley-Klein form is

$$U = \begin{vmatrix} \alpha & \beta \\ -\beta^* & \alpha^* \end{vmatrix}, \tag{56}$$

where $\alpha\alpha^* + \beta\beta^* = 1$. The Caley-Klein parameters are connected to the rotation parameters by

$$\alpha = \cos(\varphi/2) - jr_1\sin(\varphi/2), \qquad \beta = -(r_3 + jr_2)\sin(\varphi/2), \tag{57}$$

which is easily verified by expanding Eq. (53). Finally we mention the diagonal form

$$U = QDQ^\dagger, \tag{58}$$

where the diagonal matrix exhibits the eigenvalues, i.e.,

$$D = \begin{vmatrix} e^{-j\varphi/2} & 0 \\ 0 & e^{j\varphi/2} \end{vmatrix}, \tag{59}$$

and the column vectors of Q are the eigenvectors,

$$Q = (\mid r\rangle \mid r_-\rangle). \tag{60}$$

References

[1] Poole, C.D. and J.A. Nagel, "Polarization effects in lightwave systems," *Optical Fiber Telecom. IIIA*, I. P. Kaminow and T. L. Koch, eds. (Academic Press, San Diego, 1997).

[2] Kogelnik, H., L.E. Nelson, and R.M. Jopson, "Polarization mode dispersion," *Optical Fiber Telecommunications IVB*, I. P. Kaminow and T. Li, eds. (Academic Press, San Diego, 2002).

[3] Shieh, W. "Principal states of polarization for an optical pulse," IEEE Photon. Technol. Lett., **11**, 677–679 (1999).

[4] Eyal, A., W. Marshall, M. Tur, and A. Yariv, "Representation of second-order PMD," Elect. Lett., **35**,1658–1659 (1999).

[5] Li, Y., A. Eyal, P.-O. Hedekvist, and A. Yariv, "Measurement of High-Order Polarization Mode Dispersion," IEEE Photon. Technol. Lett., **12**, 861–863 (2000).
Eyal, A., Y. Li, W. K. Marshall and A. Yariv, " Statistical determination of the length dependence of high-order polarization mode dispersion," Opt. Lett., **25**, 875–877 (2000).

[6] Leppla, R. and W. Weiershausen, "A new point of view regarding higher order PMD pulse distortion: Pulse spreading interpreted as multi path propagation," Proc. European Conf. on Optical Communication, ECOC 2000. 3:211–212, 2000. .

[7] Bononi, A. and A. Vannucci, "Statistics of the Jones matrix of fibers affected by polarization mode dispersion," Opt. Lett., **26**, 675–677 (2001).
Vannucci, A. and A. Bononi, "Sensitivity Penalty distribution in fibers with PMD: a novel semi-analytical technique," Proc. Optical Fiber Comm. Conf., OFC 2002. 54–56, 2002. .

[8] Gordon, J. P. and H. Kogelnik, "PMD Fundamentals: Polarization mode dispersion in optical fibers," Proc. Natl. Acad. Sci. USA, **97**, 4541–4550 (2000).

[9] Heffner, B. L, "Automated measurement of polarization mode dispersion using Jones matrix eigenanalysis," IEEE Photon. Technol. Lett., **4**, 1066–1069 (1992).
Heffner, B. L, "Accurate, automated measurement of differential group delay dispersion and principal state variation using Jones matrix eigenanalysis," IEEE Photon. Technol. Lett., **5**, 814–817 (1993).

[10] Jopson, R., L. Nelson, and H. Kogelnik, "Measurement of second-order PMD vectors in optical fibers," IEEE Photon. Technol. Lett., **11**, 1153–1155 (1999).
Nelson, L., R. Jopson and H. Kogelnik, "Müller matrix method for determining PMD vectors," Proc. European Conf. on Optical Communication, ECOC'99, 2:10, 1999.

[11] Nelson, L., R. Jopson, H. Kogelnik, and J. P. Gordon, "Measurement of polarization mode dispersion vectors using the polarization-dependent signal delay method," Opt. Express, **6**, 158–167 (2000).
Nelson, L., R. Jopson, and H. Kogelnik, "Influence of measurement parameters on polarization mode dispersion measurements using the signal delay method," Proc. European Conf. Opt. Comm., ECOC 2000, 1:143–144, 2000.

[12] Bruyere, F., "Impact of first- and second-order PMD in optical digital transmission systems," Opt. Fiber Technol., **2**, 269–280 (1996).

[13] Penninckx, D. and V. Morénas, "Jones matrix of polarization mode dispersion," Opt. Lett., **24**, 875–877 (1999).

[14] Kogelnik, H., L. E. Nelson, J. P. Gordon, and R. M. Jopson, "Jones matrix for second-order polarization mode dispersion," Opt. Lett., **25**, 19–21 (2000).

[15] Coddington, E.A. and N. Levinson, *Theory of Ordinary Differential Equations* (McGraw-Hill, NY, 1955).
Gantmacher , F. R., *The Theory of Matrices* (Chelsea Publishing, NY, 1964).

[16] Forestieri, E. and L. Vincetti, "Exact Evaluation of the Jones Matrix of a Fiber in The Presence of Polarization Mode Dispersion of Any Order," J. Lightwave Technol., **19**, 1898–1909 (2001).

[17] Orlandini, A. and L. Vincetti, "A Simple and Useful Model for Jones Matrix to Evaluate Higher Order PMD Effects," IEEE Photon. Technol. Lett., **13**, 1176–1178 (2001).

[18] Möller, L. and H. Kogelnik, "PMD emulator restricted to first- and second-order PMD generation," Proc. European Conf. Opt. Comm., ECOC'99, 2:64–65, 1999.
Shtaif, M., A. Mecozzi, M. Tur and J. A. Nagel, "A compensator for the effects of high-order PMD in optical fibers," IEEE Photon. Technol. Lett., **12**, 434–436 (2000).

[19] Patscher, J. and R. Eckhardt, "Component for second-order compensation of polarization-mode dispersion," Elect. Lett., **33**, 1157–1159 (1997).

[20] Nelson, L., R. Jopson, and H. Kogelnik, "PMD penalties associated with rotation of principal states of polarization in optical fiber," Proc. Optical Fiber Comm. Conf., OFC 2000, ThB2, 25–27, 2000.

[21] H. Kogelnik, L. E. Nelson, and J. P. Gordon, "Emulation and inversion of polarization mode dispersion," J. Lightwave Technol., **9**, 1439–1456 (2003).

DOI: 10.1007/s10297-004-0006-0
Originally published in J. Opt. Fiber Commun. Rep. 1, 141–149 (2004)

Numerical modeling of PMD

D. Chowdhury, M. Mlejnek, and Y. Mauro

SP-TD-02-8, Science & Technology, Corning Inc., Corning, NY 14831

Abstract. A procedure for evaluating system performance considering all orders of polarization mode dispersion (PMD) is presented. The method provides engineering rules for computing system penalty introduced by PMD. Computer simulation of PMD induced penalty indicates that for generating a realistic system budget, effect of higher order PMD in presence of chromatic dispersion is essential. Considering all of these penalties, in order to obtain a system-reach of 3200 km at 40 Gb/s, a fiber PMD of less than 0.04 ps/$\sqrt{\text{km}}$ is required when at the signal wavelength chromatic dispersion is 0. In presence of 18 ps/nm chromatic dispersion for the same system reach this PMD value reduces to about 0.02 ps/$\sqrt{\text{km}}$.

1. Introduction

Polarization mode dispersion (PMD) is defined as the statistical ensemble average of differential group delay (DGD) between the two principal states of polarization (PSP) [1]-[3] of an optical fiber. With increasing bit-rate, PMD is going to be a major system impairment [4]-[11]. Since PMD is a statistical parameter of the system, depending sensitively on environmental condition and deployment [3],[12],[13], evaluation of system performance degradation by PMD is not trivial experimentally. Understanding how to evaluate system performance in presence of PMD is critical in order to budget for and mitigate PMD induced penalty.

Probability density functions of DGD and its frequency derivative are well understood [3]. The worst performance of the system is associated with large values of DGD and its frequency derivative that are low probability events [14]. Simple emulators used in the lab to simulate realistic situations generally do not reproduce these distributions well. Consequently, computer modeling is intensively used to understand PMD induced penalty and its mitigation [7],[8],[10],[11].

System impact of PMD is generally simulated by time domain simulation of vector nonlinear schrödinger equation [15],[16]. Random concatenation of small birefringence sections are used to represent realistic fiber PMD [17] in a Monte-Carlo type simulation. In order to sample the rare events in the probability distribution of DGD and its frequency derivative in such simulations, multiple methods have been proposed [18]-[21]. In this paper we will discuss a method of estimating system penalty caused by PMD considering the statistical nature of it.

The paper is organized as follows. In section 2 we first describe how we compute the performance parameter Q for the system. Then in section 3 we discuss a method of computing PMD related Q degradation based on only first order PMD. Using this estimate of system performance we compute the maximum length of a system with a given performance, system reach, as a function of fiber and component PMD. In section 4 we extend the method of PMD induced penalty outlined in the section before to include all orders of PMD. Estimation of system reach is shown when all orders of PMD is considered. Finally in Section 5 we summarize our findings.

2. *Q* Estimation

System penalty is measured by the estimation of the bit-error-rate (BER). For a noise limited system with an assumed Gaussian distributed intensity noise generated by the amplified spontaneous emission of optical amplifiers one obtains a quantity Q which is related to BER as BER= $\frac{1}{2}\,\mathrm{erfc}\left(\frac{Q}{\sqrt{2}}\right)$ [22]. With the same assumptions on the noise properties one can also derive a simple expression [22]

$$Q = \frac{< i_1 > - < i_0 >}{\sigma_1 + \sigma_0}, \tag{1}$$

where $< i_n >$ $(n = 0, 1)$ is the average level of logical 0 or 1, and σ_n is the corresponding noise variance at the same level. If we make a further simplifying assumption that only signal spontaneous beat noise is responsible for the detected noise at the receiver, we can express the noise variance as $\sigma_n = \sqrt{2\rho_{ASE} < i_n >}$ $(n = 1, 0)$. With the use of the expression for noise variance in Eq. (1) we obtain

$$Q = \frac{\sqrt{< i_1 >} - \sqrt{< i_0 >}}{\sqrt{2\rho_{ASE}}}, \tag{2}$$

where ρ_{ASE} is the noise power density. If we want to only estimate the Q-penalty instead of absolute Q and assume that output Q of the optical system without PMD is Q_0 and the output Q with PMD is Q_{PMD} then PMD induced degradation is given by

$$\Delta Q\ [dB] = 10\ \log_{10}\left(\frac{Q_{PMD}}{Q_0}\right) = Q_{PMD}[dB] - Q_0[dB] =$$
$$10\ \log_{10}\left(\frac{\sqrt{< i_1 >_{PMD}} - \sqrt{< i_0 >_{PMD}}}{\sqrt{< i_1 >_0} - \sqrt{< i_0 >_0}}\right), \tag{3}$$

where the subscripts PMD or 0 indicate the average current value for system with or without PMD correspondingly. We assume that the value of ρ_{ASE} is the same for both the systems. In the rest of the paper we will use Eq. (3) to compute the dBQ (Q expressed in dB) penalty introduced by PMD.

In order to compute penalty induced by PMD, we simulate a realistic system with chromatic dispersion and PMD without *noise and nonlinearity* and use Eq. (3) to compute the penalty. We do not use any optical amplifiers in the simulation because we are only interested in the incremental degradation of the system caused by PMD [see arguements made about ρ_{ASE} in order to obtain Eq. (3)], not in the absolute Q or BER. The average values of the level 1 and 0 are computed from the computed eye diagrams for each case.

3. System Reach Considering First-order PMD only

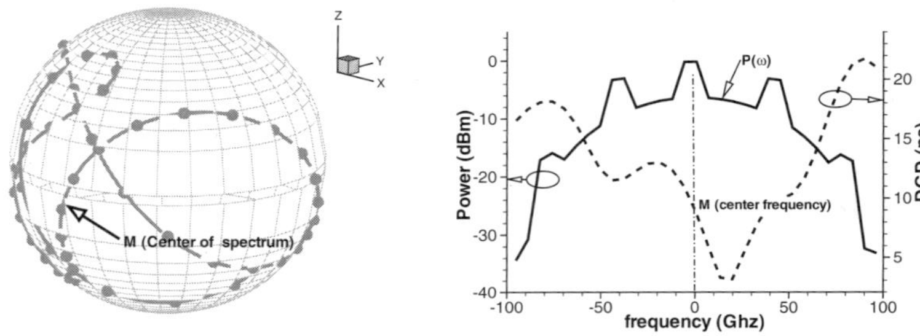

(a) (b)

Fig. 1. In (a) principal states of polarization (PSP) after propagation through a fiber is shown. In (b) the value of DGD as a function of offset from the signal center frequency is shown along with the optical power spectrum, P(ω), for a 40 Gb/s 50% duty cycle signal.

In this section we introduce the method of PMD induced penalty computation considering only the impact of first-order PMD, or DGD, and the fact that DGD distribution is Maxwellian. We ignore the impact of all other orders of PMD. Figure 1(a) shows computed normalized DGD vector (PSP) on the Poincaré sphere. The corresponding DGD amplitude as a function of frequency along with the optical spectrum for a 40 Gb/s signal is shown in Fig. 1(b). In computing system reach in this section we assume that only DGD value at the center of the frequency spectrum (denoted by M in Fig. 1) is causing the impairment.

Figure 2(a) shows the system penalty in dBQ as a function of DGD when only first order PMD is present for a 10 Gb/s system. The dots in Fig. 2(a) shows the data for a 10 Gb/s system (see [23]). However, since the x-axis is normalized by the bit-period, T_b, and PMD is a linear impairment, the results also apply for a 40 Gb/s system. Figure 2(b) shows the Maxwell DGD distribution and one minus the cumulative distribution of DGD for a communication fiber, i.e., for a randomly mode coupled fiber. In order to estimate the system reach we first decide on the power penalty that we want to associate with PMD. By choosing 0.5 dBQ of penalty for a 10 Gb/s system, from Fig. 2(a) we see that maximum tolerable DGD is $\Delta\tau_{\max} = 0.33 \times T_b = 0.33 \times 100$ ps = 33 ps. The second parameter that we have to decide is the probability level that we want to associate with $\Delta\tau_{\max}$. We will choose the probability of exceeding a targeted $\Delta\tau_{\max}$ value as our parameter. Unfortunately this choice does not translate into system outage time since the rate at which a new random value of DGD is realized in a real system

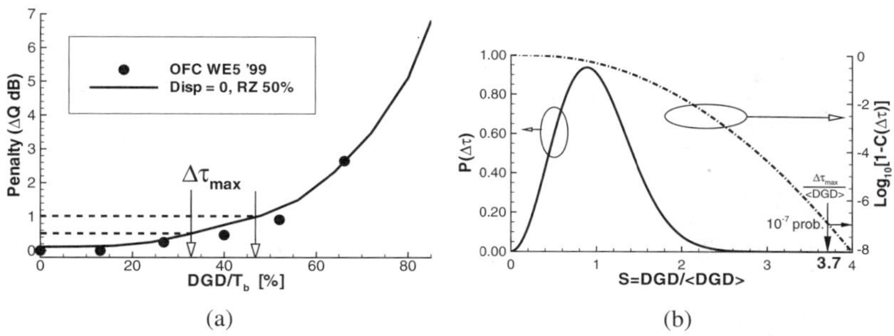

Fig. 2. In (a) power penalty in dB as a function of DGD is shown. Dots are measured results as published in[23]. Solid line is ΔQ computed as described in Section 2. x-axis is normalized by the bit-period T_b. In (b) Maxwell distribution and cumulative distribution for DGD is shown. x-axis normalized by the average DGD, $\langle\text{DGD}\rangle$ (i.e., PMD).

depends on the deployment. For example, in an undersea system the rate of change of DGD will be significantly slower than that for an aerial system. Consequently, for a chosen targeted $\Delta\tau_{\max}$ value and a probability of exceeding that value, outage time for an aerial system will be significantly higher than for an undersea system. But going forward, we will continue to use this probability as the value to be specified. This approach is consistent with current approaches taken by the standards body in the US and in Europe. Let us assume that we want to ensure that the system DGD does not exceed a value of 33 ps more than once in 10^7 instances of DGD realizations. This means that $1 - C(\Delta\tau)$ in Fig. 2(b) should be less than 10^{-7}. Since for a Maxwell distribution 10^{-7} probability occurs when DGD is 3.7 times the average DGD or PMD, we find that maximum tolerable PMD is 33 ps/3.7 = 8.92 ps. The factor 3.7, denoted by S in the following, is called the safety factor. With the use of quadrature addition of PMD for transmission fiber, dispersion compensating (DC) fiber, and network elements [24] maximum DGD for a system is computed as

$$\Delta\tau = S\sqrt{N_{\text{span}}\ (L_{\text{span}}\delta\tau^2_{\text{fiber}} + L_{\text{DC}}\,\delta\tau^2_{\text{DC}} + \Delta\tau^2_{\text{NE}})} \tag{4}$$

where L_{span} and L_{DC} are the span length and DC fiber length respectively, $\delta\tau_{\text{fiber}}$ and $\delta\tau_{\text{DC}}$ are fiber and DC fiber PMD in ps/$\sqrt{\text{km}}$ respectively, $\Delta\tau_{\text{NE}}$ is the network element DGD in ps, and N_{span} is the number of spans. Note that quadrature addition only applies to PMD and not to DGD. In order to compute DGD value at 10^{-7} probability, in Eq. (4) we multiplied the computed DGD with S, the safety factor.

In order to obtain the system reach $N_{\text{span}} \times L_{\text{span}}$, in Eq. (4) for any combination of $\delta\tau_{\text{fiber}}$ and $\Delta\tau_{\text{NE}}$ we maximize N_{span} while keeping $\Delta\tau \leq \Delta\tau_{\max}$, where $\Delta\tau_{\max}$ = 33 ps as defined before for a 10 Gb/s system. $\Delta\tau_{\max}$ is set to 0.33×25 ps = 8.25 ps for a 40 Gb/s system. L_{span} in this paper is set to 80 km. Figure 3 shows the maximum reach that can be achieved with 0.5 dBQ penalty for four different cases. These four cases are obtained by combining two different bit-rates with two different types of fiber. The two bit-rates are 10 and 40 Gb/s. Two fibers considered are standard single mode fiber (SMF) with 17 ps/(nm-km) chromatic dispersion with fully compensated span and NZDSF fiber with chromatic dispersion of 5 ps/(nm-km) with fully compensated span. As Fig. 3(a) shows, with SMF and DC system, in order to go 1600 km for a 10-

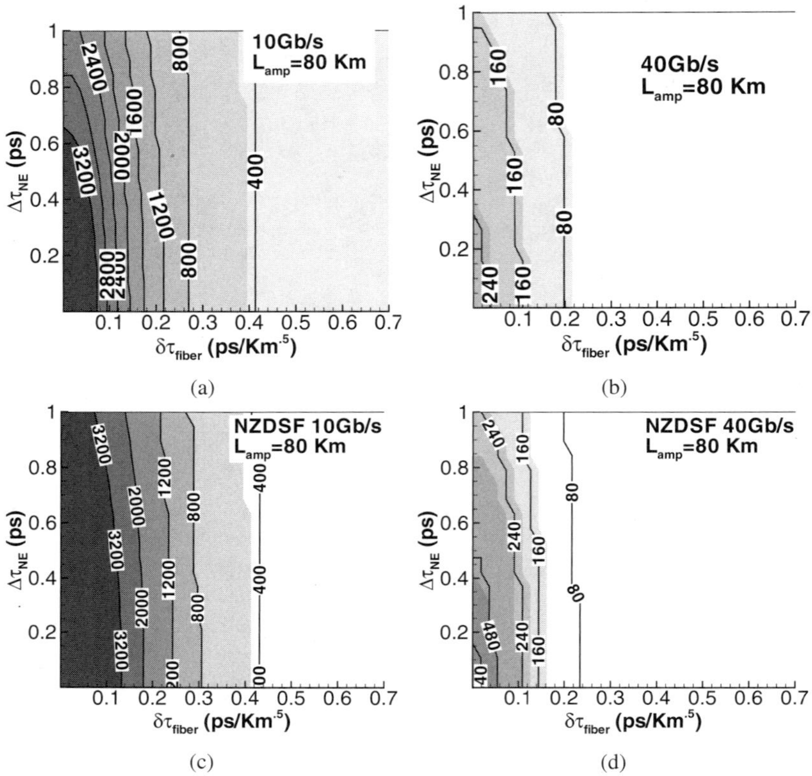

Fig. 3. System reach as a function of fiber DGD and network element DGD when 0.5 dBQ penalty is associated with PMD induced impairment. Only first-order PMD penalty is considered. In (a) and (c) system reach for a 10 Gb/s system is shown for SMF (dispersion at 1550 nm of 17 ps/nm/km) fiber and for NZDSF (dispersion at 1550 of 5 ps/nm/km) fiber respectively. In (b) and (d) bit-rate is 40 Gb/s and the rest of the parameters are the same as in (a) and (c). For both these cases $\Delta\tau_{\rm DC} = 0.35$ ps/$\sqrt{\rm km}$.

Gb/s system fiber PMD has to be less than 0.2 ps/$\sqrt{\rm km}$. On the other hand, Fig. 3(b) shows that a SMF/DC type system will not support more than three spans (240 km) for a 40-Gb/s system no matter what the network element PMD is! The main limitation for a 40-Gb/s system is caused by the DC fiber PMD. For all these computations $\Delta\tau_{\rm DC}$ = 0.35 ps/$\sqrt{\rm km}$. However, with the use of NZDSF fiber the length of DC fiber used is minimized and system reach of 3200 km at 10 Gb/s and 240 km at 40 Gb/s become feasible for reasonable values of network element PMD. In the following we will outline the process of extending this method to include all orders of PMD.

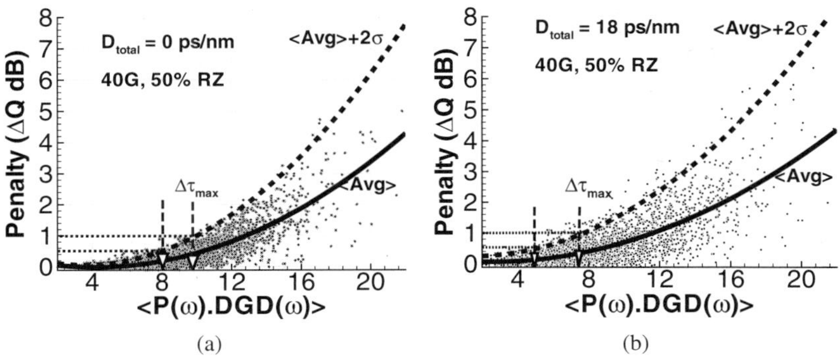

Fig. 4. Q Penalty in dB as a function of spectrally averaged DGD for 50% duty cycle RZ 40 Gb/s system. The dots are 3000 simulation with different random realization of eigen polarization angles for 900 segments in the simulation. In (a) simulation was done for no chromatic dispersion and in (b) a net 18 ps/nm dispersion is considered. Maximum tolerable DGD, $\Delta\tau_{\max}$, for 0.5 and 1 dBQ penalty are also shown in the figure.

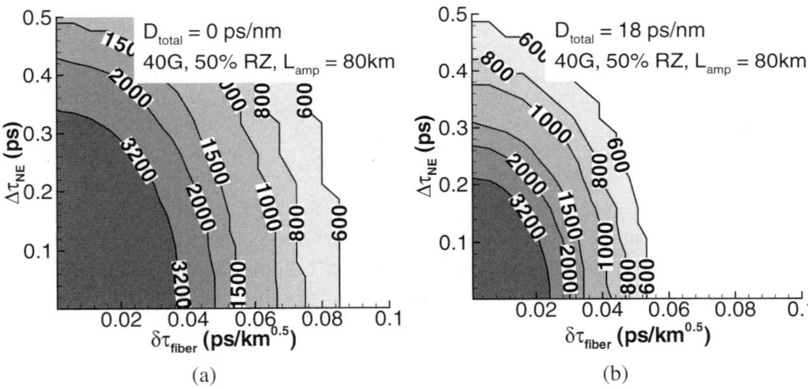

Fig. 5. System reach as a function of fiber and network element PMD is shown when all orders of DGD are considered and 0.5 dBQ penalty is associated with PMD induced impairments. In (a) simulation was done for no chromatic dispersion with the use of $\Delta\tau_{\max}$ from Fig. 4(a) and in (b) a net 18 ps/nm dispersion is considered with the use of $\Delta\tau_{\max}$ from Fig. 4(b).

4. System Reach Considering All PMD orders

In this section we extend the method outlined in Section 3 in order to incorporate penalty induced by PMD of all orders. The process involves estimating the penalty as a function of DGD for all orders of PMD included. Figure 4 shows simulated penalty vs. spectrally averaged DGD for a 40 Gb/s, 50% duty cycle RZ system. In these simulations no DC fiber is used. In Figs. 4(a) and 4(b) the end-to-end system chromatic dispersion is 0 ps/nm and 18 ps/nm respectively. Spectrally averaged DGD means that the DGD spectrum (similar to the one shown in Fig. 1) is multiplied by the optical spectral power and averaged over the spectral range of the signal. In other words, spectrally averaged DGD is

$$< \mathrm{DGD}(\omega)P(\omega) >= \frac{1}{2\Delta\omega}\int_{-\Delta\omega}^{\Delta\omega} \mathrm{DGD}(\omega)P(\omega)d\omega, \tag{5}$$

where $P(\omega)$ is the power spectrum. The simulated system was 90 km with 900 randomly concatenated fiber segments. Simulated data is then processes by computing average (marked <avg>) and standard deviation, σ, of 60 consecutive data points at a time along the x-axis. In Figure 4 we also indicate the maximum tolerable DGD $\Delta\tau_{\max}$ for 0.5 and 1 dBQ penalty from the <avg>+2σ curve. We chose the <avg>+2σ in order to obtain a conservative system design. Finally, we use Eq. (4) again with the $\Delta\tau_{\max}$ from Figure 4 to compute system reach for a 40 Gb/s system in Fig. 5. In this case no DC fiber is used, i.e., $\Delta\tau_{\mathrm{DC}}$ = 0.0 ps. Figure 5 shows the system reach as a function of fiber PMD and network element DGD for a 40 Gb/s, 50% duty cycle RZ system with and without residual chromatic dispersion when all orders of DGD are considered. Figure 5 shows that in order to achieve a system reach of 1600 km at 40 Gb/s, fiber PMD must be less than 0.05 and 0.03 ps/$\sqrt{\mathrm{km}}$ for 0 and 18 ps/(nm-km) residual chromatic dispersion, respectively. The tolerable PMD decreases steadily as network element PMD increases. From Fig. 4 we also note that 18 ps/(nm-km) chromatic dispersion adds about 0.5-1 dBQ of added penalty on top of the PMD induced penalty for zero residual dispersion.

5. Summary

A method for evaluating system performance considering all orders of PMD is presented. The method provides engineering rules for computing system penalty introduced by PMD and hence is useful for system budgeting procedure.

Results presented here indicate that if only first-order PMD penalty is considered, system reach for a SMF or NZDSF based system is limited by the DC fiber PMD rather than network element PMD. System reach for these systems is fairly independent of network element PMD. We find that at 10 Gb/s bit-rate, system reach of about 3200 km is achievable if fiber PMD is less than 0.1 ps/$\sqrt{\mathrm{km}}$. On the other hand, NZDSF fiber based system at 10 Gb/s can reach 3200 km with fiber PMD slightly greater than 0.1 ps/$\sqrt{\mathrm{km}}$. However, 40 Gb/s systems are not feasible beyond 240 km for either of the fiber choices.

When all orders of PMD is considered, system reach on SMF/DC based systems are severely limited. For a 40-Gb/s system without any chromatic dispersion, a system reach of 3200 km is obtained with a fiber PMD of less than 0.04 ps/$\sqrt{\mathrm{km}}$. This tolerance is reduced to about 0.02 ps/$\sqrt{\mathrm{km}}$ when a reasonable amount of chromatics dispersion (18 ps/nm) is present. This amount of chromatic dispersion is expected to be within the system budget specification for a 40 Gb/s system because of thermal and other fluctuations of chromatic dispersion. As can be seen from the results presented here, higher order PMD and residual chromatic dispersion add significant ΔQ penalty over first-order PMD. This added penalty suggests that for a realistic system design higher order PMD penalties in presence of chromatic dispersion must be included.

References

1. N. Gisin, R. Passy, and J. P. Von der Weid, "Definition and measurement of polarization mode dispersion: interferometric versus fixed analyzer methods," IEEE Photon. Technol. Lett. **6**, 730-732 (1994).
2. C. D. Poole and R. E. Wagner, "Phenomenological approach to polarization dispersion in long single-mode fibers," Electron. Lett. **22**, 1029-1030 (1986).
3. G. J. Foschini and C. D. Poole, "Statistical theory of polarization dispersion in single mode fibers," IEEE J. Lightwave Technol. **9**, 1439–1456 (1991).
4. C. D. Poole and J. Nagel, "Polarization effect in lightwave systems," *in Optical Fiber Telecommunications IIIA*, I. P. Kaminow and T. L. Koch, eds. (Academic Press, New York, 1997).
5. P. R. Morkel, V. Syngal, D. J. Butler, and R. Newman, "PMD-induced BER penalties in optically amplified IM/DD lightwave systems," Electron. Lett. **30**, 806-807 (1994).
6. P. Ciprut et al. "Second-order polarization mode dispersion: impact on analog and digital transmissions," IEEE J. Lightwave Technol. **16**, 757-771 (1998).
7. H. Bullow, "System outage probability due to first- and second-order PMD," IEEE Photon. Tecnol. Lett. **10**, 696-698 (1998).
8. M. Karlsson, "Polarization mode dispersion-induced pulse broadening in optical fibers," Opt. Lett. **23**, 688-690 (1998).
9. C. D. Poole and C. R. Giles, "Polarization-dependent pulse compression and broadening due to polarization dispersion in dispersion-shifted fiber," Opt. Lett. **13**, 155-157 (1988).
10. C. Vassallo, "PMD pulse deformation," Electron. Lett. **31**, 1597-1598 (1995).
11. F. Bruyere, "Impact of first- and second-order PMD in optical digital transmission systems," Opt. Fiber Technol. **2**, 269-280 (1996).
12. C. D. Poole, "Measurement of polarization-mode dispersion in single-mode fibers with random mode coupling," Opt. Lett. **14**, 523-524 (1989).
13. F. Curti, B. Daino, G. De Marchis, and D. Matera, "Statistical treatment of the evolution of principal states of polarization in single-mode fibers," IEEE J. Lightwave Technol. **8**, 1162–1165 (1990).
14. S. L. Fogal, G. Biondini, and W. L. Kath, "Multiple Importance Sampling for First- and Second-Order Polarization-Mode Dispersion," IEEE Photon. Technol. Lett. **14**, 1273-1275 (2002).
15. C. R. Menyuk,"Nonlinear pulse propagation in birefringent optical fibers," IEEE J. Quantum Electron. **QE-23**, 174-176 (1987).
16. G. P. Agarwal, *Nonlinear Fiber Optics* (Academic Press, New York, 1989).
17. F. Matera and C. G. Someda, "Random birefringence and polarization dispersion in long single-mode optical fibers," in *Anisotropic and Nonlinear Optical Waveguide*, C. G. Someda and G. Stegeman, eds. (Elsevier, New York, 1992).
18. David Yevick, "Multicanonical Communication System Modeling—Application to PMD Statistics," IEEE Photon. Technol. Lett. **14**, 1512-1514 (2002).
19. G. Biondini,W. L. Kath, and C. Menyuk, "Importance sampling for polarization mode dispersion," IEEE Photon. Technol. Lett. **14**, 310–312 (2002).
20. I. T. Lima, Jr., G. Biondini, B. S. Marks,W. L. Kath, and C. R. Menyuk, "Analysis of PMD compensators with fixed DGD using importance sampling," IEEE Photon. Technol. Lett. **14**, pp. 627–629 (2002).

21. V. Chernyak, M. Chertkov, I. Gabitov, I. Kolokolov, and V. Lebedev, "PMD induced fluctuation of bit-error-rate in optical fiber," to appear in IEEE J. Lightwave Technol. special issue on polarization mode dispersion (2004).
22. G. P. Agrawal, *Fiber-Optic Communication Systems* (Wiley, New York, 2002).
23. H. Ooi, Y. Akiyama, and G. Ishikawa, "Automatic polarization-mode dispersion compensation in 40-Gbit/s transmission," Proceedings of Optical Fiber Communication Conference, paper WE5, page 86-88 (1999).
24. D. Chowdhury, E. Brarens, S. Nipple, A. Zainul, T. Hanson, "Statistical impact of EDFA polarization mode dispersion on long-haul optical systems," Proceedings of 26th European Conference on Optical Communication, Page 147 (2000).

DOI: 10.1007/s10297-004-0007-z
Originally published in J. Opt. Fiber Commun. Rep. 1, 14–31 (2004)

Applications of importance sampling to polarization mode dispersion

Gino Biondini[1] and William L. Kath[2]

[1] Department of Mathematics, State University of New York
Buffalo, NY 14260 USA
Email: biondini@buffalo.edu
[2] Department of Engineering Sciences and Applied Mathematics
Northwestern University, 2145 Sheridan Road, Evanston, Illinois 60208 USA
Email: kath@northwestern.edu

Abstract. We describe the application of importance sampling to Monte-Carlo simulations of polarization-mode dispersion (PMD) in optical fibers. The method allows rare PMD events to be simulated much more efficiently than with standard Monte-Carlo methods, thus making it possible to assess the effect of PMD on system outage probabilities at realistic bit error ratios.

1. Introduction

Polarization-mode dispersion (PMD) has emerged as a major impairment in both terrestrial and undersea wavelength-division-multiplexed (WDM) systems. To first order in frequency, PMD splits a pulse between the fast and the slow axis of the fiber; at the same time, higher orders of PMD induce depolarization and polarization-dependent chromatic dispersion. At high bit rates, all of these effects lead to unacceptable transmission penalties. A key difficulty with PMD is its stochastic nature. PMD is caused by random variations of an optical fiber's birefringence with distance. In addition, the penalties it produces change randomly over time as the ambient temperature and other environmental parameters vary. In system design, a maximum power penalty (typically 1 dB) is usually assigned to PMD, and one demands that the outage probability (that is, the probability of the PMD-induced penalty exceeding this allowed value) is very small (typically a minute per year, 10^{-6} or less). Because of this small probability, it is very difficult to use either Monte-Carlo simulations or laboratory experiments to determine a system's outage probability, due to the extremely large number of system configurations that must be explored in order to obtain a reliable estimate.

A measure of PMD is provided by the PMD vector $\boldsymbol{\tau}$ [1, 2, 3], the magnitude of which is the differential group delay (DGD). The rare events where the DGD is significantly larger than its mean are particularly important, since they are the ones most likely to result in system outages. Of course, the PMD vector is frequency-dependent, and therefore in general the DGD at any given frequency is not the sole determiner of system outages. In particular, second-order PMD, which includes both depolarization and polarization-dependent chromatic dispersion (PCD), is known to produce additional system penalties [4, 5]. This is especially important when PMD compensation is applied, since first-order PMD is typically reduced to moderate values (or perfectly cancelled at a particular frequency); in this case, it is possible for second-order PMD to become the primary cause of system penalties [6, 7, 8, 9].

In the absence of effective tools for calculating outage probabilities, system designers have resorted to stopgap techniques. One such technique is to produce artificially large DGD values, determine the penalties at these large DGDs, and then weight the results using the DGD's well-known Maxwellian probability distribution [1]. A fundamental problem with this method, however, is that there is no direct relationship between the DGD (and/or second-order PMD) and the power penalty. In addition, it is possible for different configurations to give both the same DGD and the same second-order PMD, but not contribute equally to the penalty (because of different higher-order PMD), and therefore they should be weighted differently in calculations of outage probability.

The technique of *importance sampling* (IS) [10, 11, 12], which is one of a family of methods known as variance reduction techniques [13, 14], addresses these difficulties and provides a tool that can be used in numerical simulations — and, in principle, in experiments — to accurately estimate PMD-induced system penalties. The technique allows low probability events to be efficiently simulated by enabling one to concentrate Monte-Carlo simulations in the most significant regions of interest in sample space. We have applied IS to the numerical calculation of PMD-induced effects generated by a concatenation of birefringent sections [15, 16, 17]. In this case, the regions of interest in the sample space are the configurations that lead to large values of DGD and/or second-order PMD. This method has also been applied to numerically calculate PMD-induced transmission penalties [7, 9, 18]. Here we give a complete description of the technique and its application.

In Sections 2–4 we introduce the technique by first discussing some basic principles and by considering the simpler case in which large DGD values are the events of interest. Of course, in many practical situations it is necessary to also consider the frequency dependence of the PMD vector. Thus, in Section 5 we present a multiple biasing technique that allows one to target arbitrary amounts of first- and second-order PMD. Use of this composite biasing method in importance-sampled Monte-Carlo simulations yields a more comprehensive determination of PMD-induced system penalties than first-order biasing alone. Finally, in Sections 6–7 we discuss some technical issues that are needed to implement the method. To demonstrate the method, in Section 9 we calculate the probability density functions of the DGD, the magnitude of the second-order PMD vector and their joint pdf for a device composed by a concatenation of birefringent sections.

2. The PMD Concatenation Equations

A standard technique for simulating PMD effects is the coarse-step method [19], which approximates the continuous birefringence variations present in real fibers by a concatenation of fixed length birefringent sections. Many experimental PMD generation techniques also employ a concatenation of birefringent elements, such as high-birefringence fibers [20] or birefringent waveplates [21], connected by either polarization scramblers (e.g., polarization controllers [20]) or rotatable connectors [21].

In all such cases, the total PMD vector after the $(n+1)$-st fiber section can be obtained from the PMD concatenation equation [3],

$$\boldsymbol{\tau}^{(n+1)} = \mathsf{R}_{n+1}\boldsymbol{\tau}^{(n)} + \Delta\boldsymbol{\tau}_{n+1}\,, \tag{1}$$

illustrated in Fig. 1. Here $\boldsymbol{\tau}_n$ is the PMD vector after n sections and $\Delta\boldsymbol{\tau}_{n+1}$ is the differential contribution to the PMD vector of the $(n+1)$-st section. For fixed length sections, the magnitude of $\Delta\boldsymbol{\tau}_{n+1}$ is fixed, and only its direction varies. For linearly birefringent elements, $\Delta\boldsymbol{\tau}_{n+1}$ lies on the equatorial plane of the Poincaré sphere [22]. The Müller matrix $\mathsf{R}_{n+1} = \exp\left[\varphi_{n+1}(\hat{\mathbf{r}}_{n+1}\times)\right]$ represents a rotation through an angle φ_{n+1} about the axis $\hat{\mathbf{r}}_{n+1}$. This rotation is frequency dependent; for linearly birefringent elements, $\hat{\mathbf{r}}_{n+1} = \Delta\boldsymbol{\tau}_{n+1}/|\Delta\boldsymbol{\tau}_{n+1}|$ and $\varphi_{n+1}(\omega) = 2b'\ell_{n+1}\omega = \varphi_{n+1}(\omega_0) + 2b'\ell_{n+1}(\omega - \omega_0)$, where b' is the birefringence strength of each element and ℓ_{n+1} is its length [3]. A similar concatenation equation holds for the frequency derivative of the PMD vector, $\boldsymbol{\tau}_\omega$ [3]:

$$\boldsymbol{\tau}_\omega^{(n+1)} = \mathsf{R}_{n+1}\boldsymbol{\tau}_\omega^{(n)} + \Delta\boldsymbol{\tau}_{n+1} \times \boldsymbol{\tau}^{(n+1)}\,. \tag{2}$$

If linearly birefringent elements are used, $\mathsf{R}_{n+1}\Delta\boldsymbol{\tau}_{n+1} = \Delta\boldsymbol{\tau}_{n+1}$ and Eqs. (1)–(2) can be written as

$$\boldsymbol{\tau}^{(n+1)} = \mathsf{R}_{n+1}\big(\boldsymbol{\tau}^{(n)} + \Delta\boldsymbol{\tau}_{n+1}\big), \tag{3}$$

$$\boldsymbol{\tau}_\omega^{(n+1)} = \mathsf{R}_{n+1}\big(\boldsymbol{\tau}_\omega^{(n)} + \Delta\boldsymbol{\tau}_{n+1} \times \boldsymbol{\tau}^{(n)}\big). \tag{4}$$

When polarization controllers are present, an additional rotation matrix Q_{n+1} preceeds $\boldsymbol{\tau}^{(n)}$ and $\boldsymbol{\tau}_\omega^{(n)}$ in Eqs. (3)–(4). It is possible to factor out this rotation Q_{n+1} from the concatenation equations: the resulting equations are formally equivalent to Eqs. (3)–(4) with a new rotation matrix $\mathsf{R}'_{n+1} = \mathsf{R}_{n+1}\mathsf{Q}_{n+1}$, except that the new contributions $\Delta\boldsymbol{\tau}'_{n+1} = \mathsf{Q}_{n+1}^{-1}\Delta\boldsymbol{\tau}_{n+1}$ are now uniformly distributed on the Poincaré sphere. This property facilitates the implementation of IS with polarization scramblers. In the following, we refer to the situation where there are no polarization controllers as the case of rotatable waveplates.

3. Importance Sampling

In order to apply IS to the PMD concatenation Eqs. (1)–(2) we need to target the rare events in which the DGD and/or second-order PMD assume values much larger than their mean. Here, for brevity, we only present the implementation of importance sampling for PMD emulators which employ fixed-length birefringent sections [15, 16]. Importance sampling methods for PMD emulators with variable-length sections are

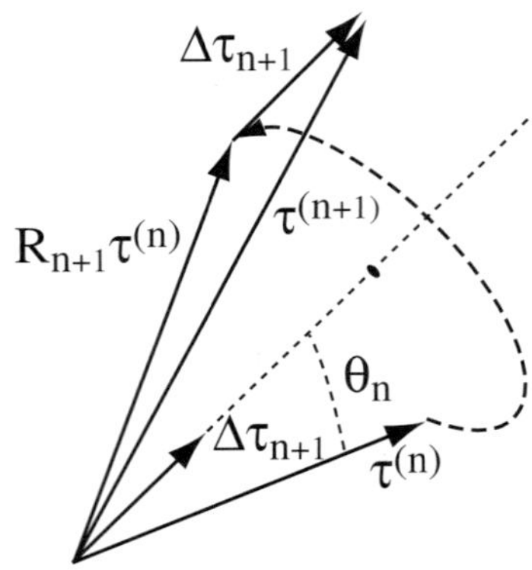

Fig. 1. Graphical representation of the first-order PMD concatentation, Eq. (1).

discussed in [17]. When fixed-length birefringent sections are employed, we have found that the appropriate variables to control are the polar angles θ_n between the PMD vector after the first n sections, $\boldsymbol{\tau}_n$, and the differential contribution of the $(n+1)$-st section, $\Delta\boldsymbol{\tau}_{n+1}$, as shown in Fig. 1. (Of course, a second set of angles, giving the orientation of $\Delta\boldsymbol{\tau}_{n+1}$ about $\boldsymbol{\tau}_{n+1}$ at fixed θ_n, is needed to uniquely specify each birefringent section. These angles do not need to be controlled and can be left uniformly distributed.) For example, the configurations that lead to these large DGD values are the ones in which the individual contributions to the PMD vector from each section tend to be aligned with each other. These configurations are the ones where the angles θ_n tend to be closer to zero than on average.

Suppose we are interested in determining the probability P that a random variable which depends upon the angles $(\theta_1, \theta_2, \dots \theta_N) \equiv \boldsymbol{\theta}$ falls in a given range, where N is the total number of sections. In our first example we will target the total DGD, $|\boldsymbol{\tau}_N|$, but later on we will consider an arbitrary combination of DGD and second-order PMD. The method can be applied to any random variable, such as the amount of pulse broadening, the power penalty, or any combination thereof. The probability P can be represented as the expectation value of an indicator function $I(\boldsymbol{\theta})$, where $I = 1$ if the random variable of interest falls in the prescribed range and $I = 0$ otherwise. That is, P is represented by the N-dimensional integral

$$P = \int_{\Theta} I(\boldsymbol{\theta})p(\boldsymbol{\theta})d\boldsymbol{\theta}\,, \tag{5}$$

where the $p(\boldsymbol{\theta})$ is the unbiased joint probability distribution function for the angles and Θ represents the configuration space. When using importance sampling, we rewrite the integrand in Eq. (5) as $I(\boldsymbol{\theta})p(\boldsymbol{\theta}) = I(\boldsymbol{\theta})L(\boldsymbol{\theta})p^*(\boldsymbol{\theta})$, where $p^*(\boldsymbol{\theta})$ is the biasing distribution, and where $L(\boldsymbol{\theta}) = p(\boldsymbol{\theta})/p^*(\boldsymbol{\theta})$ is the importance sampling likelihood ratio [10, 11, 12]. We then estimate the corresponding integral through Monte Carlo simulations; that is, we write an importance-sampled Monte-Carlo estimate of the above probability P as [10, 11, 12]

$$\hat{P} = \frac{1}{M}\sum_{m=1}^{M} I(\boldsymbol{\theta}_m)L(\boldsymbol{\theta}_m)\,, \tag{6}$$

where M is the total number of trials, and where the samples $\boldsymbol{\theta}_m$ are drawn according to the biased distribution $p^*(\boldsymbol{\theta})$. If $p^*(\boldsymbol{\theta}) \equiv p(\boldsymbol{\theta})$ (that is, for unbiased Monte Carlo simulations), Eq. (6) simply is the relative frequency of trials falling in the range of interest. The problem with this choice is that, when simulating low probability events, an exceedingly large number of samples is necessary in order for the desired events to occur, and an even larger number is required in order to obtain an accurate estimate. Using a biased probability distribution allows the desired regions of sample space to be visited much more frequently. At the same time, the likelihood ratio $L(\boldsymbol{\theta})$ automatically adjusts the results so that all of the different realizations are correctly weighted, thus contributing properly to the final probability.

Not all biasing schemes are equivalent, of course. In order to obtain a fair estimator for the desired quantity P, the biased probability distribution $p^*(\boldsymbol{\theta})$ should generate all of the configurations that contribute to the final result. The best methods are the ones that require the smallest numbers of realizations to be performed for all of the significant regions of sample space to be visited; such methods are called asymptotically efficient [23]. The algorithms that we have chosen to obtain targeted values of first- and/or second-order PMD work by orienting the sections so that the successive differential PMD vectors are preferentially aligned in certain directions with respect to the first- and second-order PMD vectors up to that point. These algorithms appear to be asymptotically efficient, as evidenced by the numerical results to be presented later, which show that the tails of the probability distribution can be reached with approximately the same amount of effort as the main part of the distribution.

4. IS for the DGD

When polarization scramblers are employed, the successive differential PMD vectors $\Delta\boldsymbol{\tau}_{n+1}$ can be regarded as fixed, while the orientations of the previous PMD vectors $\boldsymbol{\tau}_n$ at the output of the polarization scrambler vary. We therefore bias the simulations by making the scramblers preferentially orient $\boldsymbol{\tau}_n$ near the direction of $\Delta\boldsymbol{\tau}_{n+1}$. More specifically, we set the polarization scramblers so that θ_n (the angle between the PMD vector at the output of the n-th scrambler and the $(n+1)$-st differential PMD vector) is biased towards zero. This choice does not uniquely determine the orientation of the PMD vector at the scrambler output, of course, because $\boldsymbol{\tau}_n$ can still be rotated by an arbitrary amount about $\Delta\boldsymbol{\tau}_{n+1}$ while keeping θ_n constant. We assume that this additional rotational angle is uniformly distributed.

In the unbiased case, the angles θ_n are independent random variables, with $\cos\theta_n$ uniformly distributed in $[-1,1]$. When applying importance sampling, we choose $\cos\theta_n = 2x^{1/\alpha} - 1$, where x is a uniform random variable in $[0,1]$ and $\alpha \geq 1$ is a biasing parameter. Other choices are possible for the biased distribution used for the θ_n; the effectiveness of the method does not seem to be very sensitive to the particular distribution used. The above choice yields the likelihood ratio as

$$L(\boldsymbol{\theta}) = \prod_{n=1}^{N} \frac{p_1(\cos\theta_n)}{p_\alpha(\cos\theta_n)}, \tag{7}$$

where $p_\alpha(\cos\theta) = (\alpha/2)[(1+\cos\theta)/2]^{\alpha-1}$ [24]. The value $\alpha = 1$ reproduces the unbiased case (i.e., $\cos\theta$ is uniformly distributed), while increasing values of α bias the configurations towards increasingly larger values of DGD, as illustrated in Figs. 2–3. Thus, each biasing distribution allows us to sample a different range of DGDs, and simulations with different values of α can be combined in order to cover the whole range of DGD values. An efficient strategy of combining results from different biasing distributions is described in Section 6.

We emphasize that different configurations with the same value of DGD (and/or second order PMD) can have quite different likelihood ratios, and thus their relative contribution to the final result can vary substantially. As a consequence, different configurations producing the same DGD and/or second-order PMD are expected to give very different contributions to the power penalty. This point is illustrated in Fig. 4, where we plot the distribution of likelihood ratios for some specific samples. Note that the horizontal axis uses a logarithmic scale, which implies a variation of almost three orders of magnitude of the likelihood ratios.

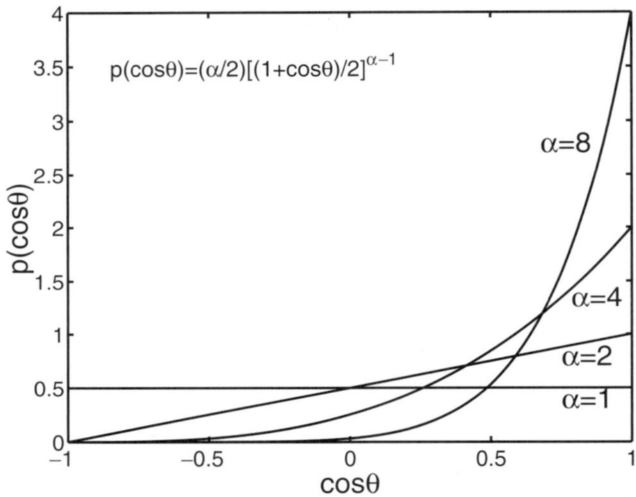

Fig. 2. The angular biasing distributions $p_\alpha(\cos\theta)$ for different values of α. Note that the distribution for the angle θ is $p_\alpha(\theta) = p_\alpha(\cos\theta)\sin\theta$.

When using rotatable waveplates, the relative orientations between sections are the primary variables determining the total DGD. The biasing toward large DGD values is done by choosing the next differential PMD vector, $\Delta\boldsymbol{\tau}_{n+1}$, to be preferentially aligned with the projection of the total PMD vector obtained thus far, $\boldsymbol{\tau}_n$, onto the equatorial plane. Specifically, we choose $\Delta\boldsymbol{\tau}_{n+1}$ so that the angle θ_n between it and the projection of $\boldsymbol{\tau}_n$ is biased toward zero (mod 2π). This is done, for example, by taking $\theta = \pi\left[1 + \operatorname{sgn}(2x-1)|2x-1|^{1/\alpha}\right]$, where x is again uniformly distributed between 0 and 1. Other choices are possible, of course. If $\alpha = 1$ (unbiased case), θ is uniformly distributed between 0 and 2π, while for $\alpha > 1$, θ is concentrated near 0 and 2π.

If frequency dependence is desired, or in the case of rotatable waveplates, the differential phase retardations φ_{n+1} in the Müller matrix R_{n+1} must also be specified. Recall that each beat length of a birefringent section generates a 2π retardation. In practice, sections with significant DGDs will be many beat lengths long, and unless

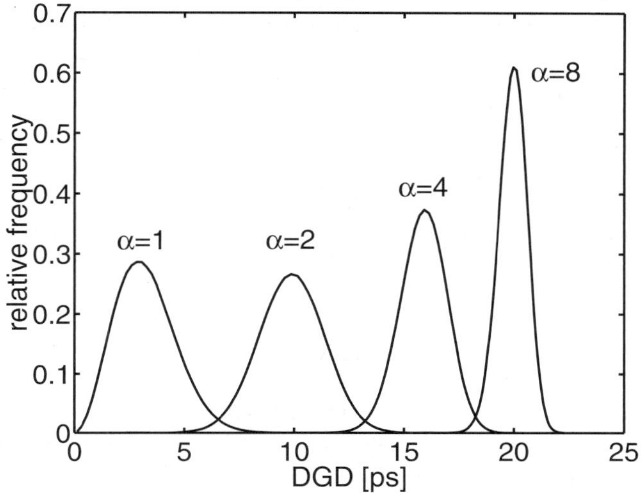

Fig. 3. The DGD frequency histograms for the biasing distributions in Fig. 2 for a concatenation of 50 birefringent elements with 0.5 ps DGD each (corresponding to a mean DGD of 3.26 ps) with polarization scramblers.

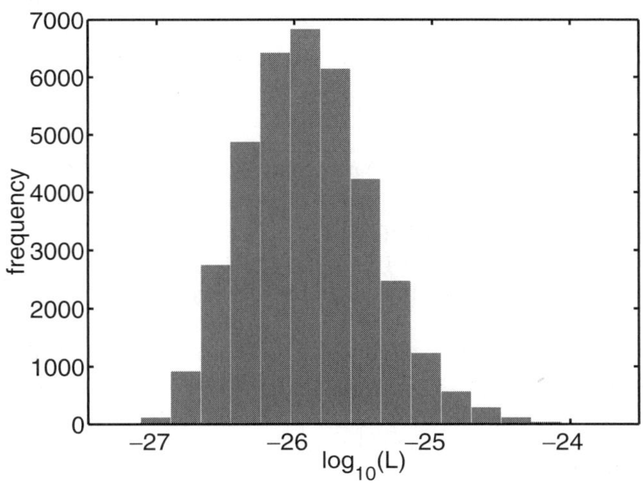

Fig. 4. The likelihood ratios and their relative frequency for samples with DGD values within a 0.2 ps interval around 20 ps and simultaneously with second-order PMD lying within a 6 ps^2 interval around 20 ps^2. A concatenation of 50 sections with 0.5 ps DGD each and polarization scramblers was used, as in Fig. 3.

the section lengths are precise to within a small fraction of a beat length, these retardations will vary from section to section. We therefore choose the retardation angles $\varphi_{n+1}(\omega_0)$ to be uniformly distributed between 0 and 2π. (Alternatively, one could assume a random distribution of lengths ℓ_{n+1}; if the variance of the lengths is large compared with the beat length, however, an approximately uniform distribution of the angles $\varphi_{n+1}(\omega_0)$ results. If, of course, the lengths are very precise, then specific phase retardations should be used.) In the case of rotatable waveplates, different configurations are then generated solely by rotating sections relative to one another. The results do not depend significantly upon the particular retardation angles used, except for certain clearly pathological cases such as identical angles with φ_{n+1} equal to 0 or π.

5. IS for Both First- and Second-order PMD

When calculating the PMD-induced outage probability of a system using Monte-Carlo simulations, it is also important to generate sufficient statistics of the frequency derivative of the PMD vector, $\boldsymbol{\tau}_\omega$, which quantifies the second-order PMD. It is possible to use importance sampling with multiple biasing distributions to generate arbitrary combinations of first- and second-order PMD, effectively targeting all regions of the $|\boldsymbol{\tau}|$-$|\boldsymbol{\tau}_\omega|$ plane. As a result, the method generates much more complete PMD statistics than first-order biasing alone.

As shown in the previous section, the appropriate variables to control when applying importance sampling are the orientations of the individual PMD vectors of each section, $\Delta\boldsymbol{\tau}_{n+1}$. IS works by biasing these vectors toward specific directions $\mathbf{b}^{(n)}$ which maximally increase the particular quantity of interest (e.g., the total DGD). In the following, we will characterize the vector $\mathbf{b}^{(n)}$ relative to the orthonormal frame of reference $\mathcal{U}$ formed by the unit vectors $\{\mathbf{u}_1^{(n)}, \mathbf{u}_2^{(n)}, \mathbf{u}_3^{(n)}\}$, where

$$\mathbf{u}_1^{(n)} = \boldsymbol{\tau}^{(n)}/|\boldsymbol{\tau}^{(n)}|, \qquad \mathbf{u}_2^{(n)} = \boldsymbol{\tau}_{\omega,\perp}^{(n)}/|\boldsymbol{\tau}_{\omega,\perp}^{(n)}|, \qquad \mathbf{u}_3^{(n)} = \mathbf{u}_1^{(n)} \times \mathbf{u}_2^{(n)}. \quad (8)$$

Here $\boldsymbol{\tau}_{\omega,\perp}^{(n)}$ is the component of $\boldsymbol{\tau}_\omega^{(n)}$ perpendicular to $\boldsymbol{\tau}^{(n)}$, as illustrated in Fig. 5. The magnitudes of $\boldsymbol{\tau}_{\omega,\parallel}$ (the component of $\boldsymbol{\tau}_\omega^{(n)}$ parallel to $\boldsymbol{\tau}^{(n)}$) and $\boldsymbol{\tau}_{\omega,\perp}$ quantify the PCD and the depolarization, respectively.

As described earlier, first-order biasing is achieved by choosing $\Delta\boldsymbol{\tau}_{n+1}$ to be preferentially aligned with the previous PMD vector $\boldsymbol{\tau}^{(n)}$, i.e., using $\mathbf{b}^{(n)} = \mathbf{u}_1^{(n)}$. With multiple biasing strengths (a combination of different α's), this choice produces region 1 in Fig. 6. First-order biasing yields the largest values of DGD. It does not, however, produce very large values of $\boldsymbol{\tau}_\omega$, because when $\Delta\boldsymbol{\tau}_{n+1}$ is parallel to $\boldsymbol{\tau}_n$ the contribution from that section to $\boldsymbol{\tau}_\omega^{(n+1)}$ is zero. Thus, if all the $\Delta\boldsymbol{\tau}_{n+1}$ were parallel to $\boldsymbol{\tau}^{(n)}$, no second-order PMD would be produced. Random fluctuations add second-order PMD, but large values are not specifically targeted. Over a single section, one can easily show that the choice that maximizes the contribution to $\boldsymbol{\tau}_\omega^{(n+1)}$ is to align $\Delta\boldsymbol{\tau}_{n+1}$ with the direction of $\mathbf{u}_3^{(n)}$. The rate at which $\boldsymbol{\tau}_\omega^{(n+1)}$ increases, however, also depends on $\boldsymbol{\tau}^{(n)}$, and thus, when optimizing over many sections, the growth of $\boldsymbol{\tau}^{(n)}$ must also be considered.

When the number of sections in the emulator is large, we have found it useful to employ a continuum approximation to find the deterministic configuration that generates

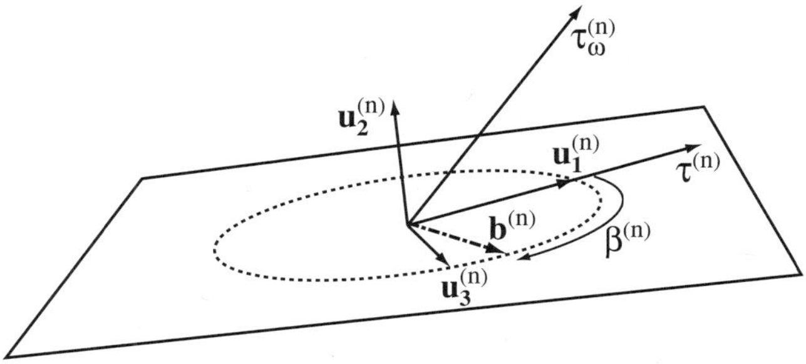

Fig. 5. Diagram showing the preferential direction $\mathbf{b}$ for biasing the simulations in the plane identified by $\boldsymbol{\tau}^{(n)}$ and $\boldsymbol{\tau}^{(n)} \times \boldsymbol{\tau}_\omega^{(n)}$.

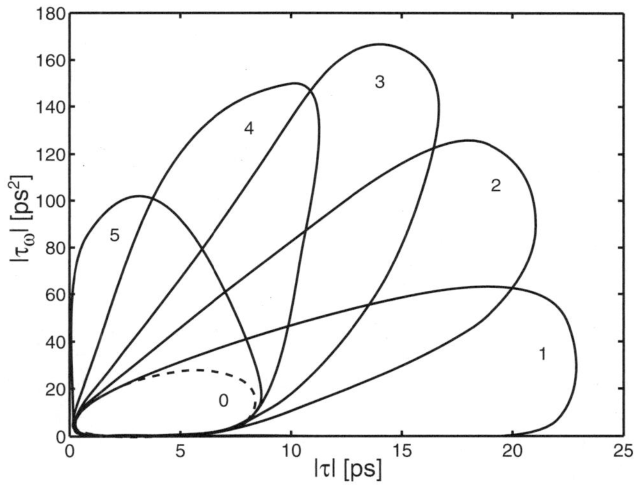

Fig. 6. The regions of the $|\boldsymbol{\tau}|$-$|\boldsymbol{\tau}_\omega|$ plane targeted by the various biasing methods. Region 1 corresponds to first-order biasing ($\beta = 0$), region 3 to pure second-order biasing ($\beta_{\max} = \pi/2$), and regions 2, 4 and 5 to $\beta_{\max} = \pi/4$, $3\pi/4$ and π, respectively. The dashed line shows the much smaller region obtained with unbiased runs. As in Figs. 3–4, 50 sections with 0.5 ps DGD each and polarization scramblers were used.

the maximum second-order PMD. Specifically, we let $\lim_{\Delta z \to 0} \Delta\boldsymbol{\tau}_{n+1}/\Delta z = \mathbf{b}(z)$. The magnitude of $\mathbf{b}(z)$ describes the relative rate at which PMD is added by the birefringent sections. For simplicity, we concentrate upon the case $|\mathbf{b}(z)| = b =$const. In the continuum limit, for the case of polarization scramblers one obtains

$$\frac{d\tau}{dz} = b_1 \,, \tag{9}$$

$$\frac{d\tau_{\omega,\|}}{dz} = b_2 \,\frac{\tau_{\omega,\perp}}{\tau} \,, \tag{10}$$

$$\frac{d\tau_{\omega,\perp}}{dz} = b_3\tau - b_2 \,\frac{\tau_{\omega,\|}}{\tau} \,, \tag{11}$$

where (b_1, b_2, b_3) are the components of $\mathbf{b}$ with respect to $\mathcal{U}$. This system of equations can be solved exactly for any $\mathbf{b}(z)$:

$$\tau(z) = \int_0^z b_1(z')\,dz'\,, \tag{12}$$

$$\tau_{\omega,\parallel} = \int_0^z b_3(z')\tau(z')\sin[B(z,z')]\,dz'\,, \tag{13}$$

$$\tau_{\omega,\perp} = \int_0^z b_3(z')\tau(z')\cos[B(z,z')]\,dz'\,, \tag{14}$$

where

$$B(z,z') = \int_{z'}^{z} \frac{b_2(z'')}{\tau(z'')}\,dz''\,. \tag{15}$$

Note that $d[|\boldsymbol{\tau}_\omega|^2]/dz = 2b_3\,\tau_{\omega,\perp}\tau$. Thus, b_2 does not directly contribute toward increasing the magnitude of second-order PMD; rather, it induces a rotation of $\boldsymbol{\tau}_\omega$ about $\mathbf{u}_3$, as seen from Eqs. (10)–(11). Since the maximum $\tau_{\omega,\perp}$ under such a rotation occurs when $\tau_{\omega,\parallel} = 0$, we expect the maximum second-order PMD to be generated for $b_2 = 0$, i.e., when $\mathbf{b}$ lies in the $\mathbf{u}_1$-$\mathbf{u}_3$ plane. Calculus of variations can then be used on Eq. (14) to show that the maximum growth of second-order PMD is obtained when

$$\mathbf{b}(z) = b\,[\mathbf{u}_1\cos\beta(z) + \mathbf{u}_3\sin\beta(z)]\,, \tag{16}$$

where $\beta(z) = \beta_{\max}\,z/z_{\max}$ and $\beta_{\max} = \pi/2$. With multiple biasing strengths, this generates region 3 in Fig. 6. We refer to this choice, which yields the largest values of $|\boldsymbol{\tau}_\omega|$, as optimal second-order biasing.

Of course, a more complete coverage of the $|\boldsymbol{\tau}|$-$|\boldsymbol{\tau}_\omega|$ plane is often needed. In this case, intermediate biasing choices must also be used in addition to optimal first- and second-order biasing. These intermediate choices can be obtained by using calculus of variations to maximize a linear combination of $|\boldsymbol{\tau}|$ and $|\boldsymbol{\tau}_\omega|$, as obtained from Eqs. (12)–(14). The resulting form of $\mathbf{b}(z)$ is the same as above, except that the value of the final angle $\beta_{\max}$ now varies between 0 and π, the particular value depending upon the specific linear combination of first- and second-order PMD being maximized. A selection of angles, together with the resulting regions in the $|\boldsymbol{\tau}|$-$|\boldsymbol{\tau}_\omega|$ plane, is shown in Fig. 6. The advantage of using multiple biasing — as opposed to just pure first- or second-order biasing or no biasing at all — is evident. When a range of biasing strengths are used, each value of $\beta_{\max}$ generates samples lying in a region that emanates in a roughly radial fashion from the location where the joint pdf is maximum. (A more careful analysis shows the samples to follow a family of parabolas.) Together, a selection of angles $\beta_{\max}$ and biasing strengths α covers the entire $|\boldsymbol{\tau}|$-$|\boldsymbol{\tau}_\omega|$ plane.

When birefringent waveplates are used, the contributions $\Delta\boldsymbol{\tau}_{n+1}$ are confined to the equatorial plane of the Poincaré sphere. In this case, the appropriate biasing directions are obtained by projecting the vectors $\mathbf{b}$ determined above onto the equatorial plane. For both models, once the biasing direction $\mathbf{b}$ has been selected, the biased distribution for the orientation of $\Delta\boldsymbol{\tau}_{n+1}$ is chosen as described previously. Furthermore, if desired, the frequency dependence of the PMD vector can be added as described as in Section 4.

6. Multiple Importance Sampling

The simultaneous use of different biasing methods is called *multiple importance sampling* [25]. When using several biasing distributions $p_j^*(\boldsymbol{\theta})$, a weight $w_j(\boldsymbol{\theta})$ is assigned to the samples from each distribution, and a multiply-importance-sampled Monte Carlo estimator for P is written as

$$\hat{P} = \sum_{j=1}^{J} \hat{P}_j = \sum_{j=1}^{J} \frac{1}{M_j} \sum_{m=1}^{M_j} w_j(\boldsymbol{\theta}_{j,m}) L_j(\boldsymbol{\theta}_{j,m}) I(\boldsymbol{\theta}_{j,m}) \,, \tag{17}$$

where M_j is the number of samples drawn from the jth distribution $p_j^*(\boldsymbol{\theta})$, and $\boldsymbol{\theta}_{j,m}$ is the mth such sample. Here, J is the number of different biasing distributions, $L_j(\boldsymbol{\theta}) = p(\boldsymbol{\theta})/p_j^*(\boldsymbol{\theta})$ is the likelihood ratio of the jth distribution and the $w_j(\boldsymbol{\theta})$ are the weights used to combined the results.

Several ways exist to choose the weights $w_j(\boldsymbol{\theta})$. The quantity $\hat{P}$ will be an unbiased estimator for P (i.e., the expectation value of $\hat{P}$ is P) for all choices such that $\sum_{j=1}^{J} w_j(\boldsymbol{\theta}) = 1$ for all $\boldsymbol{\theta}$. Each choice of weights corresponds to a different way of partitioning of the total probability. The simplest possibility, of course, is just to set $w_j(\boldsymbol{\theta}) = 1/J$ for all $\boldsymbol{\theta}$, meaning that each distribution is assigned an equal weight in all region of sample space. This choice, however, is not the most advantageous; a better strategy is to favor those biasing distributions that are expected to produce the best results in each given region.

A particularly useful choice of weights is the so-called *balance heuristics* [26]. In this case, the weights $w_j(\boldsymbol{\theta})$ are assigned according to

$$w_j(\boldsymbol{\theta}) = \frac{p_j^*(\boldsymbol{\theta})}{\sum_{j'=1}^{J} p_{j'}^*(\boldsymbol{\theta})} \,. \tag{18}$$

[Obviously, Eq. (18) can also be written in terms of likelihood ratios.] Thus, the weight of a sample obtained with the jth distribution is given by the probability of realizing that sample with the jth distribution relative to the total probability of realizing the same sample with all distributions. Thus, Eq. (18) weights more heavily those samples coming from a biasing distribution that it more likely to generate samples in that region, while samples that are unlikely to have occurred [because $p_j^*(\boldsymbol{\theta})$ is small] are weighted less heavily. Other strategies for assigning the weights are possible. However, the use of balance heuristics has been shown to become asymptotically close to optimal as the number of realizations becomes large [26]. Of course, the outputs of each Monte-Carlo realization is always adjusted for the bias through the IS likelihood ratio.

7. Variance Estimation

The variance of a Monte Carlo estimator is obviously a critical parameter, since its value determines the number of samples that will be necessary, on average, to obtain a required level of accuracy, and because its numerical estimate provides a measure of the accuracy of the results.

From probability theory we know that the variance $\sigma_{\hat{P}}^2$ of the importance-sampled Monte Carlo estimator $\hat{P}$ in Eq. (6) is $\sigma_{\hat{P}}^2 = \sigma_P^2/M$. We can then write a Monte Carlo estimate of $\sigma_{\hat{P}}^2$ as

$$\hat{\sigma}_{\hat{P}}^2 = \frac{1}{M(M-1)} \sum_{m=1}^{M} \big(I(\boldsymbol{\theta}_m) L(\boldsymbol{\theta}_m) - \hat{P} \big)^2 . \tag{19}$$

When no biasing is used (that is, for straightforward Monte Carlo simulations), we have $L(\boldsymbol{\theta}) = 1$, which yields $\sigma_P^2 = P - P^2$. If P is a rare event we also have $P \ll 1$, which means that the relative accuracy of the Monte Carlo estimator is $\hat{\sigma}_{\hat{P}}/\hat{P} \sim 1/(M\,\hat{P}^{1/2})$. Thus if $\hat{P}$ is exponentially small (as in the case of the bit error rate or the outage probability), an exceedingly large number of unbiased Monte Carlo simulations are necessary to obtain a reliable estimate. Indeed, it is because importance sampling is highly effective in reducing the variance of the Monte Carlo estimator that its use allows substantial speed ups of numerical simulations.

Of course, if $\hat{P}$ is obtained through multiply-importance-sampled Monte Carlo simulations according to Eq. (17), the variance in Eq. (19) should also be modified accordingly. The proper estimator for $\hat{\sigma}_{\hat{P}}^2$ is, in this case,

$$\hat{\sigma}_{\hat{P}}^2 = \sum_{j=1}^{J} \frac{1}{M_j(M_j-1)} \sum_{m=1}^{M_j} \big(I(\boldsymbol{\theta}_{j,m}) w_j(\boldsymbol{\theta}_{j,m}) L_j(\boldsymbol{\theta}_{j,m}) - \hat{P}_j \big)^2 . \tag{20}$$

Note also that different weighting strategies result in different values of $\hat{\sigma}_{\hat{P}}^2$, and the most effective methods are the ones that yield the smallest variances [26]. It is outside the scope of this paper to discuss the various choices; we will limit ourselves to note that the balance heuristics has been shown to give excellent results in a wide variety of situations.

8. Experimental Implementation

Experimental PMD emulators employing a concatenation of birefringent elements have been built [20, 21], and a PMD emulator that incorporates a simplified form of importance sampling which works with a small number of sections has also been built [27]. It is possible that an emulator could be constructed to use the importance-sampled algorithm described above. Essentially, in order for it to be possible to use the algorithm, it is necessary that the emulator be stable, repeatable and predictable. The biggest difficulty associated with a possible experimental implementation of the algorithm is the requirement that the orientation of the PMD vector and its frequency derivative be known after each birefringent section. One possible way of doing this would be to monitor the PMD vector after each birefringent section; obviously, such an implementation would be prohibitively cumbersome and expensive.

If the PMD vector is not monitored, then the next alternative is for each element of the emulator to be sufficiently well characterized that the PMD vector can be calculated if the orientations of the different birefringent elements is specified. Here the problem is that the differential phase retardations φ_{n+1} in the Müller matrix R_{n+1} must then be well characterized. If they are not known relatively precisely, then the Müller matrix will rotate the PMD vector to an unknown location on the Poincaré sphere, and it will then be impossible to properly determine the preferential orientation of the next differential PMD vector. As mentioned earlier, each beat length of a birefringent section generates a 2π retardation. Because sections with significant DGDs will be many beat lengths long, this means that the section lengths must precise to within a small fraction of a beat length, and that these sections lengths must be stable with respect to temperature and other environmental fluctuations. Fortunately, it has been demonstrated that it is possible to satisfy all of these constraints experimentally [28].

9. Examples

To demonstrate the ideas explained in the previous sections, we have considered a concatenation of 50 birefringent sections with 0.5 ps of DGD per section (corresponding to a mean DGD of 3.26 ps), employing either polarization scramblers or rotatable waveplates. The average DGD is given by $\langle\tau\rangle = \sqrt{8\langle\tau^2\rangle/(3\pi)}$, where $\langle\tau^2\rangle = \sum_{n=1}^{N} a_n^2$, the a_n are the individual DGDs of each section and N is the number of sections. Figure 7 shows the pdf of the DGD, obtained with optimal first-order biasing, while Fig. 8 shows the pdf of second order PMD, obtained with the optimal second-order biasing technique. In both cases, the biasing parameters $\alpha = 1, 2, \ldots, 10$ were used, with 200,000 realizations each. The individual samples are combined by sorting the values of DGD and/or second-order PMD obtained from all of the simulations into bins and adjusting the contribution *of each individual sample* for the bias using its likelihood ratio, as explained earlier. The solid lines show the analytical pdfs for real fiber [29].

When polarization scramblers are present, the evolution of the PMD vector is equivalent to a three-dimensional random walk, and an exact solution is available for the pdf of the DGD [30, 31]:

$$p_N(r) = \frac{2r}{\pi} \int_0^\infty x \sin(rx) \prod_{n=1}^{N} \frac{\sin(a_n x)}{a_n x}\, dx\,. \tag{21}$$

This integral also has Fourier and asymptotic expansions [30] which can be compared with numerical results. For moderate values of the DGD r, the pdf $p_N(r)$ is well approximated by a Maxwellian distribution, which has the pdf $p_{\mathrm{M}}(r) = 3\sqrt{6/\pi}\,\left(r^2/\langle\tau^2\rangle^{3/2}\right) e^{-3r^2/2\langle\tau^2\rangle}$. The degree of agreement, of course, improves as the number of sections is increased [31].

The numerically calculated pdf for the case of scramblers agrees extremely well with the exact solution (dashed line under the squares). For both scramblers and waveplates, the accuracy of the numerical results improves as the number of simulations (and the number of bins) is increased. It should be noted, however, that very good results are obtained with the 2,000,000 Monte-Carlo simulations employed here, as evident in Fig. 7, and reasonable results can be obtained with far fewer numbers of trials. In particular, a good approximation is achieved for probabilities below 10^{-12}; to obtain comparable accuracy at this probability with straightforward Monte-Carlo simulations, at least 10^{14} or 10^{15} trials would be required. Thus, for these cases importance sampling speeds up simulations by several orders of magnitude.

The emulator with rotatable sections yields better agreement with real fiber than the emulator with polarization scramblers. Note, however, that a concatenation of a small number of equal length, rotatable birefringent sections is not a good model for real fiber due to artificial periodicities of the PMD vector's autocorrelation function in the frequency domain [20, 32]. For both scramblers and waveplates, the pdfs deviate

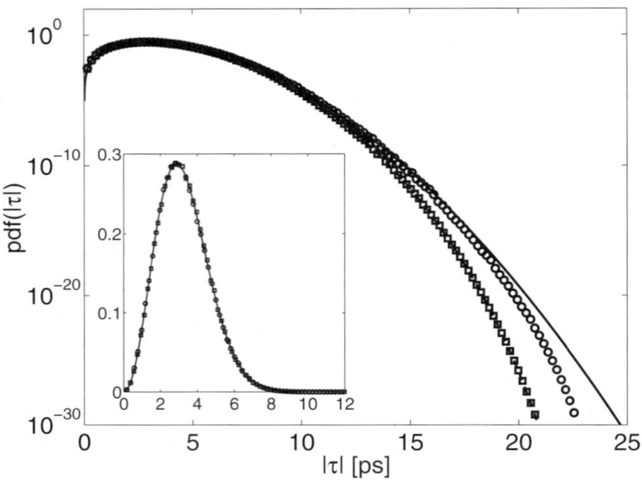

Fig. 7. Importance-sampled pdf of the DGD for 50 sections with 0.5 ps DGD each and with polarization scramblers (squares) or birefringent waveplates (circles). Solid curve: Maxwellian distribution with mean DGD of 3.26 ps; dashed line (coincident with squares): exact and asymptotic solutions from [30]. Inset: the pdfs on a linear scale. A total of 2×10^6 Monte-Carlo realizations were used.

significantly from the real fiber distributions in the tails, since here the PMD is generated by an emulator with finite number of birefringent sections, which by necessity

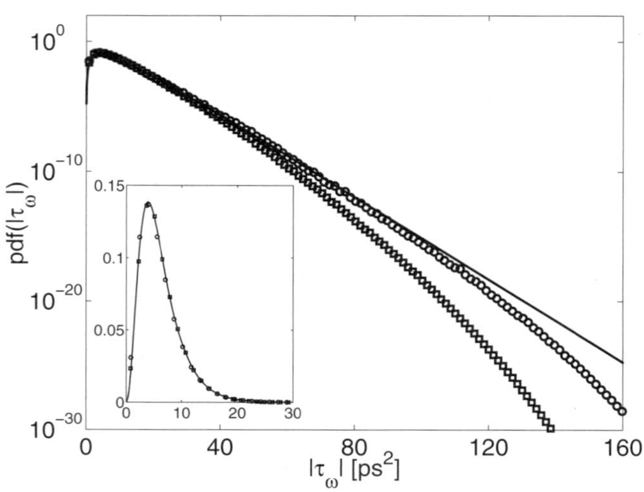

Fig. 8. The pdf of second-order PMD for a concatenation of 50 sections with 0.5 ps DGD each, using scramblers (squares) or waveplates (circles). The solid lines show the pdf for real fiber, from Ref. [29]. Inset: the pdf on a linear scale. As in Fig. 7, 2×10^6 Monte-Carlo realizations were used.

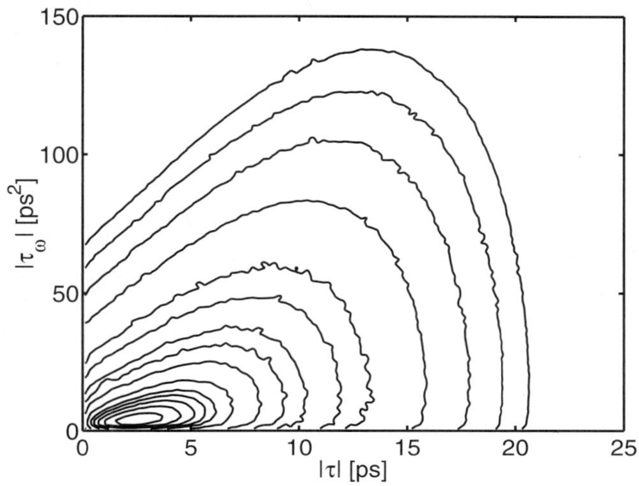

Fig. 9. Contour plots of the joint pdf for a concatenation of 50 sections with 0.5 ps DGD each and polarization scramblers. The contours are at 10^{-n} with n = 30, 25, 20, 15, 10, 8, 6, 5, 4, 3, 2.5, 2.25, 2, 1.75, and 1.5. A total of 10^6 Monte-Carlo realizations were used.

has finite maximums for both $|\boldsymbol{\tau}|$ and $|\boldsymbol{\tau}_\omega|$. As a result, large values may be inaccessible in the $|\boldsymbol{\tau}|$-$|\boldsymbol{\tau}_\omega|$ plane. (The emulator should be chosen so that these regions are unimportant in determining the outage probability of the system being tested.) For a device with equal length birefringent sections, an approximate value for the maximum second-order PMD is obtained from the exact solution of the continuum model, evaluated in the case of optimal second-order biasing: $|\boldsymbol{\tau}_\omega|_{\max} = N^2(\Delta\tau)^2/\pi$, where $\Delta\tau$ is the DGD of each section. Of course, $|\boldsymbol{\tau}|_{\max} = N\,\Delta\tau$. We also note that the finite value of the birefringence correlation length suggests that the DGD distribution in fiber might also deviate from Maxwellian in the tails.

Figure 9 shows the joint pdf of the magnitudes of first- and second-order PMD (a two-dimensional reduction of the full three-dimensional joint pdf of first- and second-order PMD [1]) for an emulator with polarization scramblers, as calculated with the multiple biasing technique. The characteristic function associated with the joint pdf of first- and second-order PMD in an optical fiber (the limit of an infinite number of infinitesimal birefringent sections) was given in [1], but, to our knowledge, no exact analytical expression exists for the joint pdf. (An approximate pdf was given in [1], however, for the case where second-order PMD is a small perturbation compared to first-order.) Similarly, the joint pdf for PMD emulators with a finite number of sections is not known analytically.

The results for the joint pdf confirm why the optimal first- and second-order biasing methods are the correct ones to use if one is only interested in first- or second-order PMD statistics, respectively. As seen from Fig. 9, for any fixed value of DGD (a vertical slice in Fig. 9), the maximum of the pdf occurs in a zone that falls within region 1 in Fig. 6. Similarly, for any given second-order PMD (a horizontal slice in Fig. 9), the maximum of the pdf occurs in a zone falling within region 3 in Fig. 6. The multiple biasing method, of course, is not limited to just these particular combinations of first- and second-order PMD, but is able to produce any combination of the two.

10. Conclusions

We have discussed the application of importance sampling to Monte-Carlo simulations of PMD generated by birefringent sections connected by either polarization scramblers or rotatable connectors. Importance sampling biases the Monte-Carlo simulations so that configurations with large values of first- and/or second-order PMD occur more frequently than they would normally. As a result, the method allows rare PMD events to be simulated much more efficiently than with standard methods. Importance-sampled Monte-Carlo techniques therefore provide a natural and effective means to assess PMD-induced impairments in optical transmission systems at realistic bit-error rates.

Acknowledgments

We thank J. N. Damask, A. Galtarossa, I. T. Lima and C. R. Menyuk for many interesting and valuable discussions. This work was supported in part by the National Science Foundation under grant numbers DMS-0101476 and DGE-9987577.

References

1. G. J. Foschini and C. D. Poole, "Statistical theory of polarization dispersion in single mode fibers", IEEE J. Lightwave Technol. **9**, 1439 (1991)
2. N. Gisin, R. Passy, J. C. Bishoff, and B. Perny, "Experimental investigations of the statistical properties of polarization mode dispersion in single-mode fibers", IEEE Photon. Technol. Lett. **5**, 819 (1993)
3. J. P. Gordon and H. Kogelnik, "PMD fundamentals: polarization-mode dispersion in optical fibers", Proc. Nat. Acad. Sci. **97**, 4541 (2000)
4. P. Ciprut, B. Gisin, N. Gisin, R. Passy, J. P. Von der Weid, F. Prieto, C. W. Zimmer, "Second-order polarization-mode dispersion: impact on analog and digital transmissions", IEEE J. Lightwave Technol. **16**, 757 (1998)
5. C. Francia, F. Bruyére, D. Penninckx, M. Chbat, "PMD second-order effects on pulse propagation in single-mode optical fibers", IEEE Photon. Technol. Lett. **10**, 1739 (1998)
6. W. Shieh, "On the Second-Order Approximation of PMD", IEEE Photon. Technol. Lett. **12**, 290 (2000)
7. S. L. Fogal, G. Biondini and W. L. Kath, "Importance-sampled pulse broadening before and after PMD compensation", OFC2002, paper ThA2
8. Q. Yu and A. E. Willner, "Performance limits of first-order PMD compensation using fixed and variable DGD elements", IEEE Photon. Technol. Lett. **14**, 304 (2002)
9. A. O. Lima, I. T. Lima, C. R. Menyuk, G. Biondini, B. S. Marks and W. L. Kath, "Statistical analysis of the performance of polarization-mode dispersion compensators with multiple importance sampling", IEEE Photon. Technol. Lett. **15**, 1716 (2003)

10. M. C. Jeruchim, "Techniques for estimating the bit error rate in the simulation of digital communication systems", IEEE J. Select. Areas Commun. **2**, 153 (1984)

11. J.-C. Chen, D. Lu, J. S. Sadowsky, and K. Yao, "On importance sampling in digital communications—Part I: Fundamentals", IEEE J. Select. Areas Commun. **11**, 289 (1993)

12. P. J. Smith, M. Shafi, and H. Gao, "Quick simulation: a review of importance sampling techniques in communications systems", IEEE J. Select. Areas Commun. **15**, 597 (1997)

13. P. Bratley, B. L. Fox and L. E. Schrage, *A guide to simulation* (Springer-Verlag, New York, 1987)

14. G. S. Fishman, *Monte Carlo: concepts, algorithms and applications* (Springer-Verlag, New York, 1996)

15. G. Biondini, W. L. Kath and C. R. Menyuk, "Importance sampling for polarization-mode dispersion", IEEE Photon. Technol. Lett. **14**, 310 (2002).

16. S. L. Fogal, G. Biondini and W. L. Kath, "Multiple importance sampling for first- and second-order polarization mode dispersion", IEEE Photon. Technol. Lett. **14**, 1273 (2002); 1487 (2002)

17. G. Biondini and W. L. Kath, "Polarization mode dispersion emulation with Maxwellian length sections and importance sampling", OFC2003, paper ThA2; IEEE Photon. Technol. Lett. **16** #3, to appear (2004)

18. I. T. Lima, G. Biondini, B. S. Marks, W. L. Kath and C. R. Menyuk, "Analysis of polarization-mode dispersion compensators using importance sampling", Photon. Technol. Lett. **14**, 627 (2002).

19. D. Marcuse, C. R. Menyuk, and P. K. A. Wai, "Application of the Manakov-PMD equation to studies of signal propagation in optical fibers with randomly varying birefringence", J. Lightwave Technol. **15**, 1735 (1997)

20. R. Khosravani, I. T. Lima, P. Ebrahimi, E. Ibragimov, A.E. Willner, and C.R. Menyuk, "Time and frequency domain characteristics of polarization-mode dispersion emulators", Phot. Technol. Lett. **13**, 127 (2001)

21. J. N. Damask, "A programmable polarization-mode dispersion emulator for systematic testing of 10 Gb/s PMD compensators", OFC2000, paper ThB3

22. A. Galtarossa, L. Palmieri, M. Schiano, and T. Tambosso, "Statistical characterization of fiber random birefringence", Opt. Lett. **25**, 1322 (2000)

23. J. S. Sadowsky and J. A. Bucklew, "On large deviations theory and asymptotically efficient Monte-Carlo estimation", IEEE Trans. Inf. Theory **36**, 579 (1990)

24. A. Papoulis, *Probability, Random Variables and Stochastic Processes* (McGraw-Hill, New York, 1965)

25. A. Owen and Y. Zhou, "Safe and effective importance sampling", J. Amer. Stat. Assoc. **95**, 135 (2000).

26. E. Veach, *Robust Monte Carlo Methods For Light Transport Simulation*, Ph.D. thesis (Stanford University, 1997).

27. L. S. Yan, M. C. Hauer, P. Ebrahimi, Y. Wang, Y. Q. Shi, X. S. Yao, A. E. Willner, and W. L. Kath, "Measurement of Q degradation due to polarisation mode dispersion using multiple importance sampling", Elec. Lett. **39**, 974–975 (2003)

28. J. N. Damask, P. Myers and T. Boschci, "Coherent PMD and programmable PMD generation", OFC2003, paper ThA6

29. G. J. Foschini, L. E. Nelson, R. M. Jopson, H. Kogelnik, "Probability densities of second-order polarization-mode dispersion including polarization-dependent chromatic dispersion", IEEE Photon. Technol. Lett. **12**, 293 (2000).

30. B. D. Hughes, *Random Walks and Random Environments* (Clarendon Press, Oxford, 1995).

31. M. Karlsson, "Probability density functions of the differential group delay in optical fiber communication systems", J. Lightwave Technol. **19**, 324 (2001)

32. I. T. Lima, R. Khoshravani, P. Ebrahimi, E. Ibragimov, A. E. Willner, and C. R. Menyuk, "Polarization mode dispersion emulator", OFC2000, paper ThB4

DOI: 10.1007/s10297-004-0008-y
Originally published in J. Opt. Fiber Commun. Rep. 1, 1–13 (2004)

PMD & PDL

Nicolas Gisin

Group of Applied Physics, University of Geneva
20 Rue de l'Ecole de Medecine
1211 Geneva 4, Switzerland
Email: Nicolas.Gisin@physics.unige.ch

Abstract. We present the formalism required to describe the combined effects of polarization mode dispersion (PMD) and polarization dependent losses (PDL) in optical fibres. The combination of PMD and PDL may lead to anomalous dispersion, which is not correctly described by a direct application of the Jones Matrix Eigenanalysis (JME) method. An extended version of this work can be found in Huttner et al.

1. Principal States in the Absence of PDL

As is well known, the polarization state at the end of an optical fiber trunk $\psi_{\text{out}}(\omega)$ depends on the optical frequency ω even if the input state ψ_{in} is independent of ω. This is because the evolution operator $U(\omega)$ depends itself on the optical frequency. Consequently, a pulse with large spectrum will be strongly depolarized by the fiber, because each frequency component corresponds to a different polarization state. How much depolarized depends on the fiber and on the spectrum, of course. However, the degree of depolarization depends also on the central output polarization state $\psi_{\text{out}}(\omega_0)$, or equivalently on the input polarization state $\psi_{\text{in}} = U(\omega_0)^{-1}\psi_{\text{out}}(\omega_0)$. Indeed, it is intuitively clear that if $\psi_{\text{out}}(\omega)$ changes rapidly for varying ω's around ω_0, then the depolarization will be larger. Hence, minimum depolarization is reached when $\psi_{\text{out}}(\omega) \approx \psi_{\text{out}}(\omega_0)$ for all optical frequencies ω in the optical pulse, where $\approx$ means "equal up to a global phase". The corresponding state ψ_{in} is called the principal (input) polarization state [2]. Let us formalize this.

We are looking for the state ψ_{in} such that $\psi_{\text{out}}(\omega_0 + \delta\omega) = e^{i\delta\omega\delta\tau/2}\psi_{\text{out}}(\omega_0)$ (where the choice of notation $\delta\tau/2$ will soon become clear). Consequently,

$$\frac{d}{d\omega}\psi_{\text{out}}(\omega) = \frac{dU(\omega)}{d\omega}\psi_{\text{in}} \tag{1}$$

$$= \frac{dU(\omega)}{d\omega}U(\omega)^{-1}\psi_{\text{out}}(\omega) \tag{2}$$

In the absence of PDL, the evolution operator is unitary: $U(\omega)^{-1} = U(\omega)^{\dagger}$. Accordingly,

$$\frac{d}{d\omega}\left(U(\omega)U(\omega)^{\dagger}\right) = 0 \tag{3}$$

$$= \frac{dU(\omega)}{d\omega}U(\omega)^{\dagger} + U(\omega)\frac{dU(\omega)^{\dagger}}{d\omega} \tag{4}$$

$$\Rightarrow \frac{dU(\omega)}{d\omega}U(\omega)^{\dagger} = -\frac{dU(\omega)^{\dagger}}{d\omega}U(\omega), \tag{5}$$

where $\Rightarrow i\frac{dU(\omega)}{d\omega}U(\omega)^{\dagger}$ is a selfadjoint operator, i.e., has real eigenvalues and two mutually orthogonal eigenvectors. Taking this into account in eq. (2) one obtains

$$\frac{d}{d\omega}\psi_{\text{out}}(\omega) = \frac{-i}{2}\delta\tau\ \psi_{\text{out}}(\omega), \tag{6}$$

where $\delta\tau$ is real and represents the Differential Group Delay (DGD).

This brief reminder should make it easier to understand the next section, in which a similar computation is presented, but without the assumption of a unitary (i.e., lossless) evolution.

2. Principal States in the Presence of PDL

2.1. The Transmission Matrix

The standard model for a single-mode telecom fiber is a concatenation of optical fiber trunks, each supporting two polarization modes. We assume that these two modes are uncoupled, all the polarization mode coupling taking place at the junctions between the trunks. Assuming furthermore that the birefringence β_j of the jth trunk is independent of wavelength, β_j is equal to the differential group delay (DGD) between the two modes. PDL is introduced through the loss coefficient α_j, defined from the differential loss between the two polarization modes by

$$e^{2\alpha_j} = \frac{I_{\max}}{I_{\min}}, \tag{7}$$

where $I_{\max}$, resp. $I_{\min}$, are the intensities of the polarization modes with lowest, resp. highest, loss. Note that $10\log_{10}(e^{2\alpha_j})$ is the jth trunk PDL expressed in dB.

In Jones formalism [3], the polarization state along the fibre is given by the complex-valued two-dimensional Jones vector ψ. The evolution of ψ along a concatenation of N trunks is described by

$$\psi_{\text{out}}(\omega) \equiv \mathbf{T}(\omega)\psi_{\text{in}} = A_N\ \mathbf{T}_N(\omega)\psi_{\text{in}}, \tag{8}$$

where

$$A_N = \exp\left(-\alpha_0 - \sum_{j=1}^{N} \alpha_j/2\right) , \tag{9}$$

with α_0 the polarization independent overall loss, and

$$\begin{aligned}\mathbf{T}_N(\omega) = \exp\left[(-i\boldsymbol{\beta}_N\omega + \boldsymbol{\alpha}_N)\cdot\ \boldsymbol{\sigma}/2\right]...\\ ...\exp\left[(-i\boldsymbol{\beta}_1\omega + \boldsymbol{\alpha}_1)\cdot\boldsymbol{\sigma}/2\right] ,\end{aligned} \tag{10}$$

where $\boldsymbol{\sigma}$ are the Pauli matrices [3]:

$$\boldsymbol{\sigma}_x = \begin{pmatrix} 0 & 1 \\ 1 & 0 \end{pmatrix}, \boldsymbol{\sigma}_y = \begin{pmatrix} 0 & -i \\ i & 0 \end{pmatrix}, \boldsymbol{\sigma}_z = \begin{pmatrix} 1 & 0 \\ 0 & -1 \end{pmatrix}.$$

Hence $\mathbf{T}_N(\omega)$ describes the polarization rotations and relative changes in intensity, whereas A_N describes the global attenuation. This global attenuation plays no role in the polarization properties, and will henceforth be omitted. The directions of the vectors $\boldsymbol{\beta}_j$ and $\boldsymbol{\alpha}_j$ define the two polarization modes of the jth trunk. For simplification, we shall assume that $\boldsymbol{\beta}_j$ and $\boldsymbol{\alpha}_j$ are parallel, though our formalism does not require this. We thus introduce the unit vector $\hat{e}_j$ defining the polarization mode of trunk j by

$$-i\boldsymbol{\beta}_j\omega + \boldsymbol{\alpha}_j \equiv b_j\hat{e}_j, \tag{11}$$

with $b_j = -i\beta_j\omega + \alpha_j$. We make the usual assumptions that the attenuation of the PDL elements and the birefringence of the HiBi fibres are both independent of the wavelength. In this case, phase and group birefringence are equal. Note that since the transmission matrix $\mathbf{T}_N(\omega)$ is made from a product of non-commuting matrices, one cannot separate directly the PMD and the PDL.

2.2. *Principal States of Polarization*

Because of the explicit frequency dependence in Eqs. (8) and (10), even when ψ_{in} is independent of the optical frequency ω, as is the case if the input polarization state is determined by a polarizer or by geometric parameters of the laser, the output polarization state $\psi_{\text{out}}(\omega)$ depends on the optical frequency ω. The principal polarization states [2] are defined as the polarization states such that the outcoming polarization is independent of the optical frequency in first order. For such states, the Jones vector $\psi_{\text{out}}(\omega)$ satisfies the following equation [4]:

$$\partial_\omega\psi_{\text{out}}(\omega) = -i\frac{\chi}{2}\psi_{\text{out}}(\omega), \tag{12}$$

where ∂_ω denotes the partial derivative with respect to ω and χ is a complex number. Indeed, when the fiber has PDL, the evolution operator $T(\omega)$ is not unitary and χ becomes complex. We shall show that its real part is still equal to the differential group delay, and its imaginary part is equal to the frequency derivative of the differential attenuation between the two principal polarization states in the fiber, similarly to Eq. (6).

From Eq. (12) we deduce that $\psi_{\text{out}}(\omega)$ is an eigenvector of the 2×2 matrix $(\partial_\omega\mathbf{T}_N)\mathbf{T}_N^{-1}$:

$$\partial_\omega\psi_{\text{out}}(\omega) = \partial_\omega\mathbf{T}_N(\omega)\psi_{\text{in}} = (\partial_\omega\mathbf{T}_N)\mathbf{T}_N^{-1}\psi_{\text{out}}(\omega), \tag{13}$$

where $\mathbf{T}_N^{-1}$ is the inverse matrix of $\mathbf{T}_N$ (it always exists, except if one α_j is infinite which would correspond to a perfect polarizer).

To solve this eigenvalue equation, we use Eq. (10) to get a recursion relation for $(\partial_\omega \mathbf{T}_N)\mathbf{T}_N^{-1}$:

$$(\partial_\omega \mathbf{T}_N)\mathbf{T}_N^{-1} = \frac{-i}{2}\beta_N \hat{e}_N \cdot \boldsymbol{\sigma} + e^{b_N \hat{e}_N \cdot \boldsymbol{\sigma}/2}(\partial_\omega \mathbf{T}_{N-1})\mathbf{T}_{N-1}^{-1} e^{-b_N \hat{e}_N \cdot \boldsymbol{\sigma}/2}. \quad (14)$$

The initial condition is: $\mathbf{T}_0 = \mathbf{1}$, the unit matrix, for all ω. Consequently, $\partial_\omega \mathbf{T}_0 = 0$ and since Eq. (14) preserves the trace, one has $\mathrm{Tr}((\partial_\omega \mathbf{T}_N)\mathbf{T}_N^{-1}) = 0$ for all N. We now use the fact that the Pauli matrices form a basis for the 2×2 matrices of zero trace, so that there exists a 3-component complex vector $\mathbf{\Omega}_N$ such that

$$(\partial_\omega \mathbf{T}_N)\mathbf{T}_N^{-1} = -\frac{i}{2}\mathbf{\Omega}_N \cdot \boldsymbol{\sigma}. \quad (15)$$

In the case of pure PMD (i.e., all $\alpha_j = 0$), the vector $\mathbf{\Omega}_N$ is real and is called the principal state vector. Physically the direction of the principal state vector is the direction of the fast PSP, while its length is the DGD. When the transmission link includes PDL, the vector $\mathbf{\Omega}$ becomes complex. The relations between the complex vector $\mathbf{\Omega}$, the PSPs on the Poincaré sphere, and the DGD are however more complicated than in the case without PDL [5]. The most important consequence is that the two PSP's are no longer orthogonal. This is easily seen from Eq. (15). When there is no PDL, since $\mathbf{\Omega}$ is real, the matrix $\mathbf{\Omega} \cdot \boldsymbol{\sigma}$ is hermitian. Therefore, it has two orthogonal eigenvectors, the PSP's. When PDL is added, since $\mathbf{\Omega}$ is a complex vector, the corresponding matrix $\mathbf{\Omega} \cdot \boldsymbol{\sigma}$ is no more hermitian, and its eigenvectors are not necessarily orthogonal. As we shall emphasize below, the non-orthogonality of the two PSP's is responsible for several peculiar features.

From Eqs. (13) and (15), we rewrite the eigenvalue equation for the PSPs as

$$\mathbf{\Omega}_N \cdot \boldsymbol{\sigma} \psi_{\mathrm{out}}(\omega) = \chi \psi_{\mathrm{out}}(\omega) \,. \quad (16)$$

This implies an important relation for χ^2:

$$\chi^2 = \mathbf{\Omega}_N \cdot \mathbf{\Omega}_N \,, \quad (17)$$

so that the two eigenvalues $\pm\chi$ have opposite signs. In the following, we will define χ as the eigenvalue with positive real part, and the two eigenvectors in Jones formalism by $\psi_\pm$, corresponding to the eigenvalues $\pm\chi$ respectively. Note that the two vectors, $\psi_\pm$ are defined at the output of the system. The DGD, $\delta\tau$ is thus simply the real part of χ:

$$\chi = \delta\tau + i\eta, \quad (18)$$

where η is the frequency derivative of the differential attenuation of the two PSP's. In analogy to the DGD, we shall call η the differential attenuation slope, or DAS.

Using Eqs. (14) and (15), the recursion equation for $\mathbf{\Omega}_N$ reads

$$\begin{aligned} \mathbf{\Omega}_N = {} & (\beta_N + \hat{e}_N \cdot \mathbf{\Omega}_{N-1}) \cdot \hat{e}_N + \\ & \cosh b_N \,(\mathbf{\Omega}_{N-1} - (\mathbf{\Omega}_{N-1} \cdot \hat{e}_N)\hat{e}_N) - i \sinh b_N \mathbf{\Omega}_{N-1} \wedge \hat{e}_N \,, \end{aligned} \quad (19)$$

where b_j is defined in Eq. (11). This equation generalizes the equation for PMD to the case where some elements along the line have PDL.

In the limit of short trunks a dynamical equation of evolution for the complex vector $\mathbf{\Omega}(z)$ can be derived [1].

The calculation of $\mathbf{\Omega}$ gives all the information about the PSP's. However, the relation between $\mathbf{\Omega}$ and the PSP's on the Poincaré sphere is more complicated than for optical links with no PDL. In particular, the decomposition of $\mathbf{\Omega}$ into real and imaginary parts, $\mathbf{\Omega} = \mathbf{\Omega}_r + i\mathbf{\Omega}_i$, is not the most interesting, as we shall see in the following. It is more relevant to define a new normalized vector, $\hat{W}$, by

$$\hat{W} \equiv \frac{\mathbf{\Omega}}{\chi}. \tag{20}$$

This vector can by decomposed into its real and imaginary parts: $\hat{W} = \mathbf{W}_r + i\mathbf{W}_i$. Note that this vector does not have the same direction as $\mathbf{\Omega}$. However, it is $\hat{W}$ which will enter into the expression of the PSP's on the Poincaré sphere. The calculation is performed in [1]. The two real vectors representing the PSP's are $\mathbf{m}_\pm$, for the two eigenvalues $\pm\chi$ respectively, and are given by

$$\mathbf{m}_\pm = \frac{\pm\mathbf{W}_r + \mathbf{W}_r \wedge \mathbf{W}_i}{W_r^2}. \tag{21}$$

The overlap between the two Jones vectors, $\gamma \equiv |\psi_+^\dagger \psi_-|$ is easily obtained from Eq. (21):

$$\gamma^2 = \frac{1 + \mathbf{m}_+ \cdot \mathbf{m}_-}{2} = \frac{W_i^2}{W_r^2} = \frac{|\mathbf{\Omega}|^2 - |\chi|^2}{|\mathbf{\Omega}|^2 + |\chi|^2}, \tag{22}$$

where $|\mathbf{\Omega}|^2 \equiv \mathbf{\Omega} \cdot \mathbf{\Omega}^\star$.

3. PDL-induced Anomalous Dispersion

In this section we show that the interaction PMD-PDL generates several peculiar effects.

3.1. Anomalous DGD

We first present a simple example [6] consisting of one pure PDL element sandwiched between two fibers with PMD. We set the axis of the first and third elements parallel to the x axis (on the Poincaré sphere) and that of the central PDL element in the z direction (i.e., orthogonal to the PMD fibers in the Poincaré sphere, or at 45° in ordinary space). Solving Eq. (19), one obtains

$$\mathbf{\Omega}(\omega) = \begin{pmatrix} \beta_3 + \beta_1 \cosh b_2 \\ \beta_2 \sin \beta_3\omega + i\beta_1 \cos \beta_3\omega \sinh b_2 \\ \beta_2 \cos \beta_3\omega - i\beta_1 \sin \beta_3\omega \sinh b_2 \end{pmatrix}, \tag{23}$$

where β_1, β_2 and β_3 are the PMD delays of the three elements, α_2 describes the PDL of the second element, and $b_2 \equiv \alpha_2 - i\beta_2\omega$.

In this case, the eigenvalue χ of Eq. (16) is complex, and the DGD corresponds to its real part. The explicit expression is cumbersome, and shall not be given here, but is rather plotted in Fig. 2 (dash-dotted curve) for the experimental system considered in the following paragraph. The effect of the PMD of the central fiber is to modulate the DGD as a function of the wavelength. The maximum value of the DGD is

$$\delta\tau_{\max} = \sqrt{\beta_1^2 + \beta_2^2 + \beta_3^2 + 2\beta_1\beta_3\cosh\alpha_2}\ . \tag{24}$$

For $\alpha_2 > 0$, i.e. in the presence of PDL, Eq. (24) shows that, when β_2 is small with respect to β_1 and β_3, $\delta\tau > \beta_1 + \beta_2 + \beta_3$. This is a surprising result: the presence of PDL in a birefringent optical fiber can increase the DGD above the sum of the DGD of all the elements. In addition to its counterintuitive aspect, this point could have concrete implications for systems that combine birefringent fibers and components with PDL, since the distortion and BER may be larger than expected from simple considerations.

In order to verify this result experimentally, we measured the DGD of the above concatenation. The experimental setup is described in Fig. 1. The system we used consisted of a PDL fiber, with PMD $\beta_2 = 1.01$ ps and PDL of 18 dB, corresponding to $\alpha_2 = 2.07$, sandwiched between two Hibi fibers, with PMD $\beta_1 = \beta_2 = 5.65$ ps. Our results are presented in Fig. 2. We first aligned the axes of the three fibers, to get the sum of the DGD for each component. This corresponds to the full curve of Fig. 2, and is in excellent agreement with the theoretical prediction:

$$\delta\tau = \beta_1 + \beta_2 + \beta_3 = 12.3\text{ps}. \tag{25}$$

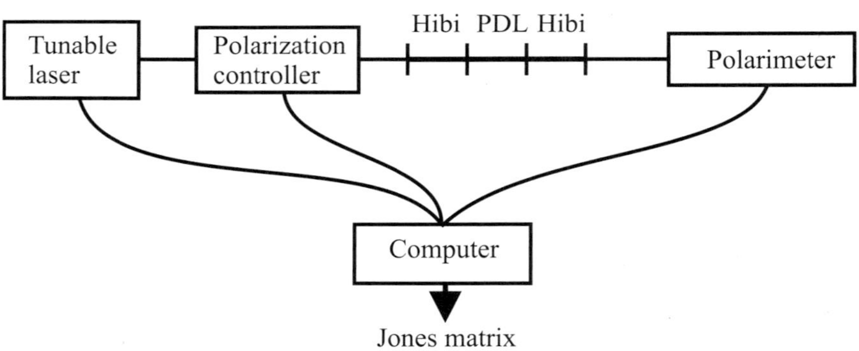

Fig. 1. Experimental setup for DGD measurements. This experimental setup implements Jones matrix eigenanalysis (JME). It measures the Jones matrix of the fiber under test as a function of the wavelength, which enables to obtain the DGD (see [4] for more details). In our case, the fiber under test is a concatenation of three trunks, one trunk with PDL being sandwiched between two Hibi trunks.

The calculated average is 12.36 ps, and is represented by a dotted line. We then rotated the axis of the PDL fiber by 45° with respect to both Hibi fibers. This is represented by the dashed curve in Fig. 2. The theoretical calculation corresponding to this setup is represented by the dashed-dotted curve, and is also in excellent agreement with the experimental curve. With the experimental values used, the maximum DGD obtained from Eq. (24) was $\delta\tau_{\max} = 17.94$ ps, which is much higher than the maximum we get with all the fibers aligned.

Our theoretical analysis, combined with the above experimental result, already show that special care has to be taken to analyze the effects of combined PMD and PDL. The usual intuition about about combining the various elements in the concatenation has to be considered cautiously.

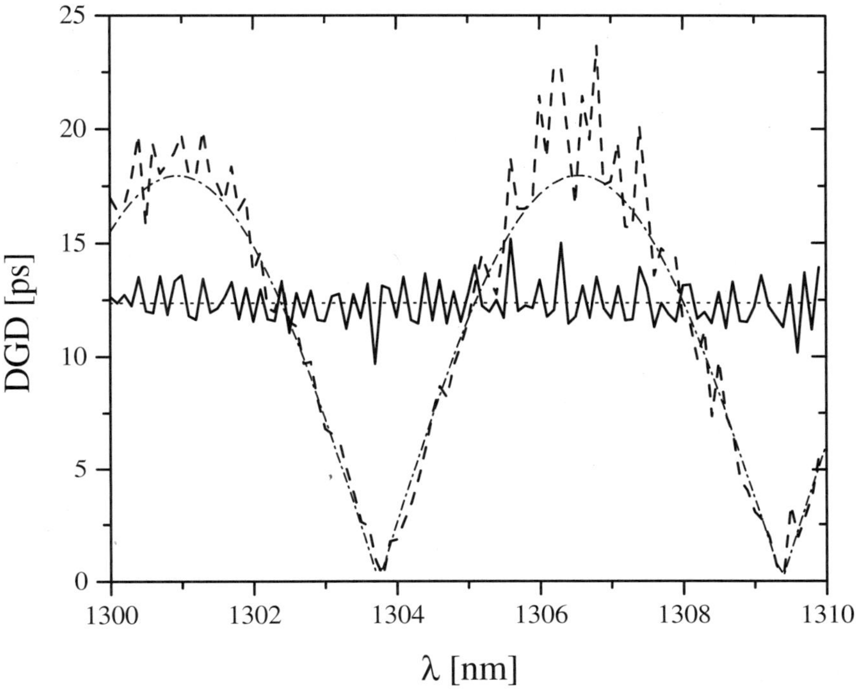

Fig. 2. Anomalous DGD. The fiber under test is a concatenation of three elements: one HiBi fiber with PMD 5.65 ps, followed by one PDL fiber with differential attenuation 18 dB and PMD 1.01 ps, and terminated with another HiBi fiber with PMD 5.65 ps. The full curve represents the DGD when all three axes are aligned, the dotted curve being the theoretical prediction (sum of the three DGD's). The dashed curve represents the anomalous DGD, obtained when the PDL fiber is rotated by 45° with respect to both HiBi fibers. In this case, the maximum DGD can be larger than the sum of all DGD's. The dashed-dotted curve is the corresponding theoretical prediction.

3.2. Anomalous Dispersion

In the above section, we have seen that the DGD of a concatenation of fibers with PMD and PDL may be higher than the sum of the DGDs of the elements. This surprising result may be complemented by the explicit calculation of the dispersion of a pulse. In general the output pulse can be written as a superposition of the two PSP's with some coefficients c_+ and c_-. Such a description is sufficient as long as the spectral spread $\Delta\omega$ is small enough for to validate a first order development in ω. This is expected to be the case whenever the coherence time is larger than the DGD $\delta\tau$. This is in fact the interesting domain for optical transmission. In this work, we only analyse the case of a Gaussian pulse, to show the combined effect of PMD-PDL. The influence of chirp could be added to the formalism, but this will not be done here. In the frequency domain, the outgoing field reads

$$E(\omega) = c_+ e^{-\frac{(\omega-\omega_0)^2}{2\Delta\omega^2}} e^{i(\omega-\omega_0)\frac{X}{2}} \psi_+(\omega_0)$$

$$+ c_- e^{-\frac{(\omega-\omega_0)^2}{2\Delta\omega^2}} e^{-i(\omega-\omega_0)\frac{\chi}{2}} \psi_-(\omega_0). \tag{26}$$

Equation (26) explains the physical content of χ. If we replace it by $\delta\tau + i\eta$ [Eq. (18)], we see that the time delay for each mode is given by $\pm\frac{\delta\tau}{2}$, corresponding to a DGD of $\delta\tau$, while $\pm\eta(\omega-\omega_0)$ corresponds to differential attenuation between the two modes, leading to a DAS of η. >From this, the outgoing intensity can be computed as

$$I(t) = \left| \int E(\omega) e^{i\omega t} d\omega \right|^2 = \left| \sum_{j=\pm} c_j e^{-(t+j\frac{\chi}{2})^2 \frac{\Delta\omega^2}{2}} e^{-i\omega_0 t} \psi_j(\omega_0) \right|^2. \tag{27}$$

Because of PDL, the principal polarization states $\psi_\pm$ are not orthogonal, in general. The above formula thus describes interferences between the fast and slow principal polarization modes. Their overlap has been derived in Eq. (22), and has to be included in the calculations.

Using Eq. (27) one can calculate the time of flight

$$< t > \equiv \frac{\int t I(t) dt}{\int I(t) dt}$$

as a function of $\delta\tau$ and γ. The calculation is rather lengthy, but completely straightforward, involving only integral of Gaussian functions. The result for the slowest and fastest time is

$$< t >_{\max/\min} = \frac{\pm\delta\tau}{2\sqrt{1-\gamma^2}}. \tag{28}$$

Note that this simple result is only correct for a Gaussian pulse, as defined in Eq. (26). The addition of chirp would lead to more complicated results. Surprisingly, for given DGD $\delta\tau$, this can be arbitrarily large. However, the fastest pulse will be more attenuated than a pulse polarized along one of the two principal polarization modes. The physical content of Eq. (28) is easily understood in terms of interference between the two PSP components. Indeed, because of PDL, the principal modes are no longer mutually orthogonal, as is the case for pure PMD. Hence, if both modes are excited by a pulse, the output pulse is distorted by the interference between both principal modes. As an example, assume that the slow mode has 1/4 of the fast mode intensity, as shown in Fig. 3, and that the interference is destructive. This amounts to "cut the back" of the fast pulse, resulting in an output pulse whose center is more advanced than that of the fast pulse. The same explanation, with the fast mode now being the weakest, leads to the slowest pulse. For optical transmissions, one can consider that the polarization of a train of pulses is constant for a very large number of pulses. Accordingly, the most relevant parameter is not the time of flight, which will not cause distortions if it fluctuates only very slowly with respect to the pulse repetition rate, but rather the pulse spreading, which may reduce the eye opening. It is again straightforward to compute the largest pulse spreading, which is obtained when the two coefficients, c_+ and c_- are equal, and the relative phase between the two terms in Eq. (26) is π, and reads

$$\delta\sigma_{\max} \equiv \sqrt{\sigma^2_{\max}(\text{output}) - \sigma^2(\text{input})} = \frac{1}{2}\sqrt{\frac{\delta\tau^2 + \eta^2\gamma}{1-\gamma}}. \tag{29}$$

This again can be arbitrarily larger than $\frac{\delta\tau}{2}$, which is the value obtained for pure PMD. Here again, it is easy to get the physical intuition behind this surprising phenomena in terms of interference between the PSP's. Assume that both principal modes have equal intensity, gaussian profiles and have delays of $\pm\delta\tau/2$. A destructive interference will destroy the central part of the outcoming pulse, as is illustrated on Fig. 4. Because of this destructive interference, the spread of the output pulse is larger than $\delta\tau$ (in the example of Fig. 4, it is almost twice as large). In Fig. 4, only the absolute value of the electric field is displayed, the intensity being equal to the square of this. The pulse obtained by interference is computed as the absolute value of the difference between the fast and slow pulse electric fields.

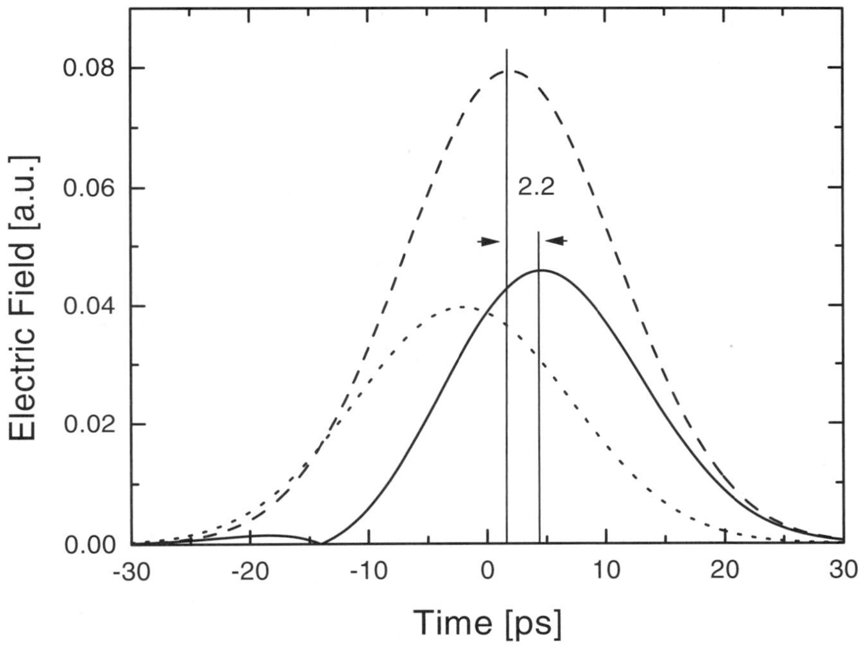

Fig. 3. Anomalous group velocity. Illustrative example of the displacement of the center of a pulse, due to the interference between the two PSP's. The dashed curve represents the fast PSP, while the dotted one represents the slow PSP. The relative intensities of the PSP's is chosen to be 4. The full curve is the electric field enveloppe of a pulse, which is the sum of the above two curves. The phase is chosen to give a destructive interference, which removes the back part of the pulse, thus displacing it in the forward direction. A similar setup, with stronger slow pulse, would displace the pulse in the backward direction.

3.3. *Pulse Spreading with Zero-DGD*

In this test [7], we analyze a slightly different concatenation, where we simply rotate the third fiber, so that its axis is now $-x$ on the Poincaré sphere, and replace the central fiber to get an element with PDL but no birefringence. Theoretical calculation shows that the DGD of such a concatenation is naught. However, a simulation of the interferometric method predicts non-zero PMD [7]. An experiment with the set of parameters of Fig. 5 confirms both predictions, with a very low value of the average

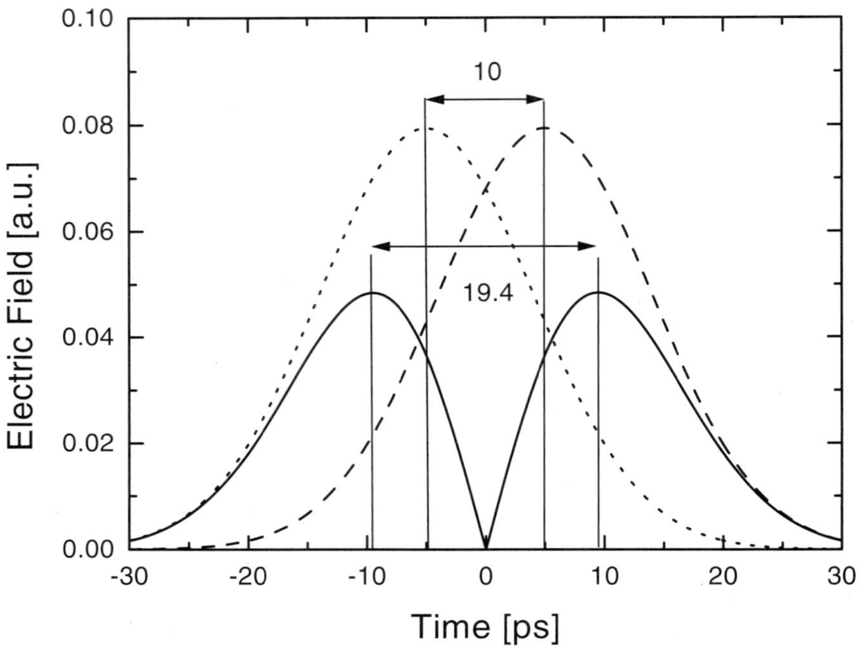

Fig. 4. Anomalous pulse spread. When the relative intensities of both PSP's are equal, the result of the destructive interference is to reduce the height of the pulse at the center, thus increasing the spread. The dashed curve represents the fast PSP, while the dotted one represents the slow PSP. The full curve is the electric field enveloppe of a pulse, which is the sum of the above two curves. Similarly, the spread could be reduced by constructive interference. Since this effect is created by an interference, it is very sensitive to minute fluctuations in the system. This would generate an extra noise in the pulse spread, which may lead to a degradation of the signal.

DGD (below 0.1 ps), and a PMD value obtained from the interferometric method, which is one order of magnitude larger (for both the simulation and the experiment).

When there is no PDL, the two eigenvalues of the PSP matrix [Eq. (16)] are real (no attenuation), so that a zero DGD really means no birefringence. However, when the fibre link has PDL, the eigenvalues becomes complex. In this case, the DGD will be zero whenever the eigenvalues are purely imaginary. As we shall see, polarization effects may still create pulse spreading. We consider a fiber section made of three trunks. The first and last trunks are purely birefringent, with birefringence β along axes x and $-x$ on the Poincaré sphere respectively. The second trunk has only PDL, given by α along axis y. We use the recursion equation, Eq. (19), to obtain the principal state vector $\mathbf{\Omega}_3$. The eigenvalue χ is then given by $\chi^2 = \mathbf{\Omega}_3 \cdot \mathbf{\Omega}_3$. After straightforward calculations, we get

$$\chi^2 = -2\beta^2(\cosh\alpha - 1). \tag{30}$$

Since this is a negative value, it corresponds to a purely imaginary χ. This corresponds to zero DGD over the full wavelength range where our approximations (i.e., wavelength-independent birefringence and PDL) are valid. However, even though the DGD is zero, the device will still cause pulse spreading. Consider a short pulse (short

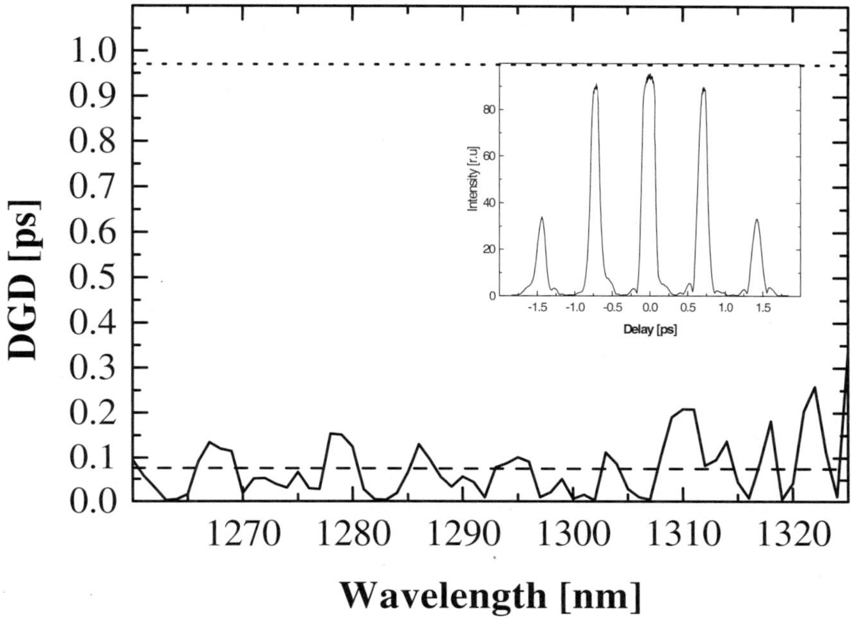

Fig. 5. Measurements of a concatenation with zero-DGD. The concatenation is built from one pure PDL element, with 7 dB differential attenuation and axis y, sandwiched between two HiBi fibers with 0.7 ps of PMD each and axes x and $-x$ respectively. The predicted DGD of such a concatenation is zero. The full curve is the DGD obtained from a JME measurement, similar to the one in Fig 1. The dashed curve is the average DGD, at 0.075 ps, while the dotted curve is the PMD obtained with the interferometric method, at 0.97 ps. The inset gives the interferogram given by the interferometric method, the variance of which gives the PMD.

with respect to β), propagating down the fibre. The first HiBi fibre separates the initial pulse into two distinct pulses, separated in time by the DGD of the first trunk, and polarized along x and $-x$ respectively. When no PDL is present, the second HiBi fibre recombines the two pulses exactly. This device therefore has zero global birefringence. The effect of a PDL element can be understood as follows: a pulse polarized along direction, say x, is decomposed along the two axes y (axis with low attenuation) and $-y$ (axis with high attenuation). The effect of the differential attenuation is to reduce the amplitude along $-y$, thus creating a rotation of the polarization towards y . Since the two components along x and $-x$ experience a different rotation, the second HiBi fibre cannot recombine the pulses fully. For the case under consideration, the output state is therefore composed of 3 pulses separated by β . This time-domain analysis clearly explains the pulse spreading, and the fact that it is predicted by the interferometric method. These theoretical considerations were also tested experimentally. The setup is quite similar to the one in Fig. 1. We first constructed an element with pure PDL and no PMD by concatenating a PDL fiber with a HiBi fiber with exactly opposite birefringence. The PDL of this device was 7 dB. This is a rather high value with respect to telecom components. The point was to choose a value, which would enable to verify easily the theoretical prediction. We then sandwiched this element between two other HiBi fibers, with birefringent axes orthogonal to one another, and at 45° of the PDL axis of the central element. The PMD of these fibers was 0.7 ps each. We measured this device with the interferometric method, using a commercial PMD analyzer. The result is shown in the inset of Fig. 5, and gives a calculated PMD of 0.97 ps. We then performed a standard DGD measurement on the whole device, by means of JME, with a HP polarimetric system. The result is presented in Fig. 5. The average DGD value is an order of magnitude lower than the interferometric PMD. Moreover, even the maximum value of the DGD between 1270 and 1320 nm is much lower than the interferometric PMD. This experimental result is in excellent agreement with the theory.

4. Conclusion

The interaction between PMD and PDL is a rather new topic, brought about by recent developments in optical networks.

In these notes we briefly recalled some of the formalism that has been developed, and some of the physical experiments that support it. Reference [1] contains also significant numerical simulations, in order to asses the potential impact on near-future optical network of the combined effects PMD+PDL. This reference presents also the concept of the most and the least attenuated polarization states and their dynamical evolution equations. Finally, in [1], it is demonstrated that any linear optics system with interleaved PMD and PDL is equivalent to another linear optic system where all the PMD comes first, followed by some PDL (however these "global PMD" and "global PDL" are not simply to sum of the ones appearing in the first system).

The physics of PMD+PDL effects is fascinating and still a research topic. For instance, in [8] we recently established a surprising connection to the quantum theory of weak-value measurements. The following are some further recent contributions to this study: [9, 10, 11, 12, 13] (and quite more can be found in the OFC and ECOC conference proceedings of the recent years).

Acknowledgments

Thanks are due to Valerio Scarani for the editorial work. Partial financial support by the Swiss OFES within the european project COST 265 and by EXFO are acknowledged.

References

1. B. Huttner, Ch. Geiser and N. Gisin, IEEE J. Select. Topics Quant. Electron., **6**, 317–329 (2000).
2. C.D. Poole and R.E. Wagner, "Phenomenological approach to polarization dispersion in long single-mode fibers", Electron. Lett. **22**, 1029–1030 (1986).
3. S. Huard, *Polarisation de la lumière* (Masson, Paris, 1994).
4. B.L Heffner, "Automated measurement of polarization mode dispersion using Jones matrix eigenanalysis", IEEE Photon. Technol. Lett. **4**, 1066–1069 (1992).

5. N. Gisin and B. Huttner, "Combined effects of polarization mode dispersion and polarization dependent losses in optical fibers", Opt. Commun. **142**, 119–125 (1997).

6. B. Huttner and N. Gisin, "Anomalous pulse spreading in birefringent optical fibers with polarization-dependent losses", Opt. Lett. **22**, 504–506 (1997).

7. B. Huttner, C. De Barros, B. Gisin and N. Gisin, "Polarisation-induced pulse spreading in birefringent optical fibres with zero differential group delay", Opt. Lett. **24**, 370–372 (1999).

8. N. Brunner, A. Acin, D.Collins, N. Gisin and V. Scarani, Phys.l Rev. Lett. **91**, 180402-1/4 (2003).

9. L. Chen, J. Cameron and X. Bao, Opt. Commun. **169**, 69 (1999).

10. L. Chen and X. Bao, IEEE J. Lightwave Technol. **18**, 665 (2000).

11. A. Bessa dos Santos, A.O. Dal Forno and J.P. Von der Weid, ECOC, #3.4.5, pp 145–146, 2000.

12. P. Lu, L. Chen and X. Bao, IEEE Photon. Technol. Lett. **13**, 451 2001.

13. Y. Li and A. Yariv, J. Opt. Soc. Am. B **17**, 1821 (2000).

DOI: 10.1007/s10297-004-0009-x
Originally published in J. Opt. Fiber. Commun. Rep. 1, 305–311 (2004)

Interaction of nonlinearity and polarization mode dispersion

Curtis R. Menyuk

University of Maryland Baltimore County
Computer Science and Electrical Engineering Department
1000 Hilltop Circle, Baltimore, MD, 21250 USA
and
PhotonEx Corporation
200 MetroWest Technology Park, Maynard, MA 01754 USA

Abstract. The derivation of the coupled nonlinear Schrödinger equation and the Manakov-PMD equation is reviewed. It is shown that the usual scalar nonlinear Schrödinger equation can be derived from the Manakov-PMD equation when polarization mode dispersion is negligible and the signal is initially in a single polarization state as a function of time. Applications of the Manakov-PMD equation to studies of the interaction of the Kerr nonlinearity with polarization mode dispersion are then discussed.

There has been a flood of recent work on polarization effects in optical fibers, most of it focused on polarization mode dispersion (PMD). There are good, practical reasons for this recent interest. As the data rate per channel increases, PMD becomes an increas-ingly important limitation in communication systems. However, my own interest in polarization issues was, at least originally, focused on more fundamental questions. Since nearly all the contributions in this volume focus on the more immediate practical issues that are a consequence of polarization effects, it seemed useful to me to focus this contribution on more fundamental issues. The nice thing about fundamental issues is that even though they attract less attention than immediate practical issues, they typically stay relevant longer and can impact practical issues 10–20 years in the future.

Throughout the 1970s, a considerable body of work established the basic polarization properties of optical fibers. Much of this work was summarized by Kaminow in 1981 [1]. It was found that the birefringence $\Delta n/n$ is in the range 10^{-4}–10^{-9}, with communication fibers in the range 10^{-6}–10^{-7}. It was found that the intrinsic birefringence is almost entirely linear, even when the fiber is twisted, because of the very small value of the electro-optic tensor in glass [2]. Remarkably, there has been no change in either the range of the birefringence values or the helicity in optical fibers in the last 20 years, although many other fiber parameters have changed considerably. The fiber-induced PMD can now be made much smaller by spinning fibers as they are drawn; the effective areas of fibers can be made much larger or smaller, affecting the strength of the nonlinearity; the dispersion can be tailored; and losses can be made smaller.

In 1980, Mollenauer et al. [3] demonstrated the propagation of solitons in optical fibers for the first time. This work began a series of studies, continuing into the present day, that explored the interaction between chromatic dispersion and nonlinearity in optical fibers and their implications for communication systems. The theoretical basis for this work was the nonlinear Schrödinger equation, which had been derived by Hasegawa and Tappert in 1973 [4]. This work took no account of the birefringence in optical fibers; however, work by Botineau and Stolen [5] had made it clear that the interaction of birefringence and nonlinearity could be quite important. In their work, they measured the nonlinear polarization rotation and demonstrated its potential for nonlinear switching. As part of this work, they demonstrated that the ratio of the cross-phase modulation and the self-phase modulation in optical fibers is 2/3. I note that this coefficient is only expected to be 2/3 if the fiber's birefringence is intrinsically linear. More generally, it equals $(2 + 2 \mid \hat{\mathbf{e}}_1 \cdot \hat{\mathbf{e}}_2 \mid)/(2+ \mid \hat{\mathbf{e}}_1 \cdot \hat{\mathbf{e}}_1 \mid)$, where $\hat{\mathbf{e}}_1$ and $\hat{\mathbf{e}}_2$ are the unit vectors for the birefringent eigenmodes [6]. Its value ranges from 2/3 for linearly birefringent fibers to 2 for circularly birefringent fibers. Thus, the nonlinear light evolution yields information about the linear properties of the fibers—a point to which I shall return.

Given the experimental evidence of the importance of birefringence, it was natural to derive an equation that takes it into account. This equation, referred to as the coupled nonlinear Schrödinger equation, was first presented in 1987 [7]. In a form that is particularly useful for studying randomly varying birefringence, it may be written as

$$i\frac{\partial \mathbf{u}}{\partial z} - ig\mathbf{u} + (\sigma_3 \cos\theta + \sigma_1 \sin\theta)\left(\Delta\beta\mathbf{u} + i\Delta\beta'\frac{\partial \mathbf{u}}{\partial t}\right) - \frac{1}{2}\beta''\frac{\partial^2 \mathbf{u}}{\partial t^2}$$
$$+\gamma\left[\mid \mathbf{u} \mid^2 \mathbf{u} - \frac{1}{3}\left(\mathbf{u}^\dagger\sigma_2\mathbf{u}\right)\sigma_2\mathbf{u}\right] + 0, \qquad (1)$$

where z and t are distance along the fiber and retarded time, respectively, while $\mathbf{u}(z,t)$ is the two-dimensional Stokes vector wave envelope. The parameter g is the gain and loss in the fiber, $\theta/2$ is the orientation angle of the axis of birefringence, $\Delta\beta$ is the birefringence, $\Delta\beta'$ is the frequency derivative of the birefringence, β'' is the dispersion, and γ is the Kerr coefficient. All parameters are evaluated at the carrier frequency. The σ_j are the standard Pauli matrices. My original motivation for deriving this equation was that for deep theoretical reasons I had hypothesized that solitons would be robust in the presence of Hamiltonian perturbations [8]–[10]. It is beyond the scope of this contribution to discuss the origin of this hypothesis and what Hamiltonian perturbations are, except to note that when the birefringence is small, it is a Hamiltonian perturbation. What does it mean for the birefringence to be small? It means that the birefringent beat length should be large compared to the nonlinear scale length and the scale length for chromatic dispersion. Under these circumstances, theory [11] and experiment [12] show that solitons are robust in the presence of birefringence, as predicted by the robustness hypothesis. However, this limit is not the usual limit that applies in optical fiber communication systems. In these systems, the beat length is short compared to the scale lengths for nonlinearity and chromatic dispersion, as shown in Fig. 1. Under these circumstances, the coupled nonlinear Schrödinger equation indicates that solitons should be ripped apart—an effect that is not observed. Clearly, there is a mystery here.

In its broad outlines, the answer to this mystery, shortly after it first became apparent in 1988, was not long in coming. The birefringence in optical fibers is randomly varying. As long as the PMD scale length is long compared to the nonlinear and dispersive scale lengths, solitons will be robust [13]–[15]. The PMD scale length is the length over which PMD will differentially change the polarization states in the bandwidth of a signal so that the polarization state is no longer constant over the signal spectrum. This scale length is the same as the scale length over which PMD will cause a linear pulse to spread significantly [16]–[18].

However, a key question remained: Why and when should the nonlinear Schrödinger equation hold? Since the early, very brief and intuitive derivation of Hasegawa and Tappert [4], numerous derivations of the nonlinear Schrödinger equation have been published. Many of these derivations contain mathematical errors. Two of the most common are neglecting the $\partial^2/\partial z^2$ term that appears in Maxwell's equations, rather than eliminating it self-consistently, and dropping the $\nabla(\nabla \cdot \mathbf{E})$ term. However, Kodama [19] has published a derivation that takes into account these issues as well as numerous other issues and is, I believe, completely correct mathematically. However, this derivation shares with all others up to my recently published derivation [18] what I consider to be a conceptual flaw. All derivations up to [18] assumed that the core of an optical fiber is perfectly round and without stress and thus neglect the birefringence entirely. In fact, a physically correct derivation that is applicable to communication systems should take into account the large, but rapidly and randomly varying, birefringence. From a conceptual standpoint, the logical chain that one must use is as follows: Starting from Maxwell's equations, one must average over the rapid oscillations in the carrier wave to obtain the coupled nonlinear Schrödinger equation, which appears in the middle group of scale lengths that I show in Fig. 1. From there, one must average over the rapidly and randomly varying birefringence to obtain the Manakov-PMD equation that applies to the largest group of length scales in Fig. 1. When PMD can be neglected, and, in addition, the initial field is in a single polarization state as a function of time, then the nonlinear Schrödinger equation holds [18].

When my colleagues and I began to tackle this question in the early 1990s, the first issue that we faced was that we did not know how to model the randomly varying birefringence in the fibers. In his analytical work, Poole [20] used a model that assumed weak off-diagonal coupling between the axes of birefringence. This model was first used to describe polarization-maintaining fiber, for which it is physically reasonable [21], but it was less obvious to my colleagues and I that it is physically reasonable for communication fibers in which we knew that the orientation of the birefringence axes had to fairly rapidly cover all possible values uniformly. How did we know that? If the opposite held, then, as noted previously, our early work had make it clear that the birefringence would rip solitons apart—something that is not observed. For numerical work, Poole et al. [22] introduced a rotating waveplate model. It is closely related to the coarse step method, which was introduced in [14] and [15] to allow users to deal effectively with systems in which nonlinearity, chromatic dispersion, and randomly varying birefringence are all present. A difficulty in principle with the rotating waveplate model that the models introduced in [14] and [15] avoid is that if the beat length

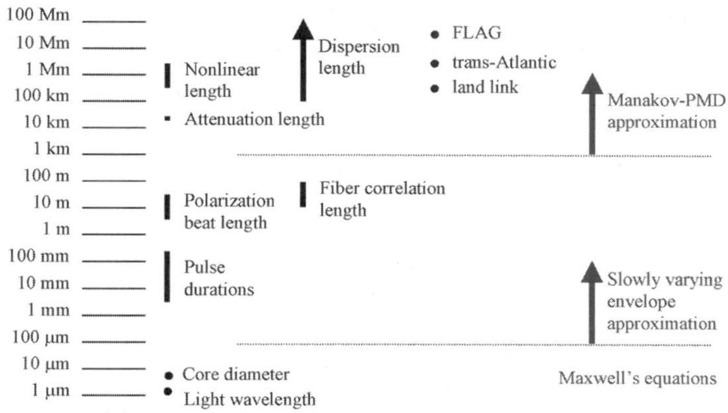

Fig. 1. Illustration of the key length scales in optical fiber transmission systems.

is resonant with birefringent fiber lengths, then a linear input polarization state can only transform to other linearly polarized states. Thus, the randomly rotating wave-plate model is not suitable for modeling broadband wavelength division multiplexed systems. By contrast, the coarse step models are widely used in both commercial and non-commercial, full-system simulators.

A key issue in deriving the Manakov-PMD equation is determining the statistics of the evolution of polarization states on the Poincaré sphere. Since the statistical variation of the birefringence in optical fibers was unknown, Wai and Menyuk [23] proposed two physical models. In the first model, the birefringence is fixed and its orientation is allowed to vary. In the second, the change in the index of refraction due to the birefringence is assumed to be Gaussian-distributed in both transverse directions of the fiber. Our original thought was that the two models might make different predictions for the field evolution that would allow one to determine experimentally which underlying fiber model is correct. In fact, the two models make nearly identical predictions. Thus, the issue remained unresolved until the recent polarization optical time domain reflectrome-try measurements by Galtarossa et al. [24] showed, in combination with earlier results indicating that the fiber birefringence is linear, that the second model is correct. Alex Wai and I found that the evolution on the Poincaré sphere is anisotropic and that it depends on two parameters, the average fiber beat length L_b and the fiber decorrelation length L_f. The field evolution on the Poincaré sphere is characterized by three parameters, an azimuthal diffusion length L_{az}, an equatorial diffusion length in a frame that is fixed $L_{\mathrm{eq,fixed}}$, and an equatorial diffusion length in a frame that rotates with the axes of birefringence, $L_{\mathrm{eq,rot}}$. Both Laz and Leq,fixed are proportional to L_f when L_b ? L_f and to L_b^2/L_f when L_f ? L_b . By contrast, $L_{\mathrm{eq,rot}}$ is proportional to L_f, regardless of L_b.

To average over the coupled nonlinear Schrödinger equation to obtain the Manakov-PMD equation, one proceeds in two steps [18], [23]. In the first step, one makes the transformation

$$\tilde{\mathbf{v}}(z,\omega) = \{\mathbf{I}\cos[\theta(z)/2] + \sigma_2 \sin[\theta(z)/2]\}\tilde{\mathbf{u}}(z,\omega), \tag{2}$$

where the tilde indicates the Fourier transform of the time domain quantities, and $\mathbf{I}$ is the 2×2 identity matrix. This transformation would diagonalize the linear portion of the coupled nonlinear Schrödinger equation in the absence of a z-varying θ. I note that ω is defined with respect to the carrier frequency. In the second step, we freeze the $\omega = 0$ motion by making the transformation $\tilde{\mathbf{v}}(z,\omega) = \mathbf{T}^{-1}\tilde{\mathbf{u}}(z,\omega) =$, where

$$i\frac{\partial \mathbf{T}(z)}{\partial z} + \{\Delta\beta\sigma_3 + [\theta_z(z)/2]\sigma_2\}\mathbf{T}(z) =); \qquad \mathbf{T}(z=0) = \mathbf{I}, \tag{3}$$

and θ_z indicates the z-derivative of θ. Returning to the time domain, we obtain the Manakov-PMD equation,

$$\begin{aligned} i\frac{\partial \mathbf{w}}{\partial z} - ig\mathbf{w} - \frac{1}{2}\beta''\frac{\partial^2 \mathbf{w}}{\partial t^2} + \frac{8}{9}\gamma \mid \mathbf{w} \mid^2 \mathbf{w} \\ = -i\Delta\beta'\bar{\sigma}_3\frac{\partial \mathbf{w}}{\partial t} + \frac{1}{3}\gamma\left[\left(\mathbf{w}^\dagger\bar{\sigma}_2\mathbf{w}\right)\bar{\sigma}_2\mathbf{w} - \frac{1}{3} \mid \mathbf{w} \mid^2 \mathbf{w}\right], \end{aligned} \tag{4}$$

where $\bar{\sigma}_j = \mathbf{T}^{-1}\sigma_j\mathbf{T}$. On the left-hand side of Eq. (4), we have the slowly varying contribution. On the right-hand side, we have the rapidly varying residual contributions. The first residual term is linear and leads to the usual linear PMD. The length scale on which averaging occurs equals $L_{\mathrm{eq,rot}} \sim L_f$, regardless of L_b. However, the accumulated differential group delay (DGD) is proportional to the birefringence; so, the expected value of the DGD is proportional to $L_f^{1/2}/L_b$ over lengths that are large compared to L_f. The second residual term is nonlinear and leads to an effect that my colleagues and I refer to as nonlinear PMD. Marcuse et al. [25] showed that this term is negligible in communication systems. However, it can be important in short-pulse systems, and Arend et al. [26] observed its effect in fiber loop mirrors. Interestingly, the strength of this effect is proportional to L_{az}, in contrast to the usual linear PMD, whose strength is proportional to $L_{\mathrm{eq,rot}}$.

While the nonlinear PMD can be neglected in present-day communication systems, the same is not true for the interaction of nonlinearity and the usual linear PMD. One example of this effect is that a nonlinearly induced pulse chirp can interact with higher-order PMD to lead to additional pulse spreading beyond the spreading that is induced by first-order PMD or can induce pulse compression. This effect was predicted by Ibragimov et al. [27] and was recently observed by Möller et al. [28]. A second example of this effect is that nonlinear polarization rotation induced by wavelength division multiplexed interactions can lead to fast variations of the polarization states that impede the ability of standard PMD compensators to mitigate signal degradation that is due to PMD. This effect was observed by Khosravani et al. [29] and by Lee et al. [30]. It is my view that much work remains to be done to explore the interaction of nonlinearity and PMD and that it is a very interesting area for future research.

In conclusion, much of my own work on polarization effects in optical fibers was motivated by a desire to understand what the basic equations are that govern light propagation in optical fibers on the length and time scales that are important for communication systems. This work led to the derivation of the coupled nonlinear Schrödinger equation that governs the evolution on the intermediate length scale of Fig. 1 and to the Manakov-PMD equation that governs the evolution on the long length scale of Fig. 1. From this work, we find that the usual scalar nonlinear Schrödinger equation and its variants will hold when PMD is negligible and the signal is launched in a single polarization state. Establishing the Manakov-PMD equation was important step forward. It includes all the physical phenomena of importance in optical fiber communications except the Brillouin and Raman nonlinearities, device effects like polarization dependent loss, and amplified spontaneous emission noise from the amplifiers. The Manakov-PMD equation can easily be modified to include these effects. It contains a rich set of phenomena, many of practical importance, which we have only begun to explore.

References

[1] I. P. Kaminow, "Polarization in optical fibers," IEEE J. Quantum Electron., **17**, 15–22 (1981).

[2] S. C. Rashleigh, "Origins and control of polarization effects in single-mode fibers," J. Lightwave Technol., **1**, 312–331 (1983).

[3] L. F. Mollenauer, R. H. Stolen, and J. P. Gordon, "Experimental observation of picosecond pulse narrowing and solitons in optical fibers," Phys. Rev. Lett., **45**, 1095–1098 (1980).

[4] A. Hasegawa and F. Tappert, "Transmission of stationary nonlinear optical pulses in dispersive dielectric fibers. I. Anomalous dispersion," Appl. Phys. Lett., **23**, 142–144 (1973).

[5] J. Botineau and R. H. Stolen, "Effect of polarization on spectral broadening in optical fibers," J. Opt. Soc. Am., **72**, 1592–1596 (1982). See also, R. H. Stolen, J. Botineau, and A. Ashkin, "Intensity discrimination of optical pulses with birefringent fibers," Opt. Lett., **7**, 512–514 (1982).

[6] C. R. Menyuk, "Pulse propagation in an elliptically birefringent Kerr medium," IEEE J. Quantum Electron., **25**, 2674–2682 (1989).

[7] C. R. Menyuk, "Nonlinear pulse propagation in birefringent optical fibers," IEEE J. Quantum Electron., **23**, 174–176 (1987).

[8] C. R. Menyuk, "Origin of solitons in the 'real' world," Phys. Rev. A, **33**, 4367–4374 (1986).

[9] C. R. Menyuk, "Why are solitons robust in optical fibers," *Guided Wave Nonlinear Optics*, D. Ostrowsky and R. Reinisch, eds. (Kluwer, Dordrecht, Netherlands, 1992), pp.457–499.

[10] C. R. Menyuk, "Soliton robustness in optical fibers," J. Opt. Soc. Am. B, **10**, 1585–1591 (1993).

[11] C. R. Menyuk, "Stability of solitons in birefringent optical fibers. I: Equal propagation amplitudes," Opt. Lett., **12**, 614–616 (1987); "Stability of solitons in birefringent optical fibers. II. Arbitrary amplitudes," J. Opt. Soc. Am. B, **5**, 392–402 (1988).

[12] M. N. Islam, C. D. Poole, and J. P. Gordon, "Soliton trapping in birefringent optical fibers," Opt. Lett., **14**, 1011–1013 (1989).

[13] L. F. Mollenauer, K. Smith, J. P. Gordon, and C. R. Menyuk, "Resistance of solitons to the effects of polarization dispersion in optical fibers," Opt. Lett., **14**, 1219–1221 (1989).

[14] P. K. A. Wai, C. R. Menyuk, and H. H. Chen, "Stability of Solitons in Randomly Varying Birefringent Fibers," Opt. Lett., **16**, 1231–1233 (1991).

[15] S. G. Evangelides, L. F. Mollenauer, and J. P. Gordon, "Polarization multiplexing with solitons," J. Lightwave Technol., **10**, 28-34, 1992.

[16] C. D. Poole and R. E. Wagner, "Phenomenological approach to polarisation dispersion in long single-mode fibres," Electron. Lett., **22**, 1029–1030 (1986).

[17] M. Karlsson, "Polarization mode dispersion-induced pulse broadening in optical fibers," Opt. Lett., **23**, 688–690 (1998).

[18] C. R. Menyuk, "Application of multiple length-scale methods to the study of optical fiber transmission," J. Engin. Math., **36**, 113–136 (1999).

[19] Y. Kodama, "Optical solitons in a monomode fiber," J. Stat. Phys., **39**, 597–614 (1985). See, also, A. Hasegawa and Y. Kodama, *Solitons in Optical Communications* (Clarendon, Oxford, UK, 1995), Chap. 1.

[20] C. D. Poole, "Statistical treatment of polarization dispersion in single-mode fiber," Opt. Lett., **13**, 687–689 (1988).

[21] S. C. Rashleigh, W. K. Burns, R. P. Moeller, and R. Ulrich, "Polarization holding in birefringent single-mode fibers," Opt. Lett., **7**, 40–42 (1982).

[22] C. D. Poole, J. H. Winters, and J. A. Nagel, "Dynamical equation for polarization dispersion," Opt. Lett., **16**, 372–374 (1991).

[23] P. K. A. Wai and C. R. Menyuk, "Polarization mode dispersion, decorrelation, and diffusion in optical fibers with randomly varying birefringence," J. Lightwave Technol., **14**, 148–157 (1996).

[24] F. Corsi, A. Galtarossa, and L. Palmieri, "Beat-length characterization based on backscattering analysis in randomly perturbed single-mode fibers," J. Lightwave Technol., **17**, 1172–1183 (1999); A. Galtarossa, L. Palmieri, M. Schiano, and T. Tambosso, "Statistical characterization of fiber random birefringence," Opt. Lett., **25**, 1322–1324 (2000).

[25] D. Marcuse, C. R. Menyuk, and P. K. A. Wai, "Application of the Manakov-PMD equation to studies of signal propagation in optical fibers with randomly varying birefringence," J. Lightwave Technol., **15**, 1735–1746 (1997).

[26] M. F. Arend, M. L. Dennis, I. N. Duling, E. A. Golovchenko, A. N. Pilipetskii, and C. R. Menyuk, "Nonlinear-optical loop mirror demultiplexer using a random birefringence fiber: Comparisons between theory and experiments," Opt. Lett., **22**, 886–888 (1997).

[27] E. Ibragimov, C. R. Menyuk, and W. Kath, "PMD-induced reduction of nonlinear penalties in terrestrial optical fiber transmission," Optical Fiber Communication Conference 2000, Baltimore, MD, paper WL3.

[28] L. Möller, Y. Su, G. Raybon, S. Chandrasekhar, and L. L. Buhl, "Penalty interference of nonlinear intra-channel effects and PMD in ultra high-speed TDM systems," Electron. Lett., **38**, 281–283 (2002). Note that Eqs. (1) and (2) contain small errors.

[29] R. Khosravani, Y. Xie, L. S. Yan, Y. W. Song, A. E. Willner, and C. R. Menyuk, "Limitations to First-Order PMD Compensation in WDM Systems Due to XPM-Induced PSP Changes," Optical Fiber Communications Conference 2001, Anaheim, CA, paper WAA5.

[30] J. H. Lee, K. J. Park, C. H. Kim, and Y. C. Chung, "Impact of nonlinear crosstalk on optical PMD compensation," Optical Fiber Communications Conference 2002, Anaheim, CA, paper ThI2.

DOI: 10.1007/s10297-004-0010-4
Originally published in J. Opt. Fiber. Commun. Rep. 1, 84–105 (2004)

PMD measurement techniques and how to avoid the pitfalls

Paul Williams

National Institute of Standards and Technology
Optical Fiber and Components Group
Optoelectronics Division
325 Broadway, Boulder, CO 80305 USA
Email: pwilliam@boulder.nist.gov

Abstract. Since polarization-mode dispersion (PMD) can often be so confusing, it is easy to see how its measurement can be complicated as well. Many different techniques for PMD measurement are available, and often many user-selectable parameters are associated with each measurement. Provided here is a description of the various measurement techniques available along with a discussion of the "best practices" for PMD measurement using these described techniques. Section 1 gives some definitions and lists the parameters that must be measured to fully characterize PMD and introduces the statistical uncertainty inherent in polarization-mode coupled devices. Section 2 describes the various measurement techniques, classifying them as either frequency-domain or time-domain techniques. Section 3 lists several useful practices to reduce measurement errors when measuring PMD. Section 4 discusses the concept of spectral efficiency as a useful figure of merit for describing measurement uncertainty normalized to spectral bandwidth. Finally, Section 5 describes the tradeoffs associated with wavelength step size (for frequency-domain techniques) and provides some "rules-of-thumb" for choosing appropriate wavelength steps in these measurements.

1. What Does it Take to Describe PMD?

1.1. Measurement Parameters

The PMD of a device is completely described when the Differential Group Delay (DGD) and the Principal States of Polarization (PSP) are characterized as functions of wavelength. Often, the terms PMD and DGD are used interchangeably, but here we will use the term PMD to describe the phenomenon and DGD to describe its magnitude. Both the DGD and the PSP are contained in the three-dimensional polarization dispersion vector $\boldsymbol{\Omega}(\lambda)$. The DGD is given by the magnitude $\Delta\tau = |\ \boldsymbol{\Omega}\ |$, and the PSP are given by the direction of $\boldsymbol{\Omega}$. For many applications, only $\Delta\tau(\lambda)$ is needed, and often even the wavelength dependence is not reported, in which case PMD is described by the wavelength-averaged DGD $\langle\Delta\tau\rangle_\lambda$ or the root-mean-square (RMS) average $\langle\Delta\tau^2\rangle_\lambda^{1/2}$.

In describing the various measurement techniques, we will discuss the two general cases of PMD-devices with and without polarization-mode coupling. A "non-mode-coupled" device is a simple birefringent element such as a single birefringent crystal. In this case, the polarization eigenaxes of the device coincide with the PSP and are independent of wavelength, and $\Delta\tau$ depends only weakly on wavelength. For example, in quartz, $\Delta\tau$ changes by less than 3% over a 1300–1800 nm wavelength range (Fig. 1) [1]. In "mode-coupled" devices (such as optical fiber), the eigenaxes do not necessarily coincide with the PSPs. The PSPs are independent of wavelength only to first order, and $\Delta\tau$ can be strongly dependent on wavelength (Fig. 2).

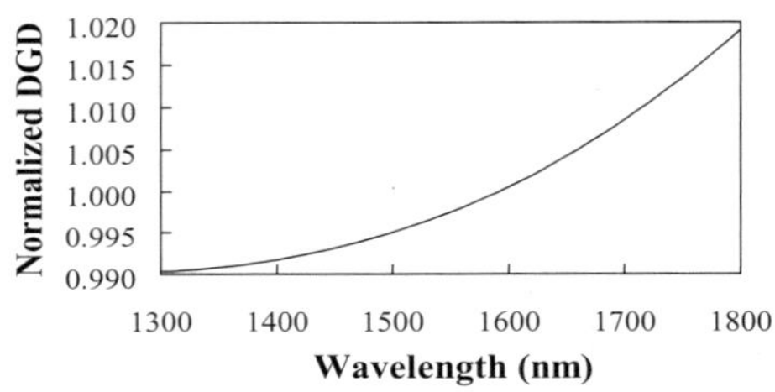

Fig. 1. Normalized DGD spectrum for quartz.

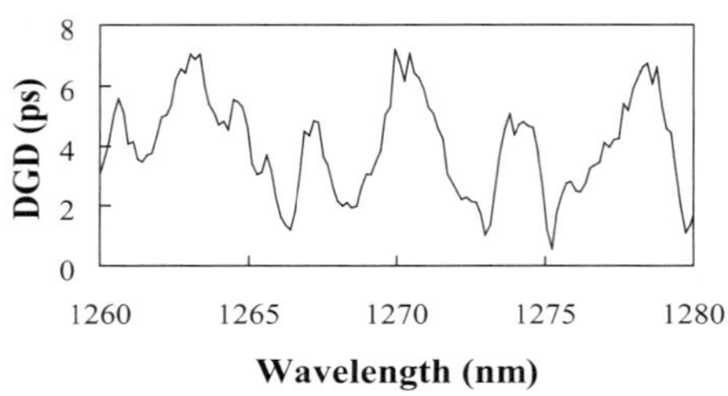

Fig. 2. Sample DGD spectrum of a mode-coupled device.

Exactly what must be measured in order to characterize the PMD of a device depends on the degree of mode coupling of the device and on what the measurement will be used for. Often, for non-mode-coupled devices, the mean DGD is well-approximated by the DGD at some particular wavelength λ_0, $\langle\Delta\tau\rangle \approx \Delta\tau(\lambda_0)$, and so the DGD need not be resolved as a function of wavelength. On the other hand, for mode-coupled devices, such as long lengths of fiber, $\langle\Delta\tau\rangle$ and $\Delta\tau(\lambda)$ can be very different. However, if only the average behavior of the device is of concern, then the mean (or RMS) DGD may suffice.

1.2. Inherent Uncertainty

A useful measurement of PMD must also report its uncertainty. Uncertainty comes not only from equipment inaccuracy, but from the environmental stability of the PMD itself. In highly mode-coupled fibers, the DGD can vary greatly with wavelength, or as temperature, stress, fiber position, or other environmental parameters change. Long, mode-coupled fiber exhibits a Maxwellian distribution of its DGD as the fiber is changed by these environmental parameters. Therefore, regardless of the quality of a measurement technique, there is an inherent uncertainty associated with measurements of DGD. So, in order to accurately report the mean DGD of a mode-coupled device, it is necessary to also report the uncertainty due to this variance of $\Delta\tau$. Gisin et al. have demonstrated that four major classes of PMD measurement techniques are subject to the same level of uncertainty due to this statistical variation of the DGD in mode-coupled devices [2]. Since all techniques essentially measure DGD (with different spectral resolutions), this variance applies to all techniques. If a fiber of mean DGD $\langle\Delta\tau\rangle$ is measured over a bandwidth of $\Delta\omega_{\mathrm{span}}$, then the standard deviation σ, normalized to $\langle\Delta\tau\rangle$, is given by [2]

$$\frac{\sigma}{\langle\Delta\tau\rangle} \approx \frac{0.9}{\sqrt{\langle\Delta\tau\rangle\Delta\omega_{\mathrm{span}}}}. \tag{1}$$

For highly mode-coupled devices, $\sigma/\langle\Delta\tau\rangle$ decreases when the average is made over a wider spectral bandwidth (or equivalently over more statistically independent samples) or when measuring a device with a larger mean DGD.

2. Measurement Techniques

The various measurement techniques can be classified as either time-domain or frequency-domain techniques. The clearest separation between the two is seen in the relationship between the coherence time T_c of the measurement light and $\Delta\tau$ (the DGD being measured). A technique is considered to be in the time domain if $T_c < \Delta\tau$, and in the frequency domain if $T_c > \Delta\tau$. Following are brief descriptions of the unique aspects of the major measurement techniques. Useful descriptions can also be found in [3].

2.1. Time-Domain Measurements

Time-of-flight Technique

The time-of-flight measurement is the most intuitive, so we consider it first. Figure 3 illustrates narrow pulses of light transmitted through a non-mode-coupled device of DGD $\Delta\tau = \tau_s - \tau_f$. The propagation delay through the device will be either τ_f (if the input pulse is polarized along the fast PSP), or τ_s (if it is polarized along the slow PSP); or if the polarization state of the pulse lies between the fast and slow PSP, the pulse will be broken into two pulses with delays τ_f and τ_s and relative intensities weighted according to the projection of the input polarization state on the PSP. So, in this intuitive measurement technique, short optical pulses are launched into a test device and detected at the output. A fast oscilloscope plots the arrival time of the pulses as the input polarization state is changed. For launch polarizations between the two PSP of non-mode-coupled devices, two received pulses are seen separated in time by the mean DGD (averaged over the spectral bandwidth of the pulses). This technique is well known [4–6], but is often impractical since the pulse width limits the temporal resolution (requiring narrow pulse widths on the order of the desired DGD resolution).

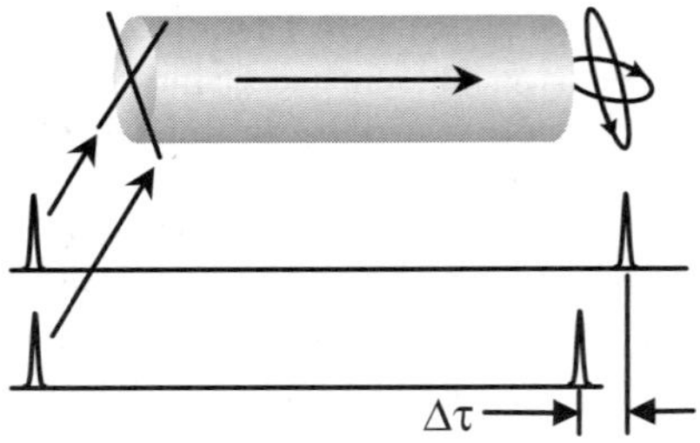

Fig. 3. Diagram of input pulses launched simultaneously down the fast and slow PSP of a device. Output pulses emerge at different times corresponding to the difference in group delay.

Low-Coherence Interferometry

A related but more practical approach to time-domain measurement is low-coherence interferometry (Fig. 4). A spectrally broad (low-coherence) source sends light through the test device and into an interferometer. As the moveable arm of the interferometer is translated, interference fringes are seen at the detector only if the time-delay difference between the two arms matches a delay generated in the test device to within the coherence time of the source. Ignoring possible phase effects, this condition can be written simplistically as

$$| \, (\tau_{\text{arm},1} - \tau_{\text{arm},2}) - (\tau_i - \tau_j) \, | < T_c, \qquad (2)$$

where $\tau_{\text{arm},1}$ and $\tau_{\text{arm},2}$ are the time delays associated with propagation along each arm of the interferometer, and τ_i and τ_j are two possible propagation times experienced by light traveling along the ith and jth polarization paths through the device. T_c is the coherence time of the source (e.g., for a Gaussian source of spectral width $\Delta\lambda$ and center wavelength λ, $T_c = 0.664\lambda^2/(c\Delta\lambda)$, where c is the speed of light) [7]. Equation (2) gives an intuitive picture of the shape of the interferogram. For the non-mode-coupled case, there are only two nondegenerate paths through the device—light traveling along the fast axis or along the slow axis. So, the only possible values of the difference $\tau_i - \tau_j$ are 0 or $\pm\langle\Delta\tau\rangle_\lambda$, where $\langle\Delta\tau\rangle_\lambda$ is the result of an average of the DGD over the spectrum of the source, weighted by the intensity of the source at each wavelength.

Plotting the envelope of interference fringes as the moveable arm of the interferometer is scanned gives a delay histogram similar to Fig. 5 for a non-mode-coupled device. The central peak of the delay histogram is the autocorrelation of the source, which gives no information about the strength of the PMD ($\tau_i - \tau_j = 0$). The two side lobes are separated from the autocorrelation peak by an amount $\langle\Delta\tau\rangle_\lambda$. So, for non-mode-coupled devices, measuring the separation of the side lobes gives $2\langle\Delta\tau\rangle_\lambda$.

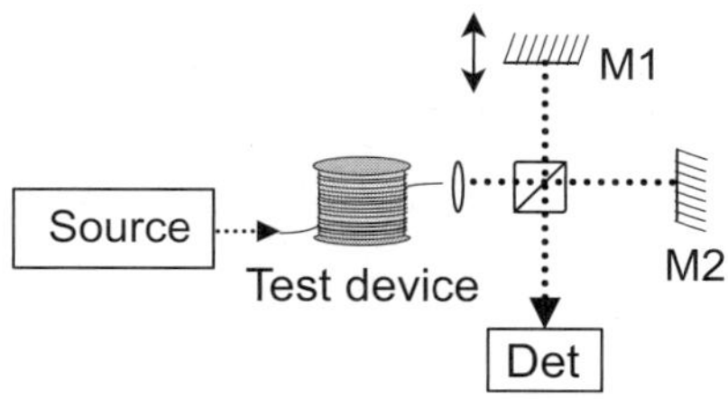

Fig. 4. Schematic diagram of low-coherence interferometer: M1 is the movable mirror.

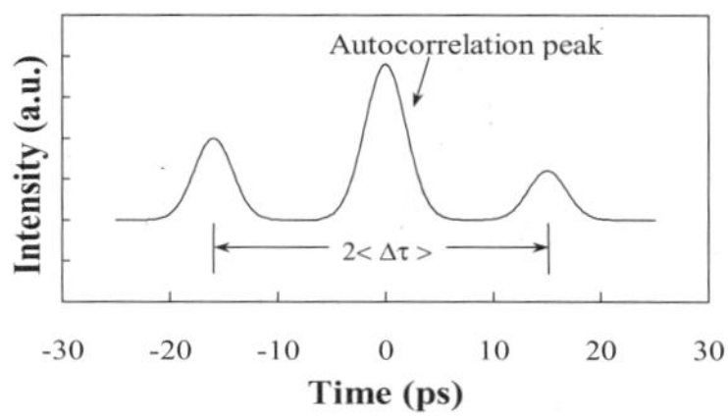

Fig. 5. Sample interferogram envelope for a non-mode coupled device.

The peak at 0 is the coherence function of the source (in the absence of chromatic dispersion or interference from the side lobes) and its width gives the source coherence time, which provides the temporal resolution limit. In other words, if the PMD is too low, the side-lobe separation will be on the order of the coherence time of the source, causing the side lobes to add coherently with the central peak and making it difficult to identify their position. This illustrates the tradeoff between bandwidth and DGD resolution; a broader source better resolves the DGD, but at the expense of the spectral resolution of the DGD.

The finite width of the side lobe peaks at $\pm\langle\Delta\tau\rangle_\lambda$ has two separate causes. First, the coherence time of the source broadens the peak. Second, if the value of $\Delta\tau$ is not constant over the source spectrum, there will be some broadening of the peak due to variation of DGD with wavelength.

Low-coherence interferometry can be used to measure mode-coupled devices as well. In this case, τ_i and τ_j in Eq. (2) can take on 2^{N+1} different values (where N is the number of mode-coupling sites in the artifact). This yields a delay histogram with 2^{N+2}-1 peaks. However, the separation of adjacent peaks can easily be less than the coherence time of the source, and so the peaks are not necessarily distinguishable. The resulting interferogram envelope comes from the coherent addition of the various delays. Figure 6 gives an example delay histogram for a highly mode-coupled device. In such devices, it is customary to characterize the RMS value of the DGD over the wavelength range of the measurement. The "second moment"

$$\sigma_M = \sqrt{\frac{\int I(t)t^2 dt}{\int I(t)dt}} \tag{3}$$

of this Gaussian-shaped delay histogram will generally[1] yield the RMS DGD value ($I(t)$ is the amplitude of the delay histogram and t is the time component). The relationship between the two is [8]

$$\langle \Delta\tau^2 \rangle^{1/2} \approx \sqrt{\frac{3}{4}}\sigma_M . \quad (4)$$

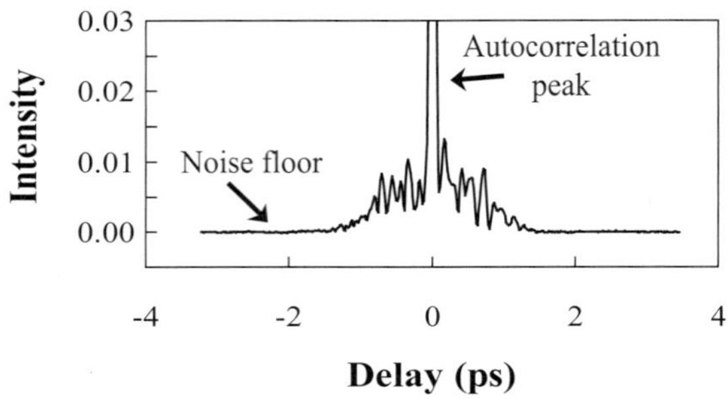

Fig. 6. Sample interferogram envelope for a mode-coupled device.

Equation (3) should be evaluated with caution. There will be non-ideal features of the delay histogram that make it deviate from a true Gaussian. The most significant features are the autocorrelation peak and the noise floor (dominant at high values of t). The integration limits for Eq. (3) must be chosen carefully in order to exclude these unwanted features, but this must be done iteratively so as not to incur a bias in the opposite direction due to excluding real data. Examples of procedures to correctly extract $\langle \Delta\tau^2 \rangle^{1/2}$ from a mode-coupled interferogram are provided in [9] and [10]. Recently, a promising approach to low-coherence interferometry was presented that uses polarimetric detection and appears to completely remove the central peak from both mode-coupled and non-mode-coupled interferograms, thus allowing the same measurement algorithm to be used regardless of the degree of mode-coupling [11].

PMD measurements using low-coherence interferometry have several advantages over other techniques. The measurements can be done quickly-the time required is essentially the travel time of the interferometer mirror over the desired scan range (a few seconds). Unlike most of the frequency—domain techniques, low-coherence interferometry does not require numerical comparison of data sampled at two different points in time. This makes interferometry less susceptible to dynamic changes in the measurement path (such as movement of the fiber leads or temperature drift). A disadvantage of low-coherence interferometry has been the need to use a spectrally broad source, which limits achievable spectral efficiency (ability to measure with fine temporal resolution in a narrow bandwidth). However, resolutions as low as 11 fs have been demonstrated using a differential technique [12], and long-haul measurements through

[1] Equation (4) assumes a large DGD-source-bandwidth product. When the magnitude of this product is small, the relationship between σ_M and the RMS DGD is a function of the exact shape of the low-coherence source [8].

multiple bandwidth-limiting optical amplifiers have demonstrated good measurement results with a bandwidth of only 20 nm [13].

Another aspect of making PMD measurements using low-coherence interferometry is the effect of multipath interference (MPI). Multiple reflections within the measurement path will cause delays in the signal transmission that are indistinguishable from delays due to PMD. While it may be useful to measure these MPI effects, the user should be aware that low-coherence interferometry does not distinguish between MPI and PMD [14,15].

2.2. *Frequency-Domain Measurements*

PMD measurements based in the frequency domain measure the same DGD as time-domain measurements but from a different perspective. The most common approach to DGD measurement in the frequency domain involves a differential method. The difference in propagation delay between light traveling on the fast and slow PSP determines the output polarization state $\hat{\mathbf{s}}_{\mathrm{out}}$ of the light. Frequency-domain measurements of PMD do not seek to find how the polarization state of monochromatic light is transformed as it passes through the device under test, but rather how the output polarization state changes as a function of optical frequency. In the absence of polarization-dependent loss (PDL), the polarization dispersion vector $\boldsymbol{\Omega}$ is related to this change in output state as [16]

$$\frac{d\hat{\mathbf{s}}_{\mathrm{out}}}{d\omega} = \boldsymbol{\Omega} \times \hat{\mathbf{s}}_{\mathrm{out}}, \tag{5}$$

where ω is the angular optical frequency of the light. The physical meaning of this expression is that PMD in a device causes the output polarization state to precess about $\boldsymbol{\Omega}$ as the optical frequency is changed (Fig. 7). From Eq. (5), the precession rate will be equal to the DGD of the device:

$$\left| \frac{d\theta}{d\omega} \right| = | \boldsymbol{\Omega} | = \Delta\tau, \tag{6}$$

where θ is defined in Fig. 7 as the angle of rotation of the output state of polarization about the precession axis $\boldsymbol{\Omega}$. The DGD can be found by measuring $\Delta\theta/\Delta\omega$ (as an approximation to $d\theta/d\omega$). The class of frequency-domain techniques that measure $\Delta\theta/\Delta\omega$ will be referred to as "polarimetric techniques."

The distinction between the various polarimetric techniques is how they measure $\Delta\theta/\Delta\omega$. The techniques all begin with launching polarized light into the test device and measuring the output state as a function of optical frequency, $\hat{\mathbf{s}}(\omega)$. But the change in $\hat{\mathbf{s}}$ with frequency yields $| \Delta\hat{\mathbf{s}}/\Delta\omega |$, not $\Delta\theta/\Delta\omega$, and this is where the variation in approaches comes in.

Three very similar techniques of polarimetric measurement of DGD will be discussed here-Jones Matrix Eigenanalysis (JME) [17], Müller Matrix Method (MMM) [18], and Poincaré Sphere Analysis (PSA) [19, 20]. All three offer means of assessing both $\Delta\tau$ and the PSP from the wavelength dependence of the output polarization state, and all three can be measured by use of the same experimental setup (Fig. 8).

Jones Matrix Eigenanalysis

The JME technique gives a measurement of $\Delta\theta/\Delta\omega$ and $\boldsymbol{\Omega}$ by turning the measurement into an eigenvalue problem [17]. A difference matrix $\boldsymbol{\Gamma}$ is defined as a 2×2 Jones

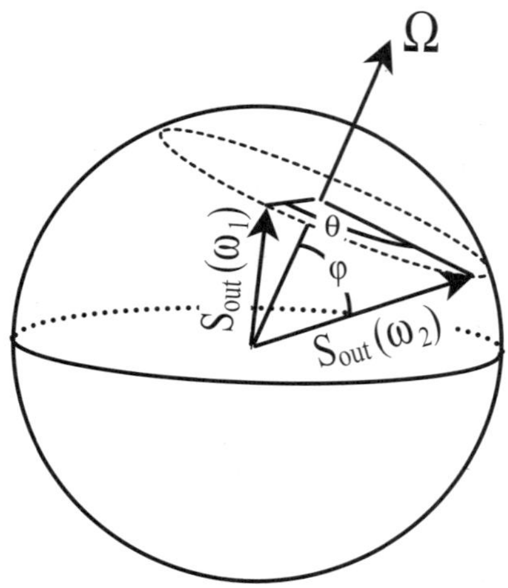

Fig. 7. Poincaré sphere representation of polarization dispersion vector $\boldsymbol{\Omega}$ and output polarization state $\hat{\mathbf{s}}_{\text{out}}$ at optical frequencies ω_1 and ω_2.

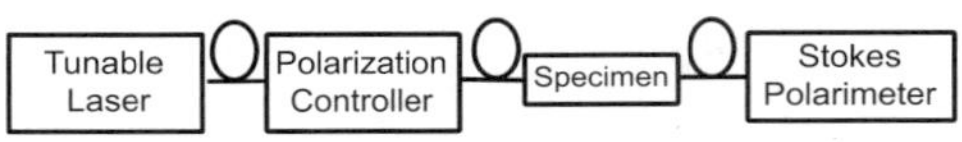

Fig. 8. General diagram of experimental setup for JME, MMM, and PSA measurement techniques.

matrix that describes the change in output polarization state as the optical frequency is changed from ω_1 to ω_2,

$$\hat{\mathbf{s}}_{\text{out}}(\omega_2) = \boldsymbol{\Gamma}(\bar{\omega}, \Delta\omega)\hat{\mathbf{s}}_{\text{out}}(\omega_1), \tag{7}$$

where $\bar{\omega} = (\omega_1 + \omega_2)/2$ is the average frequency and $\Delta\omega = \omega_2 - \omega_1$ is the frequency step size. It must be noted that $\boldsymbol{\Gamma}$ is not the transfer matrix of the DUT (it does not transform an input state to an output state). Instead, it describes the motion of the output state as the wavelength is changed. Equations (5) and (6) show that a change in wavelength causes the output polarization state to precess about the PSP with a rate equal to the DGD. Therefore, eigenstates of $\boldsymbol{\Gamma}$ are the PSP, and the DGD is derived from the eigenvalues.

At a given optical frequency ω_1, three noncollinear input polarization states (such as linear polarization with 0°, 45°, and 90° orientations) are input to the device and the corresponding output polarization states $\hat{\mathbf{s}}_{\text{out}}$ are measured. This allows the 2×2 Jones transfer matrix $\mathbf{T}(\omega_1)$ to be calculated following [21]. $\mathbf{T}(\omega_1)$ is the matrix that describes the transformation of the input polarization state to the output polarization state at the optical frequency ω_1, $\hat{\mathbf{s}}_{\text{out}}(\omega_1) = \mathbf{T}(\omega_1)\hat{\mathbf{s}}_{\text{in}}$. Then the same three states are launched at a slightly different optical frequency ω_2, and $\mathbf{T}(\omega_2)$ is calculated.

We then find $\boldsymbol{\Gamma}$ as the product $\boldsymbol{\Gamma}(\bar{\omega}, \Delta\omega) = \mathbf{T}^{-1}(\omega_2)$. As mentioned above, $\boldsymbol{\Gamma}$ describes the evolution of the output polarization state as the optical frequency is changed. When no PDL is present, the output state precesses about the PSP. So, the eigenvectors of $\boldsymbol{\Gamma}$ are the PSP, and the two eigenvalues are $\rho_q = \exp(i\tau_{g,q}\Delta\omega)$, where the index q denotes propagation along the fast or slow axis and $\tau_{g,q}$ is the associated group delay. The DGD is then

$$\Delta\tau(\bar{\omega}) = \mid \tau_{g,s} - \tau_{g,f} \mid = \left| \frac{\mathrm{Arg}(\rho_s/\rho_f)}{\Delta\omega} \right| . \tag{8}$$

Thus, JME allows the full $\boldsymbol{\Omega}(\bar{\omega})$ to be calculated, even for highly mode-coupled devices. The matrix $\boldsymbol{\Gamma}$ is found with complete generality and is correct even in the presence of PDL. However, when PDL is present, the eigenstates of $\boldsymbol{\Gamma}$ will be nonorthogonal, and the expression for DGD, Eq. (8), will no longer be exact.

Müller Matrix Method

The Müller Matrix Method (MMM) [18] measures PMD in much the same way as the JME technique, but with two slight differences. First, their calculations are carried out in different vector spaces (MMM uses Müller matrices, but JME uses Jones matrices). This simplifies the algorithm for the MMM case. Second, the entire MMM measurement assumes the absence of PDL, contrary to JME, where the absence of PDL is assumed only in the last step (as the DGD is calculated). This allows MMM to determine the difference matrix by launching only two polarization states per wavelength (as opposed to three for JME) and yields a difference matrix that describes a pure rotation θ of the output state with wavelength. The rotation angle θ substituted in Eq. (6) yields the DGD, and the rotation axis is the PSP. The tradeoff is that the JME difference matrix $\boldsymbol{\Gamma}$ is exact even in the presence of PDL, while the MMM difference matrix $\mathbf{R}_\Delta$ is not. This makes JME less susceptible than MMM to the presence of PDL. This can be demonstrated by simulation [22].

MMM first measures the Müller transfer matrix $\mathbf{R}$ for optical frequencies ω_1 and ω_2. By assuming no PDL, a 3×3 reduced Müller matrix describing $\mathbf{R}$ can be found by measuring the output state for only two input states. These two states can simply be two different linear polarizations (the angle between them is not important to the measured value, but will have an effect on the measurement noise).

The MMM technique measures the polarization transfer matrix $\mathbf{R}$ at two closely spaced optical frequencies, yielding $\mathbf{R}(\omega_1)$ and $\mathbf{R}(\omega_2)$, and calculates the difference matrix as $\mathbf{R}_\Delta = \mathbf{R}(\omega_2)\mathbf{R}^T(\omega_1)$, which describes the change in output polarization state as the optical frequency is changed from ω_1 to ω_2. In the absence of PDL, $\mathbf{R}_\Delta$ corresponds exactly to $\boldsymbol{\Gamma}$ in JME, and $\mathbf{R}$ and $\mathbf{R}_\Delta$ are pure rotation matrices, which means $\mathbf{R}^T = \mathbf{R}^{-1}$. Since $\mathbf{R}_\Delta$ is a rotation matrix, the precession angle θ of the output polarization state about $\boldsymbol{\Omega}$ is given by

$$\cos(\theta) = \frac{1}{2}(\mathrm{Tr}\mathbf{R}_\Delta - 1), \tag{9}$$

where $\mathrm{Tr}\mathbf{R}_\Delta$ is the trace of $\mathbf{R}_\Delta$. This is used in Eq. (6) to yield the DGD. The PSP is the rotation axis, and is found as the eigenvector of $\mathbf{R}_\Delta$ that corresponds to an eigenvalue of 1 [23]. This is detailed in [18].

Poincaré Sphere Analysis

The Poincaré Sphere Analysis (PSA) technique is closely related to both JME and MMM. In fact, the same measurement procedures can be used for both JME and PSA measurements; only the data analysis differs. As with MMM, PSA analysis takes place entirely in Stokes space, and assumes the absence of PDL throughout the process.

The PSA technique works with frequency derivatives of the measured output Stokes vectors rather than derivatives of the polarization transfer matrix (which JME and MMM effectively do). The relationship between the output polarization states and the PMD is as follows. Consider three output states $\hat{\mathbf{h}}$, $\hat{\mathbf{q}}$, and $\hat{\mathbf{c}}$, which are orthogonal on the Poincaré sphere. An infinitesimal change of optical frequency $d\omega$ will rotate these states through an angle $d\theta$ on the Poincaré sphere, causing changes in these states of $d\hat{\mathbf{h}}$, $d\hat{\mathbf{q}}$, and $d\hat{\mathbf{c}}$, respectively. Geometrically, the relationship between the motion of the states and the rotation angle is

$$d\theta = \sqrt{\frac{(d\hat{\mathbf{h}})^2 + (d\hat{\mathbf{q}})^2 + (d\hat{\mathbf{c}})^2}{2}}. \tag{10}$$

However, for a finite change $\Delta\omega$ in optical frequency, an approximation must be used to give a more accurate expression for $\Delta\theta$. Combining this with Eq. (6) relates the finite changes in the three orthogonal output states to the DGD as

$$\Delta\tau = \left|\frac{\Delta\theta}{\Delta\omega}\right| = \frac{2}{\Delta\omega}\arcsin\left(\frac{1}{2}\sqrt{\frac{1}{2}(\Delta\mathbf{h}^2 + \Delta\mathbf{q}^2 + \Delta\mathbf{c}^2)}\right). \tag{11}$$

The PSP is also found geometrically as the axis about which $\hat{\mathbf{h}}$, $\hat{\mathbf{q}}$, and $\hat{\mathbf{c}}$ rotate:

$$\mathrm{PSP} = \frac{\boldsymbol{\Omega}}{|\,\boldsymbol{\Omega}\,|} = \frac{\mathbf{u}}{|\,\mathbf{u}\,|}, \tag{12}$$

with $\mathbf{u} = (\mathbf{c}\cdot\Delta\mathbf{q})\mathbf{h} + (\mathbf{h}\cdot\Delta\mathbf{c})\mathbf{q} + (\mathbf{q}\cdot\Delta\mathbf{h})\mathbf{c}$.

The PSA technique finds $\hat{\mathbf{h}}$, $\hat{\mathbf{q}}$, and $\hat{\mathbf{c}}$ at two closely spaced optical frequencies, then calculates $d\hat{\mathbf{h}}$, $d\hat{\mathbf{q}}$, and $d\hat{\mathbf{c}}$ and uses Eqs. (11) and (12) to find the PMD. A variety of methods may be used, but a simple approach is to launch two linear polarization states separated by approximately 90° on the Poincaré sphere (e.g., horizontal and vertical polarizations). In the absence of PDL, the corresponding output states $\hat{\mathbf{h}}'$, and $\hat{\mathbf{q}}'$ will have the same angle between them as the input states. For nonorthogonal launch states, corrected orthogonal output states are generated as $\hat{\mathbf{h}} = \hat{\mathbf{h}}'$ and

$$\hat{\mathbf{q}} = (\hat{\mathbf{h}} \times \hat{\mathbf{q}}') \times \hat{\mathbf{h}} / \,|\,\hat{\mathbf{h}} \times \hat{\mathbf{q}}'\,|\,.$$

The third mutually orthogonal state is generated as $\hat{\mathbf{c}} = \hat{\mathbf{h}} \times \hat{\mathbf{q}}$.

PA, SOP, and PS Techniques

Various other polarimetric measurement techniques with similarities to JME, PSA, and MMM are sometimes mentioned. However, their names generally come up under the topic of "Other measurement techniques." These techniques, Poincaré Arc (PA), State of Polarization (SOP) and Poincaré Sphere (PS), are often mentioned without literature references, and so it is difficult to define a measurement procedure to associate with each name. A good generalization would be to use these somewhat generic titles to refer to techniques that measure the DGD by measuring only $\hat{\mathbf{s}}_{\mathrm{out}}/\Delta\omega$. This requires the assumption that

$$\left|\frac{d\hat{\mathbf{s}}_{\mathrm{out}}}{d\omega}\right| = \left|\frac{d\theta}{d\omega}\right|. \tag{13}$$

This is true when $\hat{\mathbf{s}}_{\text{out}}(\omega)$ lies on a great circle, and occurs when an input polarization state is launched so that both PSP are equally illuminated. This condition is difficult to maintain in a mode-coupled device such as a long fiber, but is possible in non-mode-coupled components.

Fixed-Analyzer Technique

The fixed-analyzer (FA) technique [24] offers a simplified approach to a polarimetric measurement. It is sometimes called wavelength scanning. The FA technique measures mean DGD based on Eq. (6), but indirectly. Figure 9 illustrates the basic setup. Light transmitted through a polarizer–test device–polarizer setup is detected as a function of wavelength. This can be done either with a tunable laser and detector combination or with a broadband source and an optical spectrum analyzer (or monochromator). As the output polarization vector $\hat{\mathbf{s}}_{\text{out}}(\omega)$ moves around the sphere, the normalized intensity $I_N(\omega)$ transmitted through the output polarizer of Fig. 9 is given as

$$I_N(\omega) = \frac{1}{2}(1 + \sin \Phi \cos[\theta(\omega)] \sin \varphi + \cos \Phi \cos \varphi), \tag{14}$$

where the angles are in Poincaré sphere coordinates, Φ is the angle between $\boldsymbol{\Omega}$ and the Stokes vector describing the transmission axis of the output polarizer, and φ is the angle between $\mathbf{s}_{\text{out}}$ and $\boldsymbol{\Omega}$ (Fig. 7). Φ and φ are independent of ω for non-mode-coupled devices. $\theta(\omega)$ is the azimuthal angle of the precession of $\hat{\mathbf{s}}_{\text{out}}(\omega)$ about $\boldsymbol{\Omega}$. For non-mode-coupled devices, $\theta(\omega)$ depends approximately linearly on ω and contains all of the optical frequency dependence of $I_N(\omega)$. So we can estimate $d\theta/d\omega$ (and thus average DGD $\langle \Delta\tau \rangle$) from Eq. (6) by merely counting the number of extrema (peaks and valleys) in the sinusoidal $I_N(\omega)$ curve over a given optical frequency range. That is,

$$\langle \Delta\tau \rangle_{\omega_a - \omega_b} = \frac{k N_e \pi}{\omega_b - \omega_a} = \frac{k N_e \lambda_a \lambda_b}{2(\lambda_a - \lambda_b)c}, \tag{15}$$

where the brackets indicate the average DGD measured over the frequency range from ω_a to ω_b, k is a mode-coupling constant (equal to 1 for non-mode-coupled devices), and N_e is the number of peaks and valleys measured over the frequency range from ω_a to ω_b. The right-most expression in Eq. (15) is merely the same expression as the middle one, but given in terms of wavelength instead of optical frequency.

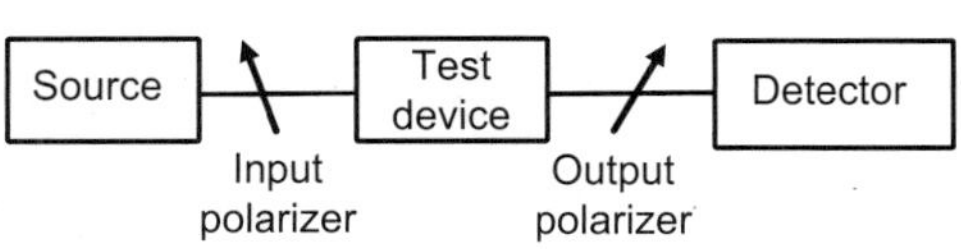

Fig. 9. Diagram of fixed analyzer setup.

If we consider the case of a mode-coupled device, the situation becomes more complicated in that $d\theta/d\omega$ can depend strongly on ω, and $\boldsymbol{\Omega}$ can also have a second-order dependence on ω ($\boldsymbol{\Omega}$ is still independent of ω to first order). This means that $I_N(\omega)$ no longer has a simple sinusoidal dependence but behaves in a quasi-random way as shown in Fig. 10. Fortunately, the mean DGD can still be estimated by counting peaks and valleys in the $I_N(\omega)$ spectrum. In this case, the coupling factor k becomes

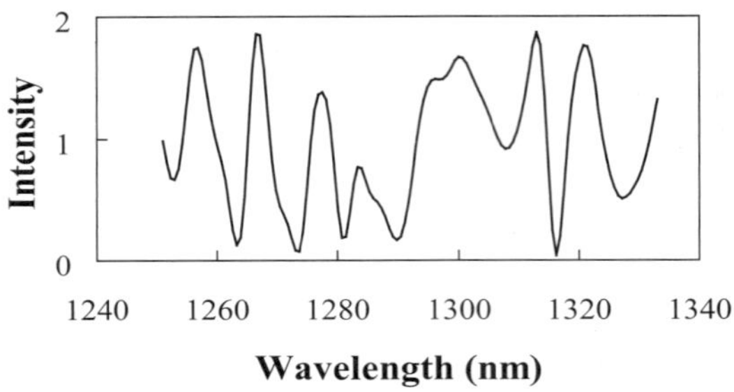

Fig. 10. Typical spectrum from fixed analyzer measurement of non-mode-coupled device.

0.805. k comes from a Monte Carlo simulation and allows mean DGD to be accurately estimated when the device is strongly mode-coupled [25].

Evaluating the mean DGD from $I_N(\omega)$ is straightforward for a non-mode-coupled device since it merely involves counting peaks and valleys of a sinusoidal signal. However, in the case of mode-coupled devices, it can become difficult to distinguish peaks and valleys from intensity noise. Often a "thresholding" algorithm is used to ignore peaks and valleys whose extent is less than some defined fraction of the full-scale excursion of $I_N(\omega)$. This practice will incur a bias due to ignoring some fraction of real peaks and can easily give errors of $\sim 10\%$. This problem is discussed in [25] and there a lookup-table correction factor is suggested to reduce this bias. FA measurements on mode-coupled devices can also be biased by sampling too coarsely (causing small peaks to be missed, thus underestimating the mean DGD). In order to sample $I_N(\omega)$ sufficiently, a device of nominal mean DGD $\langle\Delta\tau\rangle$ should be measured with at least $2\langle\Delta\tau\rangle\Delta\omega$ frequency points for a measurement spectrum of $\Delta\omega$ [25].

An alternative evaluation of the FA spectral data is often used. Rather than calculating $\langle\Delta\tau\rangle$ from an estimate of $d\theta/d\omega$ based on counting peaks and valleys, the $I_N(\omega)$ spectrum can be Fourier-transformed into the time domain. The result is a "delay histogram" very similar to that which would be seen for a low-coherence interferometric measurement with the same source spectrum as used in the FA measurement (windowing is often used to optimize the results). The mean DGD can then be evaluated in the time domain by the same means as with low-coherence interferometry [26].

RF Phase Shift Technique

This technique resembles the time-of-flight measurement. Figure 11 illustrates a typical setup. A tunable laser is intensity-modulated at a frequency ranging from several tens of megahertz to a few gigahertz. The modulated light passes through a polarization controller, then the test device, and then is detected. The RF phase φ_{RF} of the detected signal is referenced to the phase of the modulator as the polarization state of the light is changed. φ_{RF} is related to the time of flight through the device, so that as the input polarization state is changed, the maximum and minimum phases $\varphi_{\mathrm{RF,max}}$ and $\varphi_{\mathrm{RF,min}}$, can be measured and used to find the maximum and minimum polarization-dependent propagation delays through the device. Their difference gives the DGD through the test device,

$$\Delta\tau = \frac{\varphi_{\text{RF,max}} - \varphi_{\text{RF,min}}}{360^\circ \cdot f_{\text{mod}}}, \tag{16}$$

where phases are given in degrees and f_{mod} is the RF modulation frequency [27].

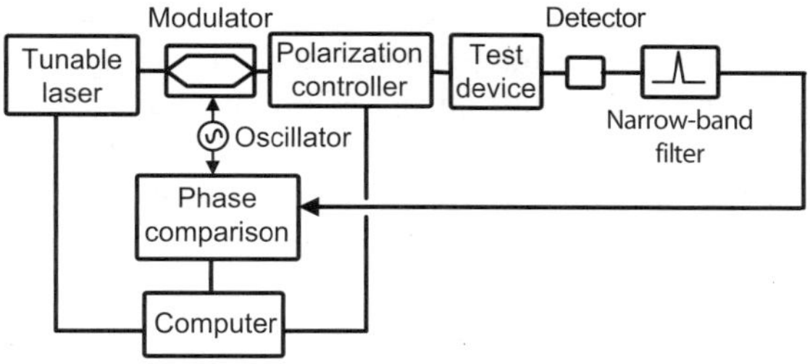

Fig. 11. Generic diagram of experimental setup for RF phase shift measurement system.

The down-side of this intuitive randomized-launch approach is that to accurately find the maximum and minimum requires many phase measurements while the polarization state is randomly varied. Over this measurement time, noise or drift in RF phase will generally increase the difference between $\varphi_{\text{RF,max}}$ and $\varphi_{\text{RF,min}}$, biasing the results toward larger DGD values. Thus, environmental drift in the optical path length of the DUT can cause significant systematic measurement errors.

This approach is improved by algorithms that allow $\Delta\tau$ and the PSP to be obtained by measuring the RF phase at just four known polarization states (launched at points orthogonal to each other on the Poincaré sphere). Two such implementations are the Modulation Phase Shift (MPS) technique [28] and Polarization-dependent Signal Delay (PSD) method [29]. By minimizing the number of phase measurements, the measurement time and resulting errors due to drift of the optical path length are significantly reduced. It has also been shown that by launching six states instead of only four, improved stability against drift can be gained [30].

RF phase shift is a promising technique in that it allows one to make narrowband measurements whose spectral width is due only to the RF modulation frequency. The classification of the RF phase-shift techniques as frequency-domain is somewhat ambiguous. The definition of time and frequency domains generally relies on the coherence length of the optical source, but in principle the MPS and PSD techniques will work with either a spectrally narrow laser source or a broadband source (though the latter negates the narrow-bandwidth advantages of RF phase-shift measurements). The choice to include RF techniques in the frequency domain is therefore based on the coherence time of the RF modulation.

Swept-wavelength Interferometry

Recently, a new class of measurement technique has been developed that finds PMD from measurement of the full optical transfer function of the DUT [31]. The "swept-wavelength interferometer" technique uses a dual Mach-Zehnder interferometer configuration with a frequency-swept laser as the source. As shown in Fig. 12, interferometers at both the input and output of the DUT enable characterization of the full magnitude and optical phase of light reflected and transmitted by the DUT. A polarization-delay element delays one polarization state of the launched light with respect to its orthogonal state. Since the source wavelength is time dependent, at a given time the two orthogonal polarization states are at different optical frequencies, causing each of the two launched polarization states to generate a different beat frequency at the detector. Filtering this frequency allows simultaneous distinguishable measurements of the two launched polarization states. Thus, the system yields the polarization-dependent magnitude and optical phase for transmission and reflection, allowing calculation of the polarization-dependent transmission and reflection matrices of the DUT at each optical frequency as the laser is swept. The data can be post-processed to measure the DGD and PSP with a variable spectral resolution (trading temporal resolution for frequency resolution).

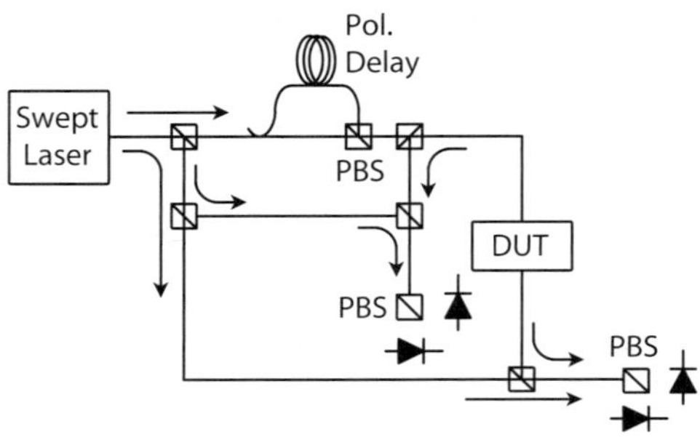

Fig. 12. Schematic of swept-wavelength interferometer. Arrows indicate light direction and PBS is a polarizing beam splitter.

This technique is attractive because of its speed and versatility, allowing it to simultaneously measure DGD, group delay, insertion loss, and polarization-dependent loss. It is also capable of simultaneous measurement of multiport components [32]. Full-scan periods are on the order of one minute. A limitation of this technique is that the source coherence limits the optical path length of the DUT to several tens of meters, making the technique unsuitable for long fibers.

3. Experimental Setup Details

Once a measurement technique is chosen, there are several precautions to be taken in making PMD measurements. Some of the most important are detailed below.

3.1. Inherent Lead PMD

When PMD is measured, it is important that there be minimal stray PMD in the measurement system or the fiber leads. The leads connecting the measurement system to the device being tested can have inherent birefringence (due to asymmetries or stresses from the manufacturing process), which leads to PMD. The DGD due to these stresses will be a few femtoseconds for lead lengths of a few meters. Selection of low-PMD leads and use of the shortest lengths possible will minimize errors due to lead birefringence.

3.2. Bend-induced PMD in Leads

Even low-PMD leads can exhibit significant PMD under bending, which causes a stress-induced birefringence. The bend-birefringence increases as $1/r^2$, where r is the radius of the fiber bend [33]. Extending this to DGD, we find the bend-induced DGD varies as $1/r$. Measurements should be made with the fiber leads as short and as straight as possible.

Once lead PMD has been minimized, it is best to measure the remaining PMD of the measurement system by removing the test device, connecting the leads together and measuring the "instrument DGD." Since PMD is a vector quantity, the device DGD and the instrument DGD will not simply add as scalars, but as vectors. The combined DGD will depend on the relative orientations between the PSP axis within the leads and the PSP axis of the test device. In other words, the final measurement cannot be corrected by simply subtracting the measured value for instrument DGD. The best solution is to measure the device several times with the leads re-oriented in between each measurement (avoiding small bend radii). The average of these multiple measurements will give the best estimate of the device DGD, and the spread in the measured values will be an estimate of the instrument DGD. The final value will still have effects due to instrument birefringence; but, they can be documented by including the instrument DGD in the uncertainty statement for the device.

3.3. Stabilizing the Measurement

Care should be taken before the measurement to stabilize both the measurement system and the test device against temperature changes or movement. Even fiber leads having only a small amount of DGD can have a big effect on measurement noise if they are moving during the measurement. This is particularly important for the polarimetric measurement techniques, which measure the output polarization state of the light in order to determine the DGD. A moving fiber lead can change the output polarization state significantly, resulting in a time-dependent output polarization state. When the polarization state is measured at different wavelengths, this effect of lead motion will falsely be identified as a change in polarization state with wavelength (DGD). The resulting DGD error will be proportional to the total DGD being measured (not just the DGD of the leads). Given the random nature of lead motion, this DGD error is likely to also be random, but could have an amplitude large enough to obscure the true DGD.

For MPS and PSD techniques, where the measured DGD comes from four measurements separated in time from each other, moving fiber leads at the input of the DUT can cause errors proportional to the total DGD. In the case of low-coherence interferometry, the polarization state is not directly measured, and the light propagating down different polarization states of the test device is detected simultaneously, so there is much less effect from moving fiber leads. In general, for all measurements, it is best to secure the fiber leads against motion during measurements.

Temperature should also be kept stable during a measurement—even for devices with relatively low temperature dependence of their DGD. The reason for this is the same as with the moving fiber leads. A change in temperature of the device will change the birefringence slightly, causing the polarization state to change with time, which produces a DGD error proportional to the total DGD of the measurement.

Generally, temperature drift is of greater concern, since as the temperature increases, the output polarization state will precess about the eigenaxis of the DUT. This means that drifting temperature will systematically bias the measured DGD (the sign of this bias is determined by the direction of the temperature drift), as opposed to the random error due to lead motion.

Figure 13 shows an example of DGD error due to temperature drift. Here, a simple quartz plate element was measured as the temperature changed from room temperature to $\sim 40°$C in about 30 min. From the graph, it is clear that when the temperature is stable, there is little difference between the DGD measured at room temperature and at the elevated temperature (the quartz plate shows a temperature coefficient of approximately 0.08 fs/°C). However, during the transition between room temperature and the elevated temperature, the DGD varied significantly. This illustrates the importance of a stable temperature environment for the measurement. It should be noted that in this example, the error appears to be random with time. This is attributable to the extreme change of temperature with time causing the polarization state to rotate on the sphere at a rate faster than the sampling time of the measurement, randomizing the sign of the bias. For slower temperature changes, a more systematic bias is observed.

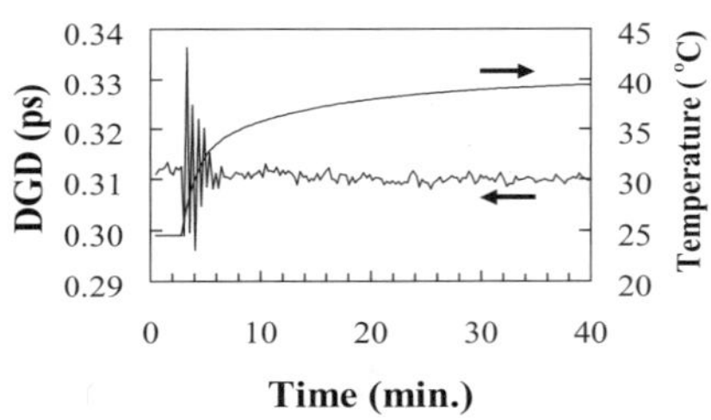

Fig. 13. Measured DGD versus time for a quartz plate as the temperature is changed as shown.

Since the magnitude of DGD error incurred by temperature drift is proportional to the rate of change of temperature, it is best if devices can be held at a constant temperature or at least insulated to allow any temperature changes to occur only slowly. Effects of temperature drift can also be reduced by careful choice of sampling order. DGD bias comes when a drifting polarization state is falsely assumed to come from the changing wavelength. This systematic bias could be randomized by sampling wavelength points in a random order, where the wavelength change is not always in the same direction. Or, when the temperature drift is linear, the problem can be virtually eliminated by measuring the polarization state (in the case of polarimetric techniques) at wavelengths λ_1, λ_2, and then at λ_1 again, and averaging the results for the two λ_1 measurements, effectively obtaining λ_1 and λ_2 measurements at the same drift position of the polarization state. This approach also works well for RF phase-shift measurements.

Again, it is expected that this effect on the DGD will be greater for frequency-domain measurements and less for interferometric measurements.

3.4. *Multipath Interference*

When multiple reflections occur within the measurement path, a cavity is set up where a fraction of the light passing through the cavity makes more than one pass, receiving extra delay. This will show up on an interferometric measurement and be indistinguishable from DGD. This will also show up as a ripple on FA spectra (in the absence of source normalization) that could be mistaken for higher DGD values. In the rest of the frequency-domain measurements, the measured DGD will not be affected by multiple reflections unless the cavity contains DGD. In that case, the DGD spectrum will have a ripple whose amplitude depends on the amount of DGD within the cavity and on the strength of the reflections [14]. The period of these multiple-reflection-induced ripples will come from the cavity spacing. When there is DGD in the cavity, there will be two spacings—one associated with the fast PSP and the other with the slow PSP. These two closely spaced periodicities will result in a fast ripple with a slow envelope modulation. The fast ripple will repeat with an angular frequency spacing of

$$\Delta\omega_0 = \frac{\pi c}{L n_g}, \tag{17}$$

and the beat note envelope will have a frequency spacing of

$$\Delta\omega_B = \frac{\pi c}{L \Delta n_g}, \tag{18}$$

where L is the physical length of the cavity, n_g is the group index and Δn_g is the group birefringence of the cavity [15].

When multipath interference is measured, it is important to be aware of the source and implications of the interference. Interferometry and un-normalized fixed-analyzer techniques do not distinguish between polarization-dependent and -independent multipath interference. This is important if general dispersion is the concern. On the other hand, polarimetric techniques report only polarization-dependent multipath interference. This is useful when only polarization-dependent dispersion is important (e.g., in polarimetric DGD compensation).

4. Spectral Efficiency

In all methods of DGD measurement, there is a trade-off between the spectral bandwidth used to make the measurement and the achievable DGD resolution. The bandwidth efficiency factor α_B relates the achievable signal-to-noise ratio (SNR) to the DGD-bandwidth product as

$$\alpha_B = \left(\frac{\text{SNR}}{\Delta\omega}\right)\frac{1}{\Delta\tau}, \tag{19}$$

where $\Delta\tau$ is the DGD of the device and $\Delta\omega$ is the angular frequency spectrum used by the measurement (i.e., the measurement bandwidth). α_B can be thought of as the achievable SNR per bandwidth normalized to the mean DGD. α_B depends on the particular measurement technique and the quality of the equipment. Equation (19) illustrates that for a given measurement (fixed α_B), the SNR can be improved by increasing the measurement bandwidth. Or, when the measurement bandwidth is fixed, the SNR improves when measuring larger DGDs. Expressing SNR as $\Delta\tau/\sigma$, where σ is the measurement standard deviation, we can also write the efficiency factor as

$$\alpha_B = \frac{1}{\sigma \cdot \Delta\omega}. \qquad (20)$$

This useful form makes calculation of α_B simple. The bandwidth efficiency is a useful figure of merit for measuring narrowband components. It describes the relationship between DGD uncertainty and spectral resolution. When comparing the performance of various measurement techniques, α_B allows comparison of performance, independent of measurement technique.

For comparison purposes, it is useful to identify "typical" values of α_B for the different techniques. Spectral efficiencies can be difficult to estimate from the literature because spectral and temporal uncertainty are often specified separately. Some "typical" values are provided here. However, it should be remembered that these values, calculated from a variety of sources, are included only to roughly compare techniques. They do not necessarily represent the state of the art.

For the FA technique (extremum counting), α_B is limited by the fact that for $\langle\Delta\tau\rangle$ to be determined, at least two extrema must be present in the measurement spectrum. From Eq. (15), this means that the minimum DGD-bandwidth product must be $\langle\Delta\tau\rangle\Delta\omega = \pi$, giving an SNR of 1, so $\alpha_B = 1/\pi$, or 0.3. This value could be improved by the use of a different evaluation technique (such as analyzing multiple level crossings, as opposed to only peaks and valleys [34]).

In the case of a low-coherence interferometric measurement, or equivalently, the FA technique (with Fourier-transform evaluation), DGD resolution is limited by the coherence time T_c of the broadband source. Reference [12] describes an interferometric technique capable of a temporal resolution of 11 fs, and from plotted results, the full-width-half-maximum of the interferogram allows an estimate of the spectral width of the source as $\Delta\omega \approx 1.4 \times 10^{13}\mathrm{s}^{-1}$ (about 20 nm). Setting σ to 11×10^{-15} s, Eq. (20) yields an α_B of 6 for low-coherence interferometry or the Fourier-transformed FA technique.

Polarimetric techniques generally have their bandwidth efficiency limited by noise in the polarization-sensitive detection. Since $\Delta\theta/\Delta\omega$ is used as the measure of DGD, the noise on the $\Delta\theta$ measurement and the noise on the $\Delta\omega$ measurement both affect α_B. However, if a wavelength meter is used in the measurement, uncertainty on $\Delta\omega$ will be small, and $\delta\Delta\theta$ (the uncertainty in $\Delta\theta$) will dominate (for small values of $\Delta\tau$). With this assumption and Eq. (6), we find $\sigma = \delta\Delta\theta/\Delta\omega$. Combining this with Eq. (20) yields $\alpha_B = 1/\delta\Delta\theta$, the inverse of the Stokes noise. For comparison purposes, achievable values of α_B for a JME technique have been found informally to be ~ 850.

For RF phase-shift techniques, the measurement bandwidth is determined by the modulation frequency, and the measurement noise is fundamentally limited by the phase resolution $\Delta\varphi$ of the phase-sensitive detector (lockin amplifier, vector voltmeter, or network analyzer). Defining the DGD resolution as the time delay corresponding to the phase resolution, we find the bandwidth efficiency factor for RF phase-shift-based techniques to be

$$\alpha_B = \frac{360^\circ}{4\pi\Delta\varphi_{\mathrm{deg}}}, \qquad (21)$$

where $\Delta\varphi_{\mathrm{deg}}$ is in degrees. Recently, an averaged MPS measurement reported an α_B of 3600 [30].

Finally, based on a specification sheet for a swept-wavelength system, the technique is expected to yield an uncertainty of 27 fs in a bandwidth of 50 pm or an uncertainty of 85 fs in a bandwidth of 10 pm. This gives an α_B of between 950 and 1500. The variation in α_B is likely due to a noise floor.

5. Wavelength Step Size

If spectral resolution and available bandwidth are not limiting factors, it is attractive to use a larger frequency step in the measurement to improve the DGD uncertainty. Since most frequency-domain measurement systems can easily vary the spectral resolution, it is important to be aware of the tradeoffs. For polarimetric techniques, which determine DGD through measurement of $\Delta\theta/\Delta\omega$, it is important to note that $\Delta\theta$ is known only to within a factor of 2π, and the determination of $\Delta\theta/\Delta\omega$ can suffer from aliasing if $\Delta\theta$ is greater than π for the given frequency step $\Delta\omega$. This has implications in all the frequency-domain measurement techniques, and can be expressed as the requirement that

$$\Delta\tau\Delta\omega \leq \pi, \tag{22}$$

or, in terms of wavelength, measurements cannot be trusted unless

$$\Delta\tau\Delta\lambda \leq 4 \text{ ps} \cdot \text{nm}, \tag{23}$$

where, for Eq. (23) only, $\Delta\tau$ is in units of picoseconds, $\Delta\lambda$ in nanometers, and a nominal operating wavelength of 1550 nm is assumed. In order to take into account variations in DGD with wavelength, $\Delta\tau$ in Eqs. (22) and (23) should represent the maximum DGD over the measurement range (not just the average). The need to resolve spectral variation in DGD requires a further increase in the spectral sampling density. Sufficient points must be measured to resolve the variations in DGD with wavelength. Multiple simulations on mode-coupled devices of various mean DGD values show that, due to the presence of second-order effects (wavelength-dependent DGD and PSP), the sampling density must be higher than for non-mode-coupled artifacts. Figure 14 shows that for a highly mode-coupled device, increased sample density improves the accuracy of the estimated mean DGD. For example, to measure a mean DGD that is $\sim 95\%$ of the true value, the sampling density must be such that

$$\Delta\tau\Delta\lambda \leq 1.5 \text{ ps} \cdot \text{nm}, \tag{24}$$

(at 1550 nm), where again $\Delta\tau$ is in picoseconds and $\Delta\lambda$ is in nanometers. This is a stricter requirement than the common expression in Eq. (23), and agrees well with the definition of the bandwidth of the principal states of polarization given by Jopson [35] as

$$\Delta\omega_{\text{PSP}} = \frac{\pi}{4\Delta\tau}. \tag{25}$$

6. Conclusion

In measuring PMD, the first choice to be made is deciding which measurement technique to use. Here, the various techniques have been described so as to illustrate their

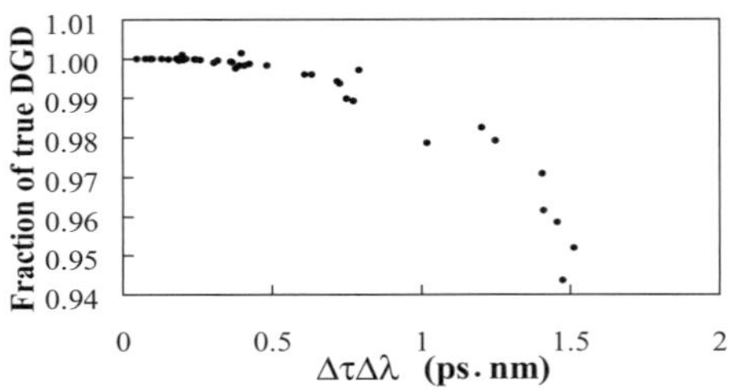

Fig. 14. Simulated results of measured DGD from a highly mode-coupled device as a function of sample density (vertical axis shows measured DGD normalized to true DGD).

relative merits. Generally, decisions are made based on measurement time, spectral resolution, and what quantities are measurable. Once a measurement technique is chosen, it is most important to understand the sources of measurement error (and to minimize them). The most significant of these sources have been described here.

References

[1] Paul A. Williams, "Mode-coupled artifact standard for polarization-mode dispersion: Design, assembly, and implementation," Appl. Opt., **38**, 6498–6507 (1999).

[2] N. Gisin, B. Gisin, J.P. Von der Weid, R. Passy, "How accurately can one measure a statistical quantity like polarization-mode dispersion?" IEEE Photon. Technol. Lett., **8**, 1671–1673 (1996).

[3] Dennis Derickson, *Fiber Optic Test and Measurement* (Prentice Hall, New Jersey, 1998).

[4] C.D. Poole, and C.R. Giles, "Polarization-dependent pulse compression and broadening due to polarization dispersion in dispersion-shifted fiber," Opt. Lett., **13**, 155–157 (1987).

[5] Yoshinori Namihira and Jun Maeda, "Polarization mode dispersion measurements in optical fibers," Technical Digest - Symposium on Optical Fiber Measurements, Boulder, 145–150 (1992).

[6] B. Bakhshi, J. Hansryd, P.A. Andrekson, J. Brentel, E. Kolltveit, B.K. Olsson, and M. Karlsson, "Measurement of the Differential Group Delay in Installed Optical Fibers Using Polarization Multiplexed Solitons," IEEE Phot. Tech. Lett., **11**, 593–595 (1999).

[7] Joseph W. Goodman, *Statistical Optics* (Wiley, New York, 1985), p. 168.

[8] B.L Heffner, "Influence of optical source characteristics on the measurement of polarization-mode dispersion of highly mode-coupled fibers," Opt. Lett., **21**, 113–115, (1996).

[9] P.A. Williams, "Accuracy issues in comparisons of time- and frequency-domain polarization mode dispersion measurements," Technical Digest—Symposium on Optical Fiber Measurements, Boulder, 125–129 (1996).

[10] TIA/EIA FOTP-124. 1999. Polarization-mode dispersion measurement for single-mode optical fibers by interferometric method. Telecommunications Industry Association, Arlington, VA.

[11] N. Cyr, "Polarization-Mode Dispersion Measurement: Generalization of the Interferometric Method to Any Coupling Regime,"J. Lightwave Technol., **22**, 794–805 (2004).

[12] Ph. Oberson, K. Julliard, N. Gisin, R. Passy, and JP Von der Weid, "Interferometric Polarization Mode Dispersion Measurements with Femtoseconds Sensitivity," Technical Digest—Symposium on Optical Fiber Measurements, Boulder, 143–146 (1996).

[13] N. Cyr, R. Roberge, J. Bradley, G. Amice, F. Audet, and G.W. Schinn, "Interferometric PMD Measurement of a Transatlantic 5512-km Fiber Link Including 119 EDFAs," Optical Fiber Communications Conference, Session MF, (2004).

[14] N. Cyr, Michel Leclerc and Bernard Ruchet, "PMD measurements in multipath components: The single waveplate example," Proceedings of Photonics North, Quebec (2002).

[15] P.A. Williams and J.D. Kofler, "Measurement and mitigation of multiple reflection effects on the Differential Group Delay Spectrum of optical components," Technical Digest—Symposium on Optical Fiber Measurement, Boulder, 173–176 (2002).

[16] C.D. Poole, N.S. Bergano, R.E. Wagner, and H.J. Schulte, "Polarization dispersion and principal states in a 147 km undersea lightwave cable," J. Lightwave Technol., **LT-7**, 1185–1190 (1989).

[17] B.L. Heffner, "Automated Measurement of Polarization Mode Dispersion Using Jones Matrix Eigenanalysis," IEEE Photon. Technol. Lett., **4**, 1066–1069 (1992).

[18] R.M. Jopson, "Measurement of Second-Order Polarization-Mode Dispersion Vectors in Optical Fibers," IEEE Photon. Technol. Lett., **11**, 1153–1155 (1999).

[19] N. Cyr, A. Girard, and G.W. Schinn, "Stokes Parameter Analysis Method, the Consolidated Test Method for PMD Measurements," Proceedings National Fiber Optics Engineers Conference, Chicago, 1999.

[20] TIA/EIA FOTP-122. 1999. Polarization-Mode Dispersion Measurement for Single-Mode Optical Fibers by Stokes Parameter Evaluation. Telecommunications Industry Association, Arlington, VA.

[21] R.C. Jones, "A New Calculus for the Treatment of Optical Systems: VI. Experimental Determination of the Matrix," J. Opt. Soc. Am., **37**, 110–112 (1946).

[22] J.K. Kofler and P.A. Williams, National Institute of Standards and Technology, unpublished.

[23] Mary L. Boas, *Mathematical Methods in the Physical Sciences* (Wiley, New York, 1983), p. 454.

[24] Craig D. Poole, "Polarization-Mode Dispersion Measurements Based on Transmission Spectra Through a Polarizer," J. Lightwave Technol., **12**, 917–929 (1994).

[25] P.A. Williams and C.M. Wang, "Corrections to Fixed Analyzer Measurements of Polarization Mode Dispersion," J. Lightwave Technol., **16**, 534–541 (1998).

[26] B.L. Heffner, "Single-mode propagation of mutual temporal coherence: equivalence of time and frequency measurement of polarization-mode dispersion," Opt. Lett., **19**, 1104–1106 (1994).

[27] P.A. Williams, A.J. Barlow, C. Mackechnie, and J.B. Schlager, "Narrowband measurements of polarization-mode dispersion using the modulation phase shift technique," Technical Digest—Symposium on Optical Fiber Measurements, Boulder, 23–26 (1998).

[28] P.A. Williams, "Modulation phase-shift measurement of PMD using only four launched polarization states: a new algorithm," Electron. Lett., **35**, 1578–1579 (1999).

[29] L.E. Nelson, R.M. Jopson, H. Kogelnik, and J.P. Gordon, "Measurement of polarization mode dispersion vectors using the polarization-dependent signal delay method," Opt. Express, **6**, 158–167 (2000).

[30] P.A. Williams and J.D. Kofler, "Narrowband Measurement of Differential Group Delay by a Six-State RF Phase-Shift Technique: 40 fs Single-Measurement Uncertainty," J. Lightwave Technol., **22**, 448–456 (2004).

[31] G.D. VanWiggeren, A.R. Motamedi, D.M. Baney, "Single-Scan Interferometric Component Analyzer," IEEE Photon. Technol. Lett., **15**, 263–265 (2003).

[32] G.D. VanWiggeren, and D.M. Baney, "Swept-Wavelength Interferometric Analysis of Multiport Components," IEEE Phtoton. Technol. Lett., **15**, 1267–1269 (2003).

[33] L.B. Jeunhomme, *Single-Mode Fiber Optics: Principles and Applications* (Marcel Dekker, New York, 1983), p. 66.

[34] A. Galtarossa, L. Palmieri, M. Schiano, and T. Tambosso, "Accuracy Enhanced PMD Measurements with Wavelength Scanning Technique," Technical Digest— Optical Fiber Measurement Conference, Nantes 45-48, (1999).

[35] R. Jopson, L. Nelson, and H. Kogelnik, "Measurement of second-order PMD vectors in optical fibers," IEEE Photon. Technol. Lett., **11**, 1153–1155 (1999).

DOI: 10.1007/s10297-004-0011-3
Originally published in J. Opt. Fiber. Commun. Rep. 1, 235–247 (2004)

PMD measurements on installed fibers and polarization sensitive components

Marco Schiano

Telecom Italia Lab
via Reiss Romoli 274, 10148, Torino, Italy
Email: marco.schiano@tilab.com

Abstract. The polarimetric PMD measurement techniques are characterized by peculiar features that make them appropriate for different measurement purposes. This section presents some relevant application examples of the polarimetric PMD measurement methods highlighting their specific characteristics and the related field of application.

1. Introduction

PMD measurement techniques have greatly evolved in the last decade under the combined pressure of research needs and industry requirements, generating a large variety of setups ranging from laboratory benches to compact and easy-to-use field instruments. This evolution has been associated with a strong research effort devoted to the clarification of the basic theory underlying the different techniques and of their mutual relationship. At present, despite a still open debate on how to interpret Interferometric measurement results, there are two basic categories of well established PMD measurement methods: 1) Interferometric Method (IM) [and its frequency domain counterpart, the Fixed Analyzers (FA) method] which make use of broadband sources, and 2) Polarimetric methods [e.g., the Jones Matrix Eigenanalysis (JME) and the Müller Matrix (MM) method] which exploit narrowband sources. This classification, based on the optical source type, characterizes also the intrinsic features of the two categories:

1. high measurement speed but lack of PMD statistics information and spectral resolution capability, for Interferometric and Fixed Analyzers methods;

2. complete PMD statistical analysis and spectral resolution capability, but low measurement speed, for Jones Matrix Eigenanalysis and Müller Matrix.

It is straightforward to note that IM and FA are more suitable for quick tests on fibers or broadband passive components (e.g., in production or field environment), while JME and MM find their natural application in the characterization of narrowband components or whenever data on PMD statistics are required.

In spite of this relatively steady picture, PMD measurement still represent a very interesting research area because the accurate characterization of PMD peculiar properties, such as its randomness and wavelength and time dependence, allow a deep understanding of optical links behavior and of optical components and subsystems structure.

The aim of this article is just to show some remarkable applications of the polarimetric techniques that, taking advantage of their spectral resolution and statistical analysis capability, let to characterize the transmission systems outage statistics (Section 2) and the optical components and subsystem transmission properties and design criteria (Section 3).

2. PMD Measurements on Long Optical Links

Optical fiber links are usually tested using either IM or FA method because of the set-up relative simplicity and the measurement speed. However, these techniques provide only the average Differential Group Delay (DGD) of the fiber under test, precluding some interesting analysis of the PMD statistics. When such analysis is required, the adoption of polarimetric methods is mandatory as described in the following.

In the last decade, the PMD of fiber links has assumed a fundamental importance for the related limitations of optical systems performance. This key role is due to the severe impairments that PMD can introduce in high capacity long haul transmissions, especially at 10 Gbit/s and higher bit rates. A particular aspect of PMD tightly related to system outage figures is the time evolution of the DGD in a fiber, from which the relevant PMD parameter is derived through time/spectral averaging. Many papers deal with the DGD time evolution [1–5], and in particular [4] presents a very comprehensive study of this topic. However, some aspects, such as the evaluation of the outage probability and outage mean time still remain unclear and need further experimental investigation. This has been done performing measurements of PMD time evolution on a buried cable, and carrying out a complete statistical analysis of the collected data. The 37-km-long cable, installed along the Torino-Ivrea motorway, in the northwest of Italy, has a loose tube structure, containing 12 fibers of the G.653 type (1 fiber per tube). Measurements were performed using the JME technique over the 1530–1565 nm wavelength range, in 0.1-nm steps, and they were repeated every 30 min for about 10 days. 477 DGD spectra $\Delta\tau(t, \lambda)$ (each composed of 351 DGD versus wavelength samples) have been collected and analyzed as described below. First of all, the histogram of all the collected DGD data matches very well the Maxwellian Probability Density Function (PDF) with a mean value of 9.44 ps. Then, the time variations of the DGD spectra correlation coefficient were investigated. This was accomplished with the calculation of the following covariance function as a measurement of the average degree of correlation of the DGD spectra in function of time:

$$R(T) = E_\lambda\{[\Delta\tau(t,\lambda) - m_\lambda(t)][\Delta\tau(t+T,\lambda) - m_\lambda(t+T)], \quad (1)$$

where $E_\lambda[.]$ is the expectation calculated as a wavelength average. The results illustrated in Fig. 1 show that the covariance function presents strong peaks located at 24 h from the time 0. The strong correlation of the DGD data sets measured at 24-h intervals implies very similar DGD spectral behavior. This is shown in Fig. 2 where three DGD spectra corresponding to three measurements performed at 24-h intervals are drawn.

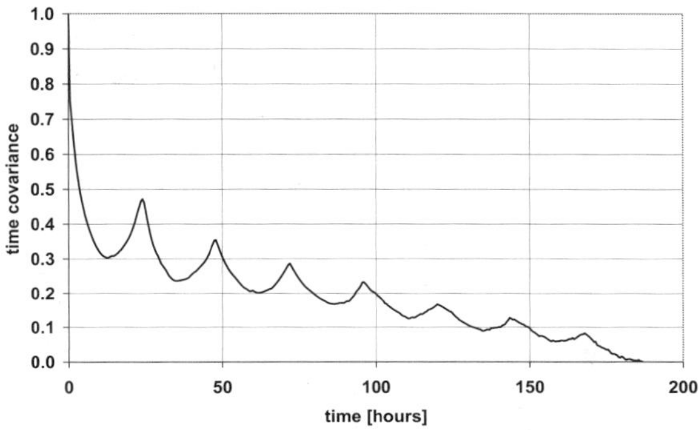

Fig. 1. Covariance function of the DGD spectra.

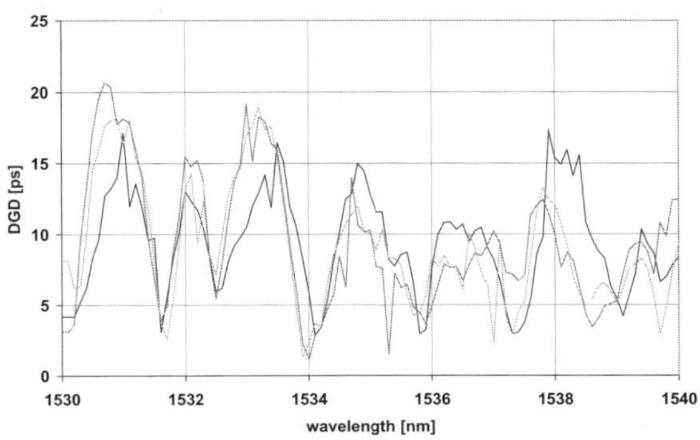

Fig. 2. Three DGD spectra measured at 24-h intervals.

The repetitive performance (in a statistical sense) of the DGD spectra at time intervals of 24 h suggests that, even if the cable is buried, its PMD behavior is strongly influenced by the environmental changes with a periodicity equal to full day duration. Hence, favorable or detrimental PMD conditions (depending on the system operating wavelength as shown in Fig. 2) can recur periodically every 24 h. A possible explanation of this behavior may be found in the plant structure. As stated in [1, 4], the buried sections of the cable probably behave as very time-stable PMD elements connected together by short cable sections strongly exposed to environmental changes (e.g., when the cable is embedded in a bridge structure). When the temperature changes, these short cables sections act as polarization rotators and vary the polarization coupling conditions between two neighboring buried cable sections, leading to different DGD spectra for the whole link. When similar temperature conditions occur (i.e., at 24-h intervals) the polarization rotators perform almost the same polarization transformation and the DGD spectra of the link are strongly correlated.

Another interesting point, addressed by means of appropriate elaborations of the measured data, is the calculation of system outage. The system impact of PMD is often expressed in terms of outage probability, Pout, which is the probability that the DGD is higher of a prescribed value tmax , and that is usually expressed in minutes per year {the approach of this paper is the same adopted in [7] where the variable excitation of the Principal States of Polarization (PSP) is neglected}. It is also interesting to evaluate the Mean Outage Rate Rout, i.e., the mean number of outage events per unit time, and the Mean Outage Duration T_{out}, i.e., the mean duration of an outage event. The calculation of the Mean Outage Rate is a simple application of the "Level Crossing" problem [6] and can be performed by the following integral evaluated for $\Delta\tau = t_{\max}$:

$$R_{\text{out}} = \frac{1}{2} f_\tau(\Delta\tau) \int_{-\infty}^{\infty} f_{\tau'|\tau}(\Delta\tau' \mid \Delta\tau) \mid \Delta\tau' \mid d\Delta\tau', \tag{2}$$

where $\Delta\tau'$ is the time derivative of $\Delta\tau$, and $f_{\tau'|\tau}(\Delta\tau' \mid \Delta\tau = t)$ is the conditional PDF of $\Delta\tau'$ assuming $\Delta\tau = t$. The values of $\Delta\tau'$ were calculated from measurements and, in the same way, the joint PDF was numerically evaluated. In order to calculate the previous integral the assumption that the DGD and its derivative are statistically independent was made. This is empirically justified by the independence of the conditional PDFs $f_{\tau'|\tau}(\Delta\tau' \mid \Delta\tau = t)$ (calculated from the measured data) of the values of t (see Fig. 3). At the end, the Mean Outage Rate is given by

$$R_{\text{out}} = \frac{1}{2} f_\tau(t_{\max}) \int_{-\infty}^{\infty} f_{\tau'}(\Delta\tau') \mid \Delta\tau' \mid d\Delta\tau', \tag{3}$$

where a Maxwellian with the same mean value of the measured data is assumed as the DGD PDF. The integral in the previous expression is evaluated numerically. Then, T_{out} is easily calculated as: $T_{\text{out}} = P_{\text{out}}/R_{\text{out}}$. The results of these calculations are shown in Fig. 4, where the three parameters are plotted in function of the ratio $t_{\max}/\langle\Delta\tau\rangle$.

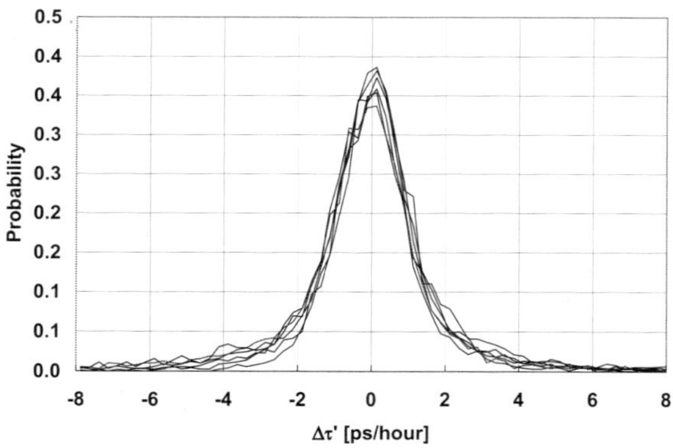

Fig. 3. Conditional PDFs $f_{\tau'|\tau}(\Delta\tau' \mid \Delta\tau = t)$, for different values of t.

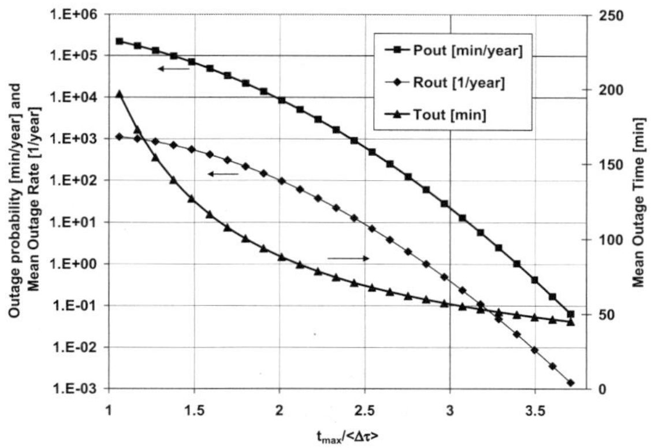

Fig. 4. $T_{\rm out}$, $P_{\rm out}$, and $R_{\rm out}$ calculated as a function of $t_{\rm max}/\langle\Delta\tau\rangle$.

For example, considering an outage probability of about 20 minutes per year (i.e., a maximum tolerable DGD of $3\langle\Delta\tau\rangle$ [7]), it results: $R_{\rm out} = 0.39$ yr^{-1}, $T_{\rm out} = 56$ min. Hence, adopting the usual design rule of [7]: PMD $\leq 1/10$ of the bit period, the transmission system should be unavailable, on average, for 56 minutes once every 2.5 years. The Mean Outage Rate and the Mean Outage Time estimated before figure out two important parameters for the system availability estimation that can be calculated from $\Delta\tau(t, \lambda)$ data sets, measured by JME technique.

3. PMD measurements on Optical Components and Subsystems

In this section, the JME measurement method will be applied to the characterization of narrowband optical components and optical amplifiers. The JME spectral resolution is pushed to its lower limit with the characterization of DGD features of fiber Bragg gratings on a wavelength scale as low as 5 pm. This will lead to important considerations on the gratings structure and to some useful design criteria. In the same way, different families of optical amplifiers are characterized to examine the statistical nature of their PMD. How components characteristics affect the whole subsystem PMD is also investigated.

3.1. PMD measurements on Chirped Bragg Gratings

Fiber Bragg gratings are optical filters based on the Bragg reflection principle. They are written exploiting the photosensitive effect, within the core of optical fibers creating a periodic sequence of high and low refractive index regions. They behave as notch filters reflecting light around the Bragg wavelength and transmitting all other wavelengths, but important additional features can be obtained by a careful design of the longitudinal refractive index profile. In particular, the so called Chirped Bragg Gratings (CBGs) are characterized by nonuniform periods (e.g., varying linearly along the grating length), exhibit high chromatic dispersion, and can be used as chromatic dispersion compensators.

PMD is an important issue in CBGs used as fixed or variable chromatic dispersion compensators in high-speed optical links. Despite their short length, average DGD of the order of 20 ps was measured in gratings written on special fibers [8]. Moreover, a strong DGD wavelength dependence, on a scale of few pm, was observed on several components. Two important issues worthy of a deeper investigation emerge from these results. The first issue is to understand the origin of PMD in components like CBGs, which are characterized on one side by a relatively short length, but on the other by a complex refractive index longitudinal structure. The second one is to clarify the statistical features of PMD in CBGs, whose knowledge is fundamental for the system outage calculation. These topics are investigated by means of very high resolution JME measurements. A simple phenomenological model of the PMD of a chirped Bragg grating can be developed assuming that the grating fiber exhibits a uniform linear birefringence characterized by an effective group index difference Δn. This birefringence can be due to intrinsic core ellipticity or to fiber bending, but could be also determined by the writing process, as will be discussed later. The result is a shift, $\Delta\lambda_B$, in the Bragg wavelength for the two fields polarized along the fiber birefringence axes:

$$\Delta\lambda_B = 2\Delta n P, \tag{4}$$

where P is the local grating period. $\Delta\lambda_B$ can be considered constant and equal to $2\Delta n\langle P\rangle$, because, in ordinary chirped gratings, the period relative variations are negligible ($\sim 1 \times 10^{-3}$). As a consequence, both reflectivity and Group Delay (GD) characteristics are polarization dependent. In particular the GD curve of the field polarized along one of the birefringence axes, $\tau_y(\lambda)$, is shifted by $\Delta\lambda_B$ with respect to the orthogonal one $\tau_x(\lambda)$:

$$\tau_y(\lambda) = \tau_x(\lambda - \Delta\lambda_B) \tag{5}$$

At a given wavelength, the DGD $\Delta\tau$ between these two orthogonal states of polarization can be written as

$$\Delta\tau(\lambda) = \mid \tau_x(\lambda - 2\Delta n P) - \tau_x(\lambda) \mid . \tag{6}$$

In practice, for small $\Delta\lambda_B$ and moderate GD ripples, the GD curve $\tau_x(\lambda)$ can be approximated by the GD curve $\tau(\lambda)$ averaged over all input states of polarization. In the case of an ideal grating, whose GD is a linear function of wavelength with slope D, it is straightforward to calculate:

$$\Delta\tau = \mid D \mid \Delta\lambda_B = \mid D \mid 2\Delta n P. \tag{7}$$

The DGD is constant and equals the product of the chromatic dispersion of the grating, $\mid D \mid$, and the Bragg wavelength shift $\Delta\lambda_B$, which is itself proportional to the fiber birefringence. It is interesting to note that CBGs can exhibit a high PMD even in presence of small fiber birefringence because of the high chromatic dispersion. For example, a small fiber birefringence $\Delta n = 2 \times 10^{-6}$, combined with a chromatic dispersion $\mid D \mid = 1500$ ps/nm (typical of many CBGs) in a grating operating around 1550 nm wavelength, determines a DGD of 3.2 ps. Furthermore, the real GD characteristic of a fiber grating should be taken into account. Ripples in the GD may lead to spikes in the DGD with peaks in the order of tens of ps, which contribute non-negligibly to the component average DGD and introduce second-order PMD, as reported in the following. Reflectivity and group delay have been measured on a commercially available

CBG with the phase shift technique [8]. PMD and Polarization Dependent Loss (PDL) have been measured by JME. In both cases the light source was an external cavity laser, tuned in 5-pm steps. This value was chosen as a tradeoff between the desired measurement resolution and the laser tuning accuracy. Fig. 5 illustrates the reflectivity and GD measurements, while Fig. 6 shows an example of DGD measurement of the grating under test.

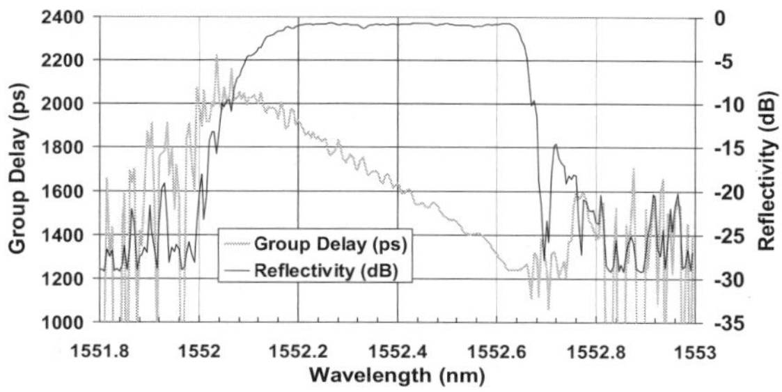

Fig. 5. CBG reflectivity and group delay.

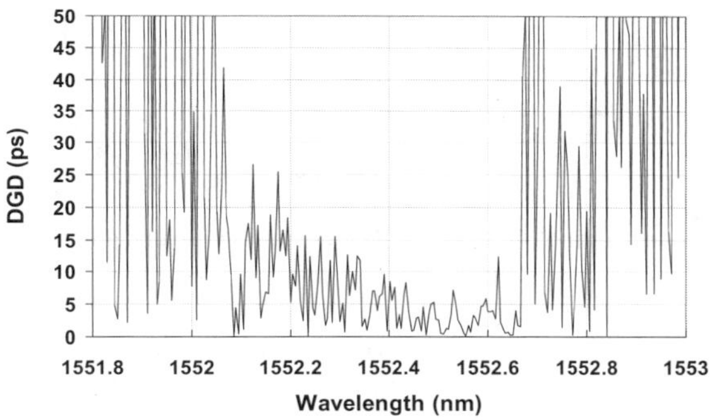

Fig. 6. DGD of CBG.

A quantitative comparison between the two curves can be carried out in the following way. An average birefringence of 2×10^{-6} was measured by JME at 1540 nm, outside the grating reflectivity band. According to Eq. (1), this birefringence causes a wavelength shift of about 2 pm between the GD curves of fields components polarized along the birefringence axes. After a spline interpolation of the measured GD curve, the GD increment on a wavelength step of 2 pm was calculated and compared with the measured DGD. Results are shown in Figs. 7 and 8, which refer to the lower and the upper section of the grating band respectively.

A very satisfactory agreement is evident from Fig. 8, which corresponds to the wavelength region where the GD curve is more regular. Also in the lower band region (Fig 7), where ripples in the GD curve are larger, the two DGD curves show similar behaviors and exhibit peaks almost in the same positions. This demonstrates the validity of Eq. (3) and puts in evidence the effects of GD ripples on DGD. An average DGD of about 3 ps is expected for an ideal grating [Eq. (4)], due to the measured fiber birefringence (2×10^{-6}) and chromatic dispersion (-1443 ps/nm), but the measured

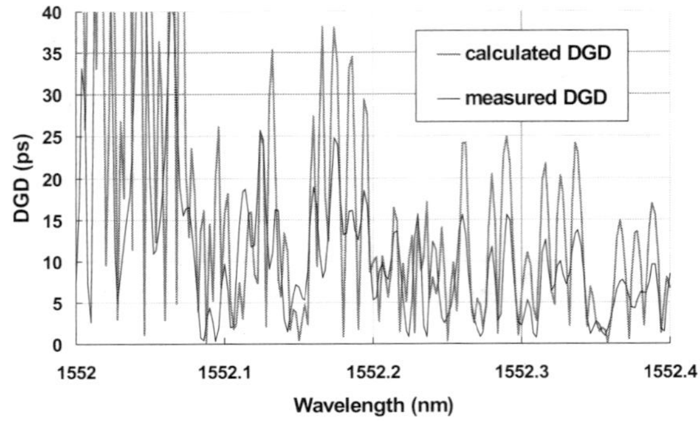

Fig. 7. Comparison of measured and calculated DGD (lower band section).

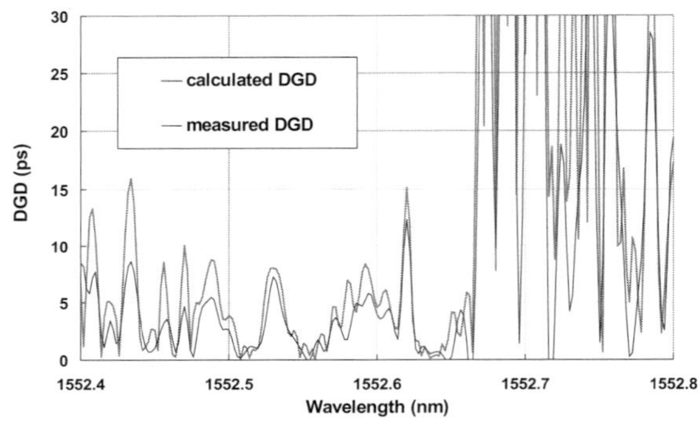

Fig. 8. comparison of measured and calculated DGD (upper band section).

value (averaged over the operating bandwidth) is twice as large (about 6.1 ps) because of the presence of DGD peaks. It is also important to note that the birefringence of 2×10^{-6}, measured outside the reflectivity band, takes into account both the contribution of intrinsic birefringence, due to core ellipticity, and of extrinsic birefringence, due, in this case, only to the writing process (as the 10-cm-long grating was enclosed in a straight package, thus avoiding curvature birefringence). In order to separate out the two effects, the same birefringence measurement was made on a short portion of the grating pigtail, leading to a value of about 0.7×10^{-6}. This indicates that the writing process is responsible for an increase in grating birefringence. These results, together with others already published data [9, 10, 11], demonstrate that the grating PMD can be kept under control acting on the following three parameters:

1. Intrinsic birefringence of the grating fiber, which should be low (of the order of 10^{-7} or less): this feature is easily achievable using standard telecommunication fibers, while special care should be taken in the case of highly doped, small core special fibers.

2. Extrinsic fiber birefringence: the writing process should be improved in order to reduce its impact on fiber birefringence [9]; this can be achieved for instance using

a proper alignment of the UV laser polarization state with respect to the fiber axis [10].

3. Group Delay Ripples: a tight control of the whole technological process is required to reduce both average DGD values and strong wavelength fluctuations responsible for second order PMD.

3.2. PMD measurements on optical amplifiers

Recently, new classes of Optical Fiber Amplifiers (OFAs) have been introduced besides the well known Conventional-band EDFA (C-EDFA). In particular Long-band Erbium-Doped Fiber Amplifiers (L-EDFAs) and Erbium-Doped Tellurite Fiber Amplifiers (EDTFAs) have been developed to extend the operation spectral window of WDM transmission systems over the whole 1530–1615 nm wavelength interval. L-EDFAs for optical amplification in the L-band (1570–1615 nm) are made by very long active fibers (up to 200 m and more), in order to transfer to longer signal wavelengths the pump energy as well as the energy of Amplified Spontaneous Emission (ASE) generated in the C-band (1530–1565 nm). It has already been shown that PMD in C-EDFAs and Erbium-Doped Fluoride Fiber Amplifiers (EDFFAs) is very small as well as deterministic [12]. This means that, even in a long-haul, high-bit-rate optically amplified link, the OFA contribution to the total PMD is small (order of 0.05–0.3 ps deterministic Differential Group Delay DGD per amplifier [13]) and can be characterized by a well defined DGD maximum value over the gain bandwidth. However, it is arguable that L-EDFAs, featuring very long active fibers, could show more severe PMD limitation and may not have a deterministic behavior. In this section the PMD features of C-band, L-band and extended band OFAs are compared both in magnitude and in the deterministic vs. stochastic behavior of the Differential Group Delay (DGD) of the light signal at the output of the device. It has been found that both EDTFAs and L-EDFAs show a deterministic PMD, even though the latter exhibits a higher mean DGD (0.6–0.8 ps). The stochastic behavior of the DGD of long active fibers is dramatically reduced by tightly wounding the fiber on a small-diameter bobbin. In general, an Optical Fiber Amplifier (OFA) consists of an input set of lumped components (WDM coupler, isolator), the active fiber and an output set of lumped components similar to the input one. In the PMD measurements presented below, the input and output components were tested separately from the active fibers, in order to discriminate between the expected deterministic and possible stochastic contributions to the OFA PMD. The measurements were performed by means of the JME technique because, due to its high immunity to the OFA ASE noise (the JME uses only the polarized part of the optical signal), it is the most appropriate method for measurements on working optical amplifiers [12].

The input and output coupling components were characterized by placing a polarization scrambler in between them, to evaluate their total PMD under different (stochastic) coupling, simulating the possible behavior of the active fiber. The resulting average DGD was quite low (< 0.25 ps) over the whole measurement range. By using the two coupling sets mentioned above, three extended band amplifiers have been made in sequence using three different active fibers. The PMD measurements have been performed with the OFAs in operating conditions (bidirectional 1480 nm pumping). Two commercial Er:silica active fibers of different length and comparable design (*Fiber 1*, 100 m, peak absorption 6.5 dB/m; *Fiber 2*, 180 m, peak absorption 7 dB/m) were tested for PMD. The first amplifier (OFA1) was made using Fiber 1 loosely wounded on its original drum (24-cm diameter). The results of two groups of three PMD measurements taken along two days are shown in Fig. 9. Differences can be observed, even though the mean value (0.6 ps) is almost the same. The measurements have been performed up to 1595 nm because of tunable laser source limitations.

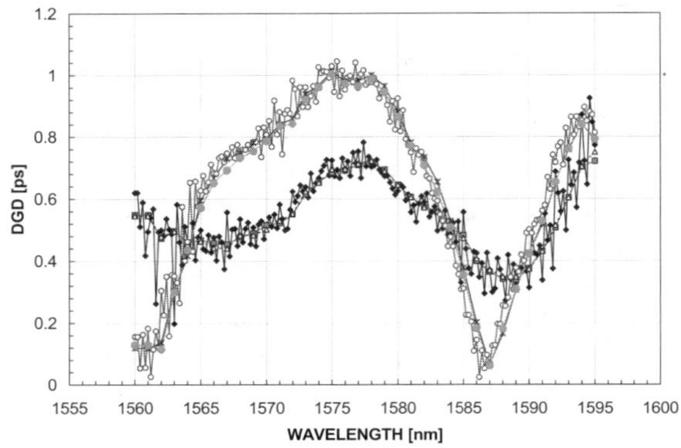

Fig. 9. DGD curves for OFA1 (Fiber 1 loosely wounded on a 12-cm-radius drum).

Measurements in the 1600–1700 nm range made on the unpumped OFA1 using the wavelength scanning technique showed the same average DGD (Fig. 10). This suggests that DGD values averaged in the 1560–1595 nm range closely agree with those averaged over the whole L-band.

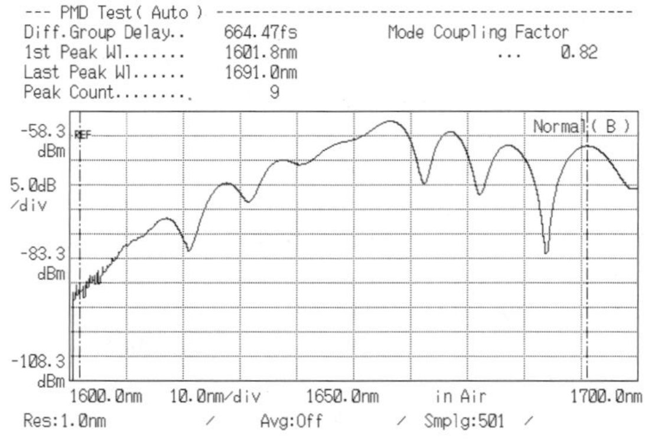

Fig. 10. PMD measurement for OFA1 (unpumped) with wavelength scanning technique.

The PMD differences in the two measurements of Fig. 9 may be ascribed to different environmental conditions. This behavior was further evidenced by measurements performed under thermal cycles, as shown in Fig. 11.

As can be seen, during the thermal cycles the mean DGD ranges from less than 0.2 ps to a value of about 0.7 ps (quite close to that of Fig. 9). This behavior could be traced back to the combined effect of polarization mode coupling variations due to

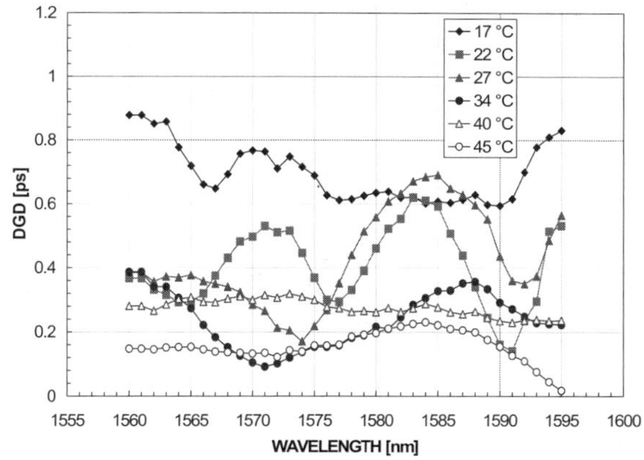

Fig. 11. DGD curves for OFA1 (Fiber 1 loosely wounded on a 12-cm-radius drum under thermal cycles).

temperature induced stress and random combination of the deterministic DGD of the input and output stages. However, in commercial OFAs' packages the fiber is clearly not deployed this way. The dependence of PMD on the fiber wounding tension was investigated in OFA2. The spectral DGD was measured in two different conditions: i) Fiber 2 loosely wounded on a 12-cm-radius drum (Fig. 12); ii) Fiber 2 tightly wounded on a 3.5-cm-radius drum (Fig. 13).

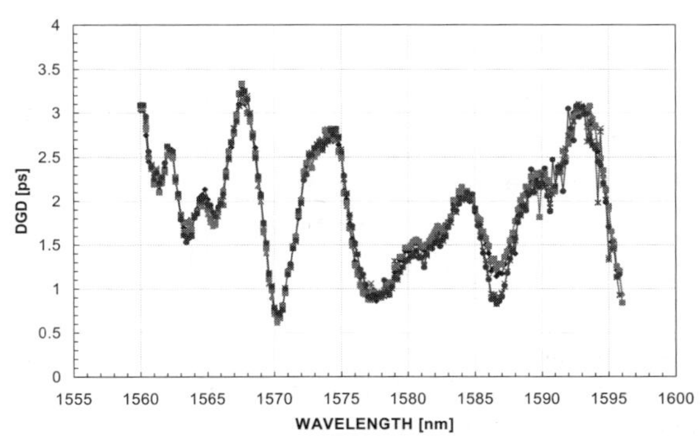

Fig. 12. DGD curves for OFA2 (Fiber 2 loosely wounded on a 12-cm-radius drum).

While no dependence upon the active fiber deployment was observed in OFA2 Gain and Noise Figure, the PMD varies considerably. PMD is markedly lower in the case of the smaller bobbin (and of the stronger tension), as shown in Fig. 13. The mean DGD value dropped down from 2 to 0.7 ps, due to the smaller contribution of the fiber to the overall DGD value. The measurement repeatability has been verified over four days and several environmental laboratory conditions. As can be seen, the DGD seems to behave in a deterministic way over wavelength.

Finally, OFA3 was made using a commercial EDTFA module. The measured DGD curves of OFA3 are shown in Fig. 14. As it is apparent, the PMD is very small and

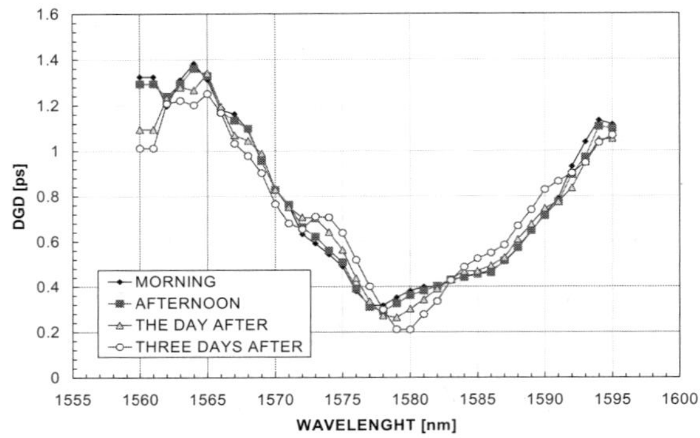

Fig. 13. DGD curves for OFA2 (Fiber 2 tightly wounded on a 3.5-cm-radius bobbin).

deterministic, since at each wavelength the DGD variance is much smaller than the mean value. In contrast with the previous two cases, the active fiber is only 5 m long and it resembles the case of C-EDFA [12].

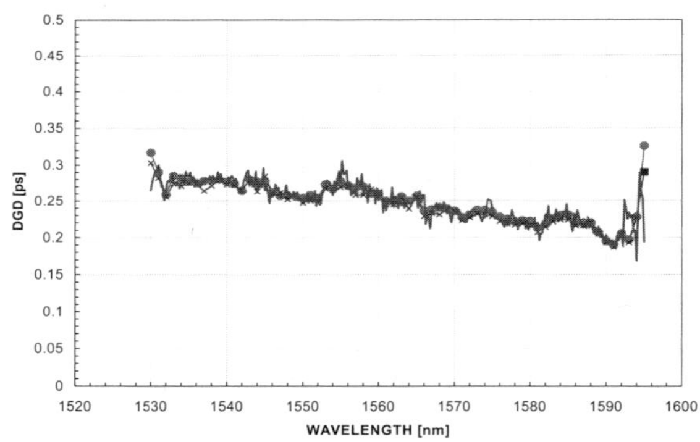

Fig. 14. DGD curves for the EDTFA.

These results demonstrate that, although the active medium is very long, PMD of L-EDFAs can be effectively controlled in the packaged OFA device. However, DGD values are higher than those of conventional EDFAs by at least a factor of two. The DGD values of EDTFAs, are on the contrary very low and similar to those of C-EDFAs, due to the short length of active fiber. In conclusion, no particular PMD system penalty is expected from the use of extended-band OFAs, but, possibly, for extremely long amplifier chains.

Acknowledgments

The author gratefully acknowledges the fundamental contributions of several colleagues: M. Artiglia, R. Caponi, A. Pagano, A. Percelsi, M. Potenza, B. Riposati, A. Rossaro, B. Sordo, T. Tambosso, G. Zaffiro.

References

[1] C. De Angelis, A. Galtarossa, G. Gianello, F. Matera, and M. Schiano, "Time evolution of polarization mode dispersion in long terrestrial links," IEEE J. Lightwave Technol.,**10**, 552–555 (1992).

[2] B. W. Hakki, "Polarization mode dispersion in a single mode fiber," IEEE J. Lightwave Technol., **14**, 2202–2208 (1996).

[3] J. Cameron, L. Chen, X. Bao, and J. Stears, "Time evolution of polarization mode dispersion in optical fibers," IEEE J. Lightwave Technol., **10**, 1265–1267 (1998).

[4] M. Karlsson, J. Brentel, and P. A. Andrekson, "Long-term measurement of PMD and polarization drift in installed fibers," IEEE J. Lightwave Technol.,**18**, 941–951 (2000).

[5] D. Waddy, P. Lu, L. Chen, and X. Bao, "The Measurement of Fast State of Polarization Changes in Aerial Fiber," proceedings OFC2001, paper TH A3, Anaheim, 2001.

[6] A Papoulis, *Probability, Random Variables, and Stochastic Processes*, Third Edition (MacGraw-Hill, New York, 1991).

[7] ITU-T Recommendation G.691, "Optical interfaces for single-channel STM-64, STM-256 and other SDH systems with optical amplifiers," Appendix I.

[8] S. Bonino et al., "Measurement of polarisation properties of chirped fibre gratings," proceedings OFMC '97, pp.10–13, Teddington, September 1997.

[9] T. Erdogan et al, "Characterization of UV-induced birefringence in photosensitive Ge-doped silica optical fibers," J. Opt. Soc. Am., **11** (10), 2100–2105 (October 1994).

[10] T. Meyer et al., "Reversibility of photoinduced birefringence in ultralow-birefringence fibers," Opt. Lett., **21** (20), 1661–1663 (October 1996).

[11] K. Dossou et al., "Contribution of the transverse asymmetry of the index change to the birefringence of fiber Bragg gratings: a numerical calculation," proceedings ECOC'01, pp. 390–391, Amsterdam, September 1991.

[12] R. Caponi et al., "Deterministic nature of polarisation mode dispersion in fibre amplifiers," proceedings ECOC'98, pp. 543–544, Madrid, September 1998.

[13] N. Jolley, IEE Coll. on new developments in optical amplifiers, London, 1998.

DOI: 10.1007/s10297-004-0012-2
Originally published in J. Opt. Fiber Commun. Rep. 1, 150–179 (2004)

Reflectometric measurements of polarization properties in optical-fiber links

Andrea Galtarossa[1] and Luca Palmieri[2]

[1] DEI, University of Padova, V. Gradenigo 6/A, Padova, Italy
andrea.galtarossa@unipd.it
[2] DEI, University of Padova, V. Gradenigo 6/A, Padova, Italy
luca.palmieri@dei.unipd.it

Abstract. Polarization sensitive reflectometric techniques can be effectively used to perform spatially resolved measurements of polarization properties of fiber optic link — such as birefringence, polarization mode dispersion and polarization dependent loss. In particular, this contribution is focused on polarization sensitive OTDR and provides a survey of its theory and main applications. Special emphasis is given to the characterization of fiber birefringence, that allows to inspect the fiber while cabled and, consequently, to test and improve the cabling process. In addition, the analysis of the birefringence also allows to define reliable mathematical models, which are essential for the design of low polarization mode dispersion fibers. Reflectometric measurements of polarization mode dispersion and polarization dependent loss are also discussed.

1. Introduction

Owing to their perfectly radial symmetry, ideal single-mode fibers preserve the polarization of a propagating electromagnetic field, therefore the propagation in such waveguides can be completely described with a scalar theory [1]. Conversely, when the radial symmetry is broken, either accidentally or intentionally, a single-mode fiber behaves like a birefringent medium, where the polarization of the field evolves accordingly to the local birefringence properties [2,3]. In this case, different polarization states may propagate with different group velocities, causing polarization mode dispersion (PMD) [4,5].

The birefringence properties of a perturbed single-mode fiber can be profitably described by the *local birefringence vector*, $\boldsymbol{\beta}(z,\omega)$, defined in the three-dimensional space of Stokes vectors [6]. The birefringence vector is in general a function of both the position along the fiber, z, and the optical frequency, ω, and contains all the information about perturbations and defects acting on the fiber. Actually, $\boldsymbol{\beta}(z,\omega)$ completely describes all the PMD sources.

More than twenty years ago, pioneering studies [3], [7] established precise relationships between the kind of perturbation acting on the fiber and the properties of the corresponding induced birefringence. However, later on the interest in fiber birefringence waned somewhat, mainly because it was found that, in the case of the randomly perturbed telecommunication fibers, the knowledge of the mean differential group delay (DGD) is enough to characterize the transmission properties of fiber links [8–13]. Roughly speaking, any couple of fibers with the same mean DGD have the same PMD statistical properties, independently of the nature of their birefringence, provided that they are sufficiently long and their birefringence is sufficiently random. Therefore, the research activity focused mostly on the study of the overall effects of PMD.

Although this approach widely proved successful in the analysis of telecommunication system performances, it is not suitable for many other PMD-related problems. Probably, the most meaningful example of the importance of birefringence properties is represented by the design of optimized spin profile. The spinning process is by far the most effective way to reduce PMD, and consists in rotating the fiber during the drawing, while the silica is still in the viscous state [14–16]. In this way the intrinsic birefringence is forced to rotate and, if the spin parameters are properly tuned, the mean DGD of the fiber can be reduced by even two orders of magnitude. Nevertheless, it has been recently shown that the effectiveness of the spin strongly depends on birefringence properties of the fiber, to the extent that a spin design based on an inaccurate birefringence model may overestimate the DGD reduction by orders of magnitude [17, 18].

But spin is not the only effect influenced by the birefringence. Another example is given by optical non-linearities, which affect high bit-rate telecommunication systems, and whose interaction with fiber birefringence does depend on birefringence properties. Therefore, in this case the knowledge of the mean DGD would no longer suffice for a complete characterization of system performances [19, 20].

From the discussion made so far, it appears that birefringence measurements are needed to characterize real fibers, in order to define accurate mathematical models for the birefringence. Besides these theoretical aims, birefringence measurements may also provide an effective tool for inspecting and testing fiber production process, from the drawing to the cabling.

Any measurement technique intended to characterize fiber birefringence has to be able to perform a local analysis of fiber properties. Several proposals have been published in literature [3, 21–47], but the most effective techniques are those based on the analysis of the polarization state of the backscattered field. Among them, two main reflectometric techniques may be distinguished.

The first one is the polarization sensitive optical time domain reflectometry (P-OTDR). Proposed more than twenty years ago [21–24], it is based on the analysis of the field backscattered by a probe pulse, sent into the fiber under test. At the beginning, this technique was applied only on uniformly birefringent fiber samples, few tens of meters long. More recently, the dynamical range of P-OTDR has been extended up to several kilometers, with a spatial resolution of roughly half a meter, so that

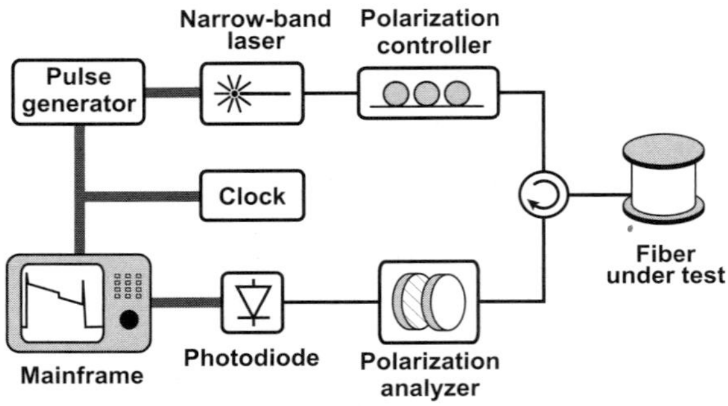

Fig. 1. Schematic of the P-OTDR set-up.

the birefringence of fibers either wound on bobbin or installed in cables have been successfully characterized [25–37].

The second technique, similar to P-OTDR, is the polarization sensitive optical frequency domain reflectometry (P-OFDR) [38–41]. In this case, a frequency-modulated continuous-wave probe field is launched into the fiber, and the polarization state the of corresponding backscattered field is analyzed. At the moment, P-OFDR cannot be applied to fibers longer than a couple of kilometers; nevertheless, on short samples of fiber (few meters long), P-OFDR may reach a spatial resolution as high as few millimeters.

This contribution reviews the theory and the applications of P-OTDR. In section 2 the experimental set-up and its principle of operation are described. Note that P-OTDR essentially measures the evolution of the backscattered-field polarization state. Once this evolution has been measured, information about the birefringence may be extracted with appropriate data analyses, described in detail in sections 3 and 4. In particular, the measurement of beat length and correlation length is considered, as well as the measurement of circular birefringence in twisted fibers. Preliminary results about the measurement of the spin profile in spun fibers are also described.

Section 5 is devoted to the analysis of reflectometric techniques as a tool for measuring the evolution of the mean DGD accumulated along a fiber optic link [48–54]. In this case, besides providing a picture of the local properties of the link, these techniques represent an alternative to standard measurement methods — such as the Jones matrix eigenanalysis [55] and the Müller matrix method [56] — that need both fiber ends to perform the measurement. Actually, by using only one end of the fiber link, reflectometric techniques may simplify the management of the measurement, although they require a somewhat more complicated experimental set-up.

Finally, in section 6 reflectometric measurements applied to links affected also by polarization dependent loss (PDL) are discussed. In this case there might be problems in measuring the local properties, whereas the characterization of the mean PDL accumulated along the link is still achievable.

2. The P-OTDR Experimental Set-up

Basically, a P-OTDR is an instrument for the measurement of the state of polarization of the backscattered field. It has the same structure of a standard OTDR, the main difference being at the source stage and at the detection stage, as shown in Fig. 1.

Actually, differently from standard OTDR's, a P-OTDR needs a narrow-band optical source, so to increase the robustness of the optical pulses with respect to the frequency dependent distortions due to PMD and PDL. Moreover, the source has to be polarized and its polarization state has to be controlled. Because of this, the optical source of P-OTDR is usually made of an external cavity or DFB laser (eventually amplified and properly filtered) followed by a polarization controller [27, 29, 30, 57]. The typical pulse-width is about 3 to 5 ns, and the peak power is set in the order of hundred milliwatts and so to avoid non-linearities.

Of course, the detection stage of P-OTDR has to be polarization sensitive, whereas that of standard OTDR should not. Two different solutions have been proposed to achieve this requirement. The first solution (the one represented in Fig. 1) consists in routing the backscattered field to a polarization analyzer, made of a quarter wave-plate followed by a linear polarizer [30]. The former may be rotated at any angle, while the latter is fixed. Owing to the polarizer, polarization evolutions of the backscattered field are converted in power fluctuations, that are detected and recorded by the mainframe. These traces are measured for several different orientations of the quarter wave-plate. In principle, four orientations would be enough, but it is preferable to use at least five different orientations, so that the measured data are redundant, and an estimate of the measurement accuracy can be performed. These raw data are downloaded in a personal computer and processed to determine the evolution of the state of polarization [6].

The alternative solution consists in using a Stokes analyzer instead of the polarization analyzer [27]. In the Stokes analyzer the optical power is split in four paths. Three of them end with vertical, horizontal and circular polarizers, respectively, while the forth path is not polarized. The four signals originated from each path are directly related to the Stokes parameters via a calibration matrix. Therefore, a four-channel receiver is needed to record simultaneously the four signals provided by the analyzer. This configuration allows faster measurements, however it does not allow an easy error estimate. Moreover, it introduces higher losses in the return path, that have to be compensated by an optical amplifier.

Typical P-OTDR's have a spatial resolution of roughly 0.5 m, a dynamical range of few tens of kilometers and allow to measure the backscattered state of polarization with an uncertainty of about 1%.

3. Basic Theory

Birefringence measurements based on reflectometric technique are grounded on the simple consideration that the evolution of backscattered field polarization is influenced by fiber birefringence. Therefore, information on the latter may be inferred by measuring the former.

Assuming that the optical link is not affected by PDL, and using the Müller/Stokes formalism [6], the state of polarization (SOP) of a field propagating in the fiber reads $\boldsymbol{s}(z) = \mathsf{R}(z)\boldsymbol{s}_0$, where $\mathsf{R}(z)$ is the Müller matrix representing the propagation and $\boldsymbol{s}_0$

is the input SOP. The matrix $\mathsf{R}(z)$ is orthogonal, hence $\mathsf{R}^{-1}(z) = \mathsf{R}^{\mathrm{T}}(z)$, where the T represents the transposition. Moreover, $\mathsf{R}(z)$ is related to the local birefringence of the link by

$$\frac{\mathrm{d}\mathsf{R}}{\mathrm{d}z} = \boldsymbol{\beta}(z) \times \mathsf{R}(z)\,, \tag{1}$$

where $\boldsymbol{\beta}(z) = (\beta_1,\, \beta_2,\, \beta_3)^{\mathrm{T}}$ is the *local birefringence vector* representing all the perturbations acting on the fiber link, and $(\boldsymbol{\beta}\times)$ should be considered a linear operator as specified in [58].

Similarly, the SOP of the field backscattered in a point z along the fiber, can be written as $\boldsymbol{s}_{\mathrm{R}}(z) = \mathsf{R}_{\mathrm{R}}(z)\boldsymbol{s}_0$, where $\mathsf{R}_{\mathrm{R}}(z)$ is the Müller matrix representing the *round-trip propagation*, i.e., the evolution of a field that propagates along the fiber link up to a point z, undergoes Rayleigh scattering, and then back-propagates to the fiber input. According to this definition, the matrix of the round-trip propagation can be written as $\mathsf{R}_{\mathrm{R}}(z) = \overleftarrow{\mathsf{R}}(z)\mathsf{S}\overrightarrow{\mathsf{R}}(z)$, where $\overrightarrow{\mathsf{R}}(z) = \mathsf{R}(z)$ represents the propagation up to z, S the Rayleigh backscattering, and $\overleftarrow{\mathsf{R}}(z)$ the back-propagation from z to the fiber input.

Neglecting the scalar attenuation, the Rayleigh backscattering is equivalent to an ideal reflection, so that S is equal to the identity matrix [26]. Moreover, in order to develop a general theory of reflectometric measurements, a relationship between $\overrightarrow{\mathsf{R}}(z)$ and $\overleftarrow{\mathsf{R}}(z)$ has to be established. This can be done in a general way only when the link satisfies the principle of reciprocity [1], because in this case $\overleftarrow{\mathsf{R}}(z) = \mathsf{M}\overrightarrow{\mathsf{R}}^{\mathrm{T}}(z)\mathsf{M}$, where $\mathsf{M} = \mathrm{diag}(1, 1, -1)$ is a constant diagonal matrix [26, 59]. On the contrary, when the principle of reciprocity is not satisfied, the relationship between $\overrightarrow{\mathsf{R}}(z)$ and $\overleftarrow{\mathsf{R}}(z)$ cannot be written in general, since it depends on the specific properties of the medium. Nevertheless, an optical fiber is always reciprocal, unless it is exposed to rather high magnetic fields [3]. Thus, owing to the above argumentations, it produces the result

$$\mathsf{R}_{\mathrm{R}}(z) = \mathsf{M}\mathsf{R}^{\mathrm{T}}(z)\mathsf{M}\mathsf{R}(z)\,. \tag{2}$$

Recalling that the SOP of a field backscattered from a point in z is given by $\boldsymbol{s}_{\mathrm{R}}(z) = \mathsf{R}_{\mathrm{R}}(z)\boldsymbol{s}_0$, upon differentiating this expression with respect to z, and since $\mathsf{R}_{\mathrm{R}}^{-1} = \mathsf{R}_{\mathrm{R}}^{\mathrm{T}}$ one finds

$$\frac{\mathrm{d}\boldsymbol{s}_{\mathrm{R}}}{\mathrm{d}z} = \frac{\mathrm{d}\mathsf{R}_{\mathrm{R}}}{\mathrm{d}z}\mathsf{R}_{\mathrm{R}}^{\mathrm{T}}\,\boldsymbol{s}_{\mathrm{R}} = \boldsymbol{\beta}_{\mathrm{R}} \times \boldsymbol{s}_{\mathrm{R}}\,, \tag{3}$$

where $\boldsymbol{\beta}_{\mathrm{R}}(z)$ is the *round-trip birefringence vector* [49] and, as shown by Eq. (3), governs the evolution of $\boldsymbol{s}_{\mathrm{R}}(z)$. The key point of this analysis is the relationship between $\boldsymbol{\beta}_{\mathrm{R}}(z)$ and the local birefringence vector, $\boldsymbol{\beta}(z) = (\beta_1,\, \beta_2,\, \beta_3)^{\mathrm{T}}$. As shown in appendix A, using Eqs. (1) and (2), it may be found that

$$\boldsymbol{\beta}_{\mathrm{R}}(z) = 2\mathsf{M}\mathsf{R}^{\mathrm{T}}(z)\boldsymbol{\beta}_{\mathrm{L}}(z) = 2\overleftarrow{\mathsf{R}}(z)\boldsymbol{\beta}_{\mathrm{L}}(z)\,, \tag{4}$$

where $\boldsymbol{\beta}_{\mathrm{L}} = (\beta_1,\, \beta_2,\, 0)^{\mathrm{T}}$ is the linear component of $\boldsymbol{\beta}$, and the last equality holds because $\mathsf{M}\boldsymbol{\beta}_{\mathrm{L}} = \boldsymbol{\beta}_{\mathrm{L}}$. Equation (4) states that the round-trip birefringence vector is equal to twice the local linear birefringence as "seen" through the link. Note also that the round-trip birefringence vector does not depend explicitly on the circular birefringence, $\beta_3(z)$.

Equation (4) together with Eq. (3) represent the core of all birefringence measurement methods based on P-OTDR. To understand how these equations may be used to characterize fiber birefringence, note that the evolution of $\boldsymbol{s}_B(z)$ along z is governed by the same kind of equation that describes the evolution of the output SOP as a function of the optical frequency, ω. In other words, Eq. (3) is formally equal to

$$\frac{\mathrm{d}\boldsymbol{s}}{\mathrm{d}\omega} = \boldsymbol{\Omega}(\omega) \times \boldsymbol{s}(\omega)\,,$$

where $\boldsymbol{\Omega}$ is the well known polarization dispersion vector [60]. This means that all of those techniques developed to measure PMD vector (like the *Jones matrix eigenanalysis* [61] or the *Müller matrix method* [30, 56]) may be adapted to measure the round-trip birefringence vector, as well.

Once the round-trip birefringence has been calculated, information on the local birefringence may be inferred using Eq. (4). As a matter of fact, since $\overleftarrow{\mathsf{R}}$ is an orthogonal matrix, Eq. (4) yields the fundamental result

$$\beta_{\mathrm{R}}(z) = |\boldsymbol{\beta}_{\mathrm{R}}(z)| = 2\,|\boldsymbol{\beta}_{\mathrm{L}}(z)| = 2\beta_{\mathrm{L}}(z)\,, \tag{5}$$

which means that if the round-trip birefringence is measured with a reflectometric technique, the local evolution of the linear birefringence strength can be calculated, even though the Müller matrix of the link is unknown. Conversely, as discussed in section 4.3, retrieving information about circular birefringence, β_3, is somewhat more involved.

An alternative to Eq. (5) is suggested in [29]. In this second approach it is assumed that the spatial resolution of the measurement is so high, that the fiber birefringence may be considered uniform over the length of three consecutive sampling points. In this way, it is possible to collect enough information to determine all the three components of $\boldsymbol{\beta}(z)$. The main problem of this technique is, however, the need for a rather high spatial resolution, which limits its applicability only to very low and very uniform birefringent fibers.

3.1. A Stochastic Model for the Birefringence

The theoretical analysis performed in the previous section leads to a point-wise characterization of the birefringence, and holds whether the fiber birefringence is deterministic or random. However, in the case of random birefringence — and, actually, standard telecommunication fibers are always randomly birefringent — it makes sense to perform a statistical analysis, too.

This second approach is greatly enhanced if a proper stochastic model for the birefringence is defined. Actually, once the model is specified, the measurement may consist in finding the values of the model parameters by fitting theoretical predictions with the experimental data. In this section the model is introduced by means of reasonable argumentations; in the next sections, however, it will also be supported and validated by experimental evidences.

The Random Modulus Model

When several perturbations act simultaneously on a fiber section, the total induced birefringence vector is given by the sum of all the vectors representing each perturbation, as if it were acting alone [3]. Linear birefringence has several origins: bendings, geometrical imperfections, stress induced anisotropies. All of them are likely to act randomly along the fiber. As a consequence, the linear component of the birefringence, $\boldsymbol{\beta}_{\mathrm{L}}(z)$, is the sum of several random vectors, most likely statistically independent from each other.

According to the above argumentations, it is reasonable to assume that the components of $\boldsymbol{\beta}_{\mathrm{L}}(z)$ are well described by Gaussian processes. This leads to the definition of the *random modulus model* (RMM), which assumes that $\beta_1(z)$ and $\beta_2(z)$ are independent Langevin process, so that [19]

$$\frac{\mathrm{d}\beta_i}{\mathrm{d}z} = -\rho\beta_i(z) + \sigma\eta_i(z)\,, \qquad (i = 1, 2) \tag{6}$$

where $\eta_1(z)$ and $\eta_2(z)$ are statistically independent white Gaussian noises such that

$$\langle \eta_i(z) \rangle = 0\,, \qquad \langle \eta_i(u)\eta_i(z+u) \rangle = \delta(z)\,, \qquad (i = 1, 2)$$

and where $\langle \cdot \rangle$ and $\delta(z)$ represent the mean value and the Dirac distribution, respectively. It may be shown that, with a suitable choice of the initial conditions [19], $\beta_1(z)$ and $\beta_2(z)$ are — as argued above — independent, wide-sense stationary, Gaussian processes. Moreover

$$\langle \beta_i(z) \rangle = 0\,, \qquad \langle \beta_i^2(z) \rangle = \sigma_\beta^2 = \frac{\sigma^2}{2\rho}\,, \qquad (i = 1, 2)$$

and the autocorrelation function (ACF) reads

$$r_\beta(z) = \langle \beta_i(u)\beta_i(z+u) \rangle = \sigma_\beta^2 \exp(-\rho\,|z|) \qquad (i = 1, 2)\,. \tag{7}$$

Another consequence of the RMM is that the linear birefringence strength, $|\boldsymbol{\beta}_{\mathrm{L}}(z)|$, is a Rayleigh distributed random variable [62].

The statistical properties of the model are controlled by the parameters ρ and σ. These can be related to other two parameters more physically meaningful:

1. the *beat length*, $L_B = 2\pi/\langle \beta_{\mathrm{L}}^2 \rangle^{1/2} = 2\pi\sqrt{\rho}/\sigma$, which is inversely proportional to the RMS linear birefringence strength;

2. the *correlation length*, $L_F = 1/\rho$, that is the distance over which the ACF of the birefringence decays to $1/e$ of its maximum value.

Note that the RMM assumes that $\beta_3 = 0$, i.e., that there is no circular birefringence. However, this assumption is not very restrictive. Actually, circular birefringence may be generated by an external magnetic field aligned with the axis of propagation, or by twisting the fiber. Telecommunication fibers are rather unlikely to be exposed to non-negligible magnetic fields; moreover, if this were the case, the fiber would not satisfy the reciprocity theorem, and P-OTDR theory would lose its validity [26]. Therefore, the only circular birefringence source of practical interest is the twist, and as long as a fiber is untwisted — which is the most common case — it is reasonable to assume $\beta_3 = 0$.

The Extended RMM for Twisted and Spun Fibers

As remarked above, the RMM represents only the linear component of the birefringence, while it neglects circular birefringence. Furthermore, it does not account for the rotation of the birefringence axes due either to the twist or to the spinning. Therefore, the RMM fails in describing both twisted fibers and spun fibers, yet it can be easily extended to include both cases.

Actually, let $\boldsymbol{\beta}_{\mathrm{L}} = (\beta_1,\, \beta_2,\, 0)^{\mathrm{T}}$ be the linear birefringence of an untwisted fiber. When the fiber is twisted, its local birefringence vector undergoes a rotation and circular birefringence is induced, hence the birefringence vector of the twisted fiber reads

$$\boldsymbol{\beta}(z) = \mathsf{T}(z)\boldsymbol{\beta}_{\mathrm{L}}(z) + \begin{pmatrix} 0 \\ 0 \\ g\tau'(z) \end{pmatrix} = \mathsf{T}(z) \begin{pmatrix} \beta_1(z) \\ \beta_2(z) \\ g\tau'(z) \end{pmatrix}, \tag{8}$$

where $\mathsf{T}(z)$ is the rotation matrix

$$\mathsf{T}(z) = \begin{pmatrix} \cos 2\tau(z) & -\sin 2\tau(z) & 0 \\ \sin 2\tau(z) & \cos 2\tau(z) & 0 \\ 0 & 0 & 1 \end{pmatrix},$$

$\tau(z)$ is the twist expressed in radians, and $\tau'(z)$ is its z-derivative, i.e., the twist-rate. The parameter g is a suitable physical constant representing the proportionality between the twist-rate and the induced circular birefringence; typically $g \simeq 0.14$ [3, 7, 63].

This extended model, and in particular Eq. (8), can be used to described also spun fibers, simply by setting $g = 0$. Actually, it is commonly accepted that the only local effect of the spin is to rotate the birefringence axes, without causing any stress-induced birefringence [14, 34, 47, 64].

Equations (8) and (6) define the extended RMM for a twisted or spun fiber. Nevertheless, it is convenient to make some more assumptions on the properties of $\tau(z)$. In particular, it is hard to figure out how external perturbations might induce a twist, randomly varying along the fiber. On the contrary, twist might be deterministically induced while fiber is either wind on or unwind from a bobbin. For example, circular birefringence measurement reported in [29] presents a quasi-periodicity that the authors relate to the unwinding process. Similarly, in the case of spun fibers $\tau'(z)$ is usually periodic or constant, but never random. On the ground of these considerations, it is then reasonable to assume $\tau(z)$ to be a deterministic function of z.

4. Birefringence Measurements

As explained in section 3, the round-trip birefringence vector $\boldsymbol{\beta}_{\mathrm{R}}(z)$ can be measured using techniques like the Jones matrix eigenanalysis (JME) or the Müller matrix method (MMM). When the link to be measured in not affected by PDL (as in the case considered here), the MMM results to be simpler to implement than the JME, since the latter requires a precise knowledge of the input SOP, whereas the former does not.

According to MMM, in order to measure $\boldsymbol{\beta}_B(z)$ one should first of all measure $\boldsymbol{s}_{\mathrm{R}}(z)$ for at least two different input SOPs (more than two input SOPs would reduce the measurement uncertainty at the expense of a longer measurement time). Then, the

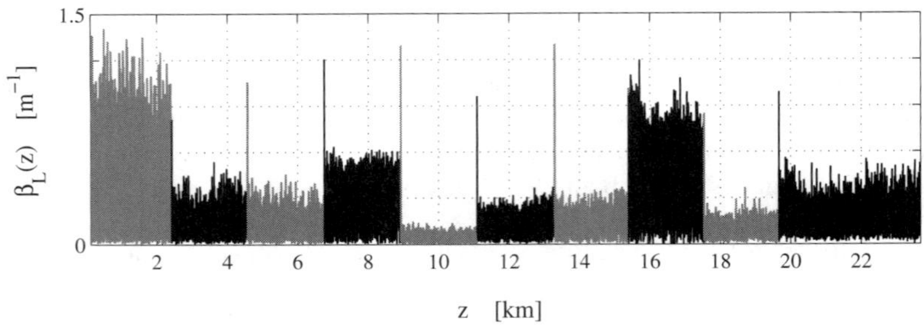

Fig. 2. Evolution of $\beta_{\mathrm{L}}(z)$ on a 24-km-long G.652 fiber, made of a concatenation of ten different fiber sections. Alternate colors are used to highlight different sections. Adapted from [57].

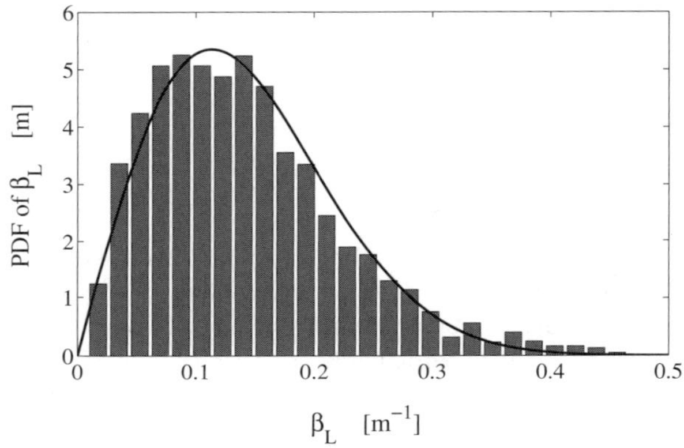

Fig. 3. Experimental histogram of the values of $\beta_{\mathrm{L}}(z)$ obtained from the second fiber section shown in Fig. 2. The solid curve is the best fitting Rayleigh PDF. Adapted from [57].

matrix $\mathsf{R}_{\mathrm{R}}(z)$ representing the round-trip can be determined [56], and hence $\mathsf{R}_{\Delta}(z) = \mathsf{R}_{\mathrm{R}}(z+\Delta z)\mathsf{R}_{\mathrm{R}}^{\mathrm{T}}(z)$ can be calculated (Δz being the sampling step). Finally, $\boldsymbol{\beta}_{\mathrm{R}}(z)$ can be found by means of the equations rearranged from [56]:

$$2\cos(\beta_{\mathrm{R}}\,\Delta z) = \mathrm{Tr}\,\mathsf{R}_{\Delta} - 1$$

$$2(\boldsymbol{\beta}_{B}\times)\frac{\sin(\beta_{\mathrm{R}}\,\Delta z)}{\beta_{\mathrm{R}}} = \mathsf{R}_{\Delta} - \mathsf{R}_{\Delta}^{\mathrm{T}},$$

where β_{R} is the modulus of $\boldsymbol{\beta}_{\mathrm{R}}$ and Tr is the trace.

An example of application of the measurement technique described so far is shown in Fig. 2, where the evolution of $\beta_{\mathrm{L}}(z)$, measured on a standard step index fiber (G.652), is reported. The fiber under test is one of the 80 contained in a 24-km-long cable, made of ten sections spliced together [32]. Within each section the evolution of $\beta_{\mathrm{L}}(z)$ is rather uniform, although random, whereas it has abrupt changes when passing from one section to the next one.

The randomness of the linear birefringence strength of each section can be characterized in terms of probability density function (PDF). For example, Fig. 3 shows the experimental histogram of the values of $\beta_{\mathrm{L}}(z)$, that are obtained from the second section of the fiber to which Fig. 2 refers. The solid curve is the best fitting Rayleigh PDF,

$$\mathrm{f}(\beta_\mathrm{L}) = \frac{2\beta_\mathrm{L}}{\langle\beta_\mathrm{L}^2\rangle} \exp\left(-\frac{\beta_\mathrm{L}^2}{\langle\beta_\mathrm{L}^2\rangle}\right), \tag{9}$$

where $\langle\beta_\mathrm{L}^2\rangle$ is the mean square value of β_L, equal to 0.027 m^{-2} in this case.

The agreement between the experimental result and the Rayleigh PDF shown in Fig. 3 is not at all a fortuitous coincidence, rather it is quite common to all measurements. This agreement does not depend on any theoretical assumption, therefore it has to be considered as an experimental evidence, a fact that should be provided for by any mathematical model for the birefringence. Actually, the agreement with the Rayleigh PDF is an experimental validation of the RMM.

4.1. *Beat Length*

Measurement of the beat length is, by definition, an immediate consequence of birefringence measurements; for instance, the beat length corresponding to the data of Fig. 3 is $L_B = 2\pi/\langle\beta_\mathrm{L}^2\rangle^{1/2} = 38.2$ m. The same kind of measurement has been repeated for each section of 30 G.652 and 28 G.653 fibers contained in the cable. Figure 4 shows the histogram of L_B measured on the G.652 fibers; similarly, Fig. 5 refers to G.653 fibers. Note that the birefringence in G.653 fibers has larger mean value but, at the same time, its standard deviation is smaller.

A simpler way to measure the beat length is based on a modified version of the *fixed polarizer* method, first proposed in [61] for the measurement of the DGD. Actually, owing to fiber birefringence, the SOP of the backscattered field evolves at random, therefore the power detected by the P-OTDR fluctuates. The spatial frequency of these fluctuations is related to the mean value of $\beta_\mathrm{R}(z)$, hence to the average of $\beta_\mathrm{L}(z)$ [see Eq. (5)]. In fact, let $T(z)$ be the normalized version of the detected power, then it has been proved that the mean linear birefringence can be expressed as [28, 30]:

$$\mathrm{E}[\beta_L] = \frac{\pi n(\upsilon)}{2\sqrt{\upsilon}},$$

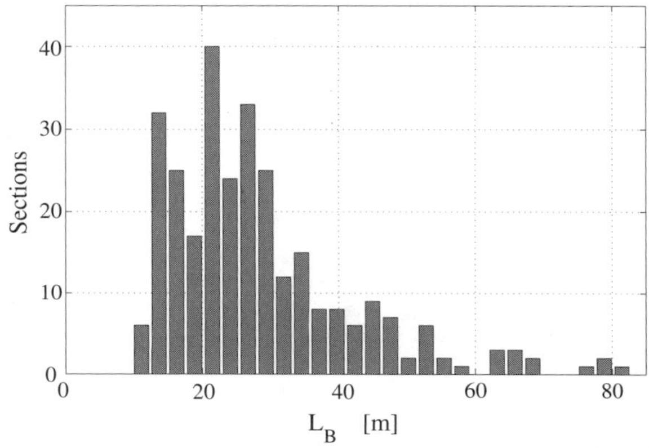

Fig. 4. Histogram of the value of L_B measured on 300 sections of G.652 fibers. The mean value of the ensemble is 31.0 m. Adapted from [57].

where v is a level between "0" and "1", and $n(v)$ is the mean number of times $T(z)$ crosses the level v per unit length. This method is also called the *level crossing rate* (LCR) analysis.

To implement the fixed polarizer method using a P-OTDR, one should measure $\boldsymbol{s}_{\mathrm{R}}(z)$ and then numerically calculate the signal $T(z)$, that is given by the formula $T(z) = [1 + \boldsymbol{s}_0 \cdot \boldsymbol{s}_{\mathrm{R}}(z)]/2$, where $\boldsymbol{s}_0$ is the input SOP. Apparently, this technique requires a complete and absolute control on the input SOP, which would drastically complicate the experiment. However, this problem can be overcome retrieving the input SOP from the ergodic spatial average of $\boldsymbol{s}_B(z)$: actually, it has been proved that $\langle \boldsymbol{s}_B(z) \rangle = \mathsf{M}\boldsymbol{s}_0/3$.

An example of application of this technique is reported in fig. 6, where it is shown the evolution of the beat length along a cascade of two G.653 fibers, spliced together. Each point of the curve $\{z_c,\ L_B(z_c)\}$ is obtained by applying the LCR analysis on a spatial "sliding window" 900 m long, centered at z_c. It is evident the difference in the values of L_B. The average values of the two subsections are represented with the dashed lines; the residual oscillations around the mean values are due both to local variation of the beat length, and to statistical error. Actually, this error can be reduced by increasing the window width at the expense, however, of a lower spatial resolution.

4.2. *Correlation Length*

Owing to the model for the birefringence introduced in section 3.1, the theoretical analysis of reflectometric measurements can be done within the framework of stochastic differential equations (SDEs) [65–67]. The theory of SDEs is a very powerful — and relatively recent — mathematical tool, which allows, among the other things, to calculate the evolution of mean values. A detailed treatment of this topic is referred to the quoted books, while a compendium may be found in [8] and [19], and in Appendix B.

Using the theory of SDE's, it can be proved that the ACF of the components of the round-trip birefringence vector reads (see Appendix C)

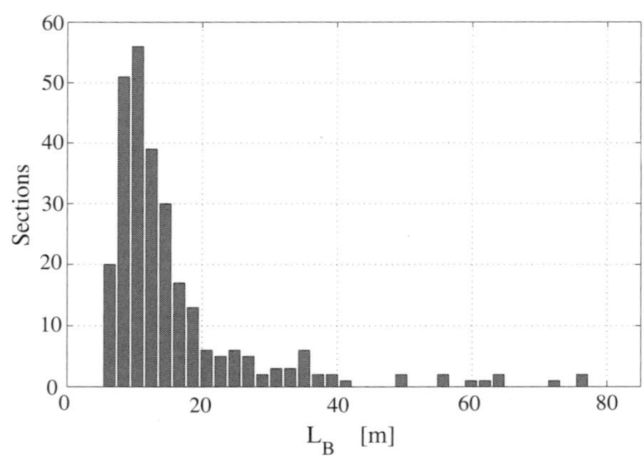

Fig. 5. Histogram of the value of L_B measured on 280 sections of G.653 fibers. The mean value of the ensemble is 16.6 m. Adapted from [57].

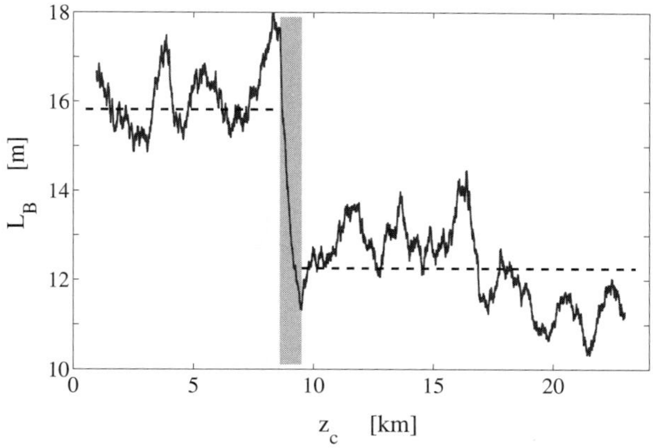

Fig. 6. Evolution of L_B along a cascade of two G.653 fibers. The shaded area indicates a window of 900 meters centered on the fiber splice. Adapted from [30].

$$r_{\beta_{\mathrm{R}}}(z) = \frac{16\pi^2}{3L_B^2} \exp\left(-\frac{|z|}{L_F}\right) . \tag{10}$$

Upon comparing Eq. (10) with Eq. (7), one sees that $\boldsymbol{\beta}_{\mathrm{R}}(z)$ has the same correlation length of $\boldsymbol{\beta}(z)$, so by experimentally estimating the ACF of the round-trip birefringence, the correlation length L_F can be measured.

Figure 7 shows the experimental ACF of the round-trip birefringence of a dispersion shifted (G.653) fiber, tightly wound on a standard shipping bobbin. The dashed curve is the best fitting bilateral exponential found from Eq. (10), corresponding to $L_B = 10.5$ m and $L_F = 12.1$ m. The fairly good agreement between experimental and theoretical results corresponds to a 10% of uncertainty.

As an example of the potential application of this kind of analysis, the ACF estimate has been repeated after the same fiber was unwound and loosely arranged on a circular

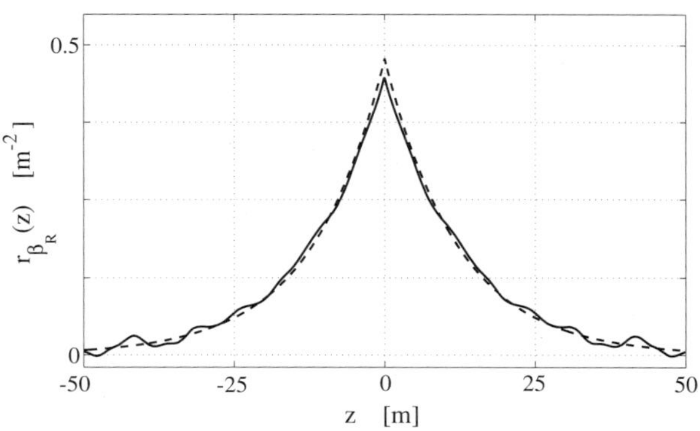

Fig. 7. Experimental ACF of round-trip birefringence (solid curve), compared with the theoretical best fit (dashed curve). The measurement refers to a 2.3-km-long G.653 fiber, tightly wound on a bobbin. Adapted from [57].

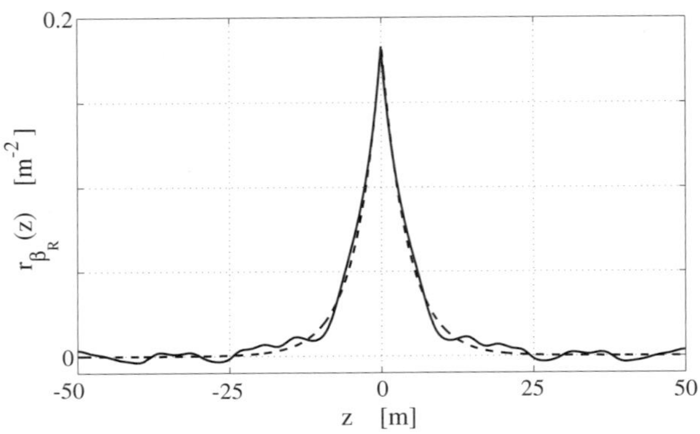

Fig. 8. Experimental ACF of round-trip birefringence (solid curve), compared with the theoretical best fit (dashed curve). The measurement refers to the same fiber of Fig. 7, loosely arranged on a 20-m-long circular path. Adapted from [57].

path, approximatively 20 m long. Figure 8 shows the experimental ACF (solid curve) with the corresponding best fit (dashed curve) obtained for $L_B = 16.9$ m and $L_F = 4.2$ m. Note that the beat length is increased (i.e., the birefringence is decreased), whereas the correlation length is decreased. Most likely this occurred because when the fiber was tightly wound on a bobbin, a rather deterministic and intense birefringence was superimposed to the intrinsic one, increasing the correlation length and shortening the beat length. Afterward, when the fiber was unwound, this extrinsic birefringence was removed.

Besides providing a way to measure the birefringence correlation length, Eq. (10) allows also to further validate the RMM. Actually, it is important to remark that Eq. (10) is strictly related to the assumption $\beta_3 = 0$ made by the RMM, and it would not hold if that assumption was not made or if the fiber was either twisted or spun. Accordingly, the most interesting outcome of the agreement shown in Figs. 7 and 8 is that the measured fiber was neither twisted nor spun.

4.3. Twisted Fibers

When the fiber is twisted, the birefringence has to be described with the extended RMM introduced in section 3.1 and, as a consequence, the ACF of the round-trip birefringence is no longer a bilateral exponential, rather it looks like in Fig. 9.

One of the most important theoretical aspects of reflectometric measurements in twisted fibers is that, in this case, the SOP evolution of the backscattered field is identical to the SOP evolution of a field backscattered by a fiber with the "equivalent birefringence vector" [36]:

$$\boldsymbol{\beta}_{\mathsf{eq}}(z) = \begin{pmatrix} \beta_1(z) \\ \beta_2(z) \\ (g-2)\tau'(z) \end{pmatrix} . \tag{11}$$

In other words, any reflectometric measurement technique sees the geometrical rotation of the fiber as an apparent circular birefringence of amplitude $-2\tau'(z)$, that adds to

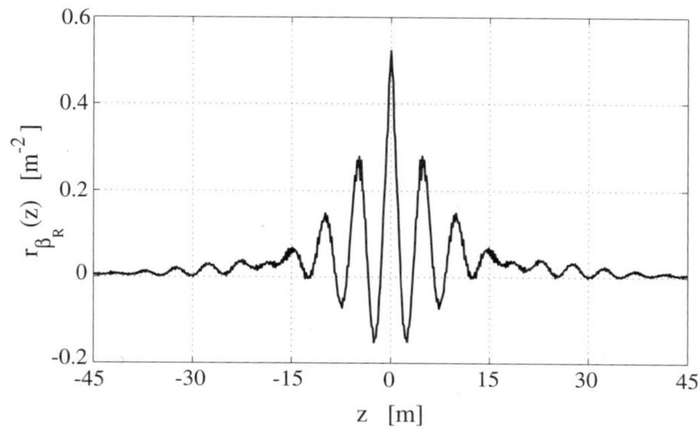

Fig. 9. Experimental ACF of round-trip birefringence, measured on a 5.8-km-long G.652 fiber, tightly wound on a bobbin. Clearly, it does not agree with the class of functions defined by Eq. (10). Adapted from [36].

the stress induced circular birefringence $g\tau'(z)$. Moreover, since $g \simeq 0.14$ is quite small compared to 2, the measurement is much more sensible to the apparent circular birefringence, rather than to the stress induced one. Note that, as reported in [40], the same phenomenon occurs also when the measurements are performed with a P-OFDR.

Using the extended RMM and the theory of SDE's, the ACF of the round-trip birefringence of a twisted fiber can be calculated. Unfortunately, the theoretical analysis is extremely involved and does not lead to an explicit expression for the ACF. Rather, it provides an infinite-dimensional system of linear differential equations (reported in Appendix D for completeness), which implicitly defines the ACF. However, this system can be numerically solved and compared with the experimental results. In particular, the case of constantly twisted fibers has been analyzed; in this case $\tau'(z) = 2\pi/p$, where p is the twist period. This choice simplifies the analysis and, most important, yields results in good agreement with the experimental data obtained so far. The complete treatment may be found in [36], while hereafter the attention is focused only on the experimental results.

Rather than the ACF $r_{\beta_\mathrm{R}}(z)$, it is profitable to consider the power spectral density (PSD) $R_{\beta_\mathrm{R}}(f)$, which is the Fourier transform of $r_{\beta_\mathrm{R}}(z)$. The PSD corresponding to the ACF reported in Fig. 9 is shown in Fig. 10 with the solid curve. The experimental PSD has been fitted with the theoretical model (dashed curve). In this way it was found that the birefringence parameters were $L_B = 14.5$ m and $L_F = 11.5$ m, and that the fiber was twisted with a period $p = 11.7$ m. Other measurements provided twist periods ranging from 10 to 80 m [36]. In all cases, there was a fairly good agreement, corroborating the reliability of the extended RMM in describing also twisted fibers. In analogy to what was observed in [29], it may be argued that the twist was induced on the fibers by the winding process. Note also that twist periods like the observed ones are too long to have any influence on fiber PMD. However, this kind of measurement is one more example of how P-OTDR may be used to inspect the fiber production process.

As Fig. 10 exemplifies, in general $R_{\beta_\mathrm{R}}(f)$ is characterized by a central peak and two symmetrical side lobes, whose position is related to the twist-rate $1/p$. As this

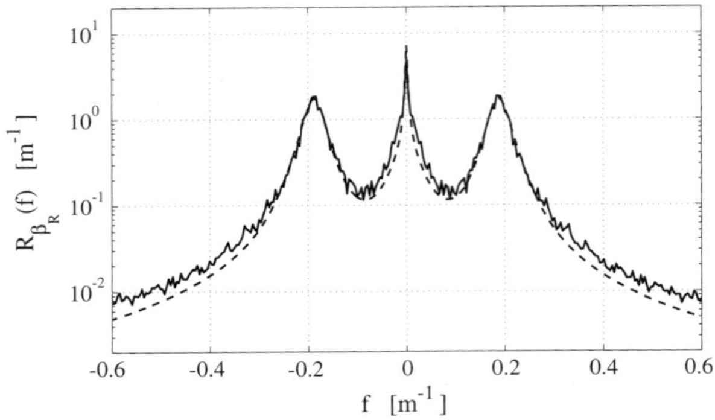

Fig. 10. Experimental PSD (solid curve) of round-trip birefringence, compared with the theoretical best fit (dashed curve). The measurement refers to the same fiber of Fig. 9.

parameter increases, the lobes move toward higher values of f. However, the position of the side lobes depends also on L_F and L_B in a non trivial way. Furthermore, $R_{\beta_\mathrm{R}}(f)$ is independent of the sign of p, hence the PSD analysis cannot distinguish between a right- or a left-handed twist.

Other circular birefringence measurement technique have been reported in literature [27, 29]; however, all of these techniques apply only to deterministic or quasi-deterministic fibers (i.e., to fibers where L_F is much longer than L_B and p).

4.4. Spun Fibers

It can be easily shown that the whole theory developed to measure twisted fibers, can be adapted to spun fibers as well, by simply setting $g = 0$. Actually, the spin consists in a torsion applied to the fiber when the silica is still at a high temperature, close to its melting point; thus, no torsional stress is induced. Accordingly, spin profiles may be measured in the same way as twist functions. Nevertheless, there is a limit to the measure. In fact, after performing a measurement like the one shown in Fig. 10, one is not able to determine whether the parameter g should be set equal to 0 (i.e. the fiber is spun) or to 0.14 (i.e., the fiber is twisted). This results in an uncertainty of roughly 7% in the measure of $\tau(z)$, which, for the moment, seems to be an intrinsic limit of reflectometric techniques.

Another issue is represented by the z-dependence of spin profiles. In the overwhelming majority of times, the spin profiles are not constant, rather they are variable and often periodic, because in this way higher DGD reductions can be achieved [15,16]. The theoretical analysis shows that when the spin profile is not constant, the round-trip birefringence vector is no longer a stationary process; in particular, when the spin profile is periodic, $\boldsymbol{\beta}_\mathrm{R}(z)$ results to be a cyclostationary process. Therefore, in the case of variable spin-rate the statistical analysis of the data becomes more complicated.

Besides these theoretical problems, there is also a serious practical issue related to the spatial resolution of the measurements. Actually, typical spin-rates are about several turns per meter, meaning that the birefringence vector undergoes a complete rotation in a distance of few centimeters. Therefore, in order to sample correctly the SOP as a function of distance, a spatial resolution of few centimeters or higher is needed. For the moment, such a resolution is in the range only of the P-OFDR that, however, has a limited dynamical range.

A different approach to P-OTDR based measurement of spin parameters has been proposed in [34], in the case of sinusoidal spin-profile. The technique consists in twisting the spun fiber at different constant rates, and for each of these twist-rates a frequency analysis of the backscattered SOP is performed. Then, from the collected data, and using a precalculated frequency map, both the spin amplitude and period can be found. However, twist can be applied only to fiber a few meters in length, thus limiting the sample length. Moreover, only slow spin-rates can be measured, unless the P-OTDR has a rather high spatial resolution.

One more technique, recently proposed in [68], enables to measure the spin period of periodically spun fiber. The technique does not require an high spatial resolution and, actually, it exploits the limited bandwidth of the P-OTDR receiver.

In summary, the characterization of spin profiles in randomly birefringent fibers by means of reflectometric measurements is still an open problem. Yet, it probably presently represents the most interesting application of reflectometric techniques.

5. DGD Measurements

As anticipated in the introduction, reflectometric techniques can be used also to measure the evolution of the DGD along a fiber optic link. Actually, the field backscattered from a point z is a function of the optical frequency too, thus $\boldsymbol{s}_{\mathrm{R}}(z,\omega) = \mathsf{R}_{\mathrm{R}}(z,\omega)\boldsymbol{s}_0$, where $\mathsf{R}_{\mathrm{R}}(z,\omega)$ is the round-trip matrix defined in section 4. From this expression one gets [26]

$$\frac{d\boldsymbol{s}_{\mathrm{R}}}{d\omega} = \frac{d\mathsf{R}_{\mathrm{R}}}{d\omega}\mathsf{R}_{\mathrm{R}}^{\mathrm{T}}\,\boldsymbol{s}_{\mathrm{R}} = \boldsymbol{\Omega}_{\mathrm{R}} \times \boldsymbol{s}_{\mathrm{R}}\,, \tag{12}$$

where $\boldsymbol{\Omega}_{\mathrm{R}}(z,\omega)$ is the *round-trip PMD* vector. Similarly to Eq. (4), the relationship between $\boldsymbol{\Omega}_{\mathrm{R}}(\omega)$ and the PMD vector $\boldsymbol{\Omega}(\omega)$ reads (see Appendix A; the dependence on z is omitted for brevity)

$$\boldsymbol{\Omega}_{\mathrm{R}}(\omega) = 2\mathsf{M}\mathsf{R}^{\mathrm{T}}(\omega)\boldsymbol{\Omega}_{\mathrm{L}}(\omega) = 2\overleftarrow{\mathsf{R}}(\omega)\boldsymbol{\Omega}_{\mathrm{L}}(\omega)\,, \tag{13}$$

where $\boldsymbol{\Omega}_{\mathrm{L}} = (\Omega_1,\ \Omega_2,\ 0)^{\mathrm{T}}$ is the linear component of $\boldsymbol{\Omega}$. Therefore, the round-trip DGD is given by

$$\Delta\tau_{\mathrm{R}}(\omega) = |\boldsymbol{\Omega}_{\mathrm{R}}(\omega)| = 2\,|\boldsymbol{\Omega}_{\mathrm{L}}(\omega)| = 2\sqrt{\Omega_1^2 + \Omega_2^2}\,. \tag{14}$$

Also in this case information about the third component of $\boldsymbol{\Omega}$ is lost, but differently from β_3, Ω_3 is quite often different from zero, especially in the case of long optical links. This impasse can be overcome resorting to the statistical properties of the PMD. Nevertheless, note that such statistical properties hold only when the fiber link is long enough, so to be in the long-length regime [5]. Therefore, in the remainder of this section it is assumed that the link is in the long-length regime.

It is well known that the components of the PMD vector of long fiber links are three statistically independent Gaussian random variables, with zero mean and same standard deviation. Consequently, the DGD, $\Delta\tau$, is a Maxwellian distributed random variable, whereas, owing to Eq. (14), the round-trip DGD, $\Delta\tau_{\mathrm{R}}$, results to be a Rayleigh

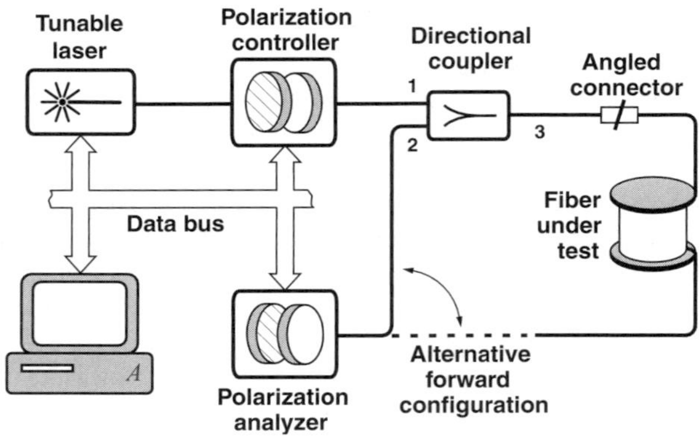

Fig. 11. Schematic of a CW-source experimental set-up. Adapted from [50].

distributed random variable [26, 50]. Furthermore, the following relationship is found between their mean values [26]:

$$\langle \Delta\tau_{\mathrm{R}} \rangle = \frac{\pi}{2} \langle \Delta\tau \rangle \,. \tag{15}$$

So, once the mean round-trip DGD is measured, the mean forward DGD can be calculated.

By means of a P-OTDR like the one shown in Fig. 1, the evolution of the round-trip SOP $\boldsymbol{s}_{\mathrm{R}}(z, \omega)$ can be measured for different frequencies ω; then, using standard techniques [55, 56], the z-evolution of the round-trip PMD vector and of the round-trip DGD can be calculated. Finally, repeating the measurement for different wavelengths and/or different environmental conditions, a proper statistical ensemble is collected, so that $\langle \Delta\tau(z) \rangle$ may be calculated from Eq. (15).

However, the measurement of $\langle \Delta\tau(z) \rangle$ with the half-a-meter resolution provided by the P-OTDR may be superfluous in many cases. More often it may be enough to sample $\langle \Delta\tau(z) \rangle$ only at the end of the link, so to measure the overall mean DGD. In this case, reflectometric measurements may be based also on a CW source [28, 51]. Actually, when a CW signal is injected into a fiber, the total backscattered power is mainly due to the Rayleigh scattering and to the Fresnel reflection of the fiber far-end face. As the fiber length increases, the Rayleigh contribution increases towards a saturation value, whereas the Fresnel contribution tends to be negligible. However, it may be shown that for a standard telecommunication fiber, as long as the fiber length is less that 40 km, the power due to the Fresnel reflection is at least 5 dB higher than the power due to the Rayleigh scattering; this difference may be further increased by placing a reflector at the fiber far-end face [50]. If these conditions are verified, then it may be assumed that only the Fresnel reflection contributes to the total backscattered power.

According to these argumentations, the PMD parameters may be measured with an experimental set-up as described in Fig. 11. The optical source is a polarization controlled, CW external cavity tunable laser, while the receiver is a polarization analyzer which is connected to port 2 of the coupler (for round-trip measurements). The fiber under test is connected to the directional coupler through a low reflection angled connector (FC-APC, return loss $>$ 60 dB), so to minimize undesired reflections. In

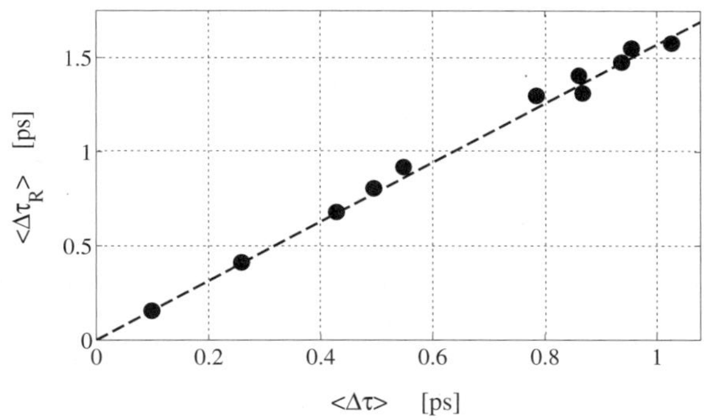

Fig. 12. Measured $\langle \Delta\tau_{\mathrm{R}} \rangle$ versus $\langle \Delta\tau \rangle$ (black dots); continuous line refers to (15). Averaging was performed over a band of 130 nm centered at 1555 nm. Adapted from [50].

order to compare the results of round-trip measurements with those of standard forward measurements, in this set-up was also possible to connect the receiver to the fiber far-end.

By means of the above set-up, a set of measurements was performed on G.652 fibers, spliced together, to obtain cascades up to 40 km long; $\Delta\tau$ and $\Delta\tau_{\mathrm{R}}$ were measured by means of JME method. Results are shown in Fig. 12, where values of $\langle \Delta\tau_{\mathrm{R}} \rangle$ are plotted versus $\langle \Delta\tau \rangle$ (black dots). Experimental data fit very well the continuous line that represents the theoretical prediction given by Eq. (15).

The technique suffers two main limitations: first, the fiber under test should not be longer than 40 km and, second, there should not be any connector or any other source of reflection between the fiber input and the fiber output. However, these limitations may be avoided improving the experimental set-up, using a pulsed source and gated receiver. This can be done simply by placing an optical switch after the CW source and another optical switch before the polarimeter. Switches should be synchronized so that the polarimeter detects only the power backscattered by the fiber far-end face, whereas all the other sources of reflection are obscured. Note that the width of the pulse generated by the transmission switch may be in the order of few microseconds — hence much longer than standard P-OTDR pulses — because the power due to the Fresnel reflection is much higher than the power due to the Rayleigh scattering occurring in few hundreds of meters of fiber (i.e., a section of fiber corresponding to the length of the pulse). Note also that, if the repetition rate of the pulse is much higher than the acquisition rate of the polarimeter, the fast light chopping due to the switches does not affect the measurement performed by the CW polarimeter.

More details about this technique may be found in the paper by Sunnerud et al. [53], where the technique was proposed and used to perform distributed measurement of the mean DGD.

Another kind of PMD measurement based on P-OTDR is proposed by Huttner et al. [52]. In that case, rather than measuring the exact evolution of the SOP, so to calculate the DGD, the authors measure the degree of polarization (DOP) of the field backscattered by a rather long probe pulse (100 ns corresponding to a length of 20 m). As a result, it is possible to establish a qualitative relationship between the DOP and the DGD of the fiber. Actually, fiber sections with lower DOP are expected to have higher DGD. This instrument provides a qualitative characterization of the fiber link, yet the use of long pulses greatly simplifies the set-up.

5.1. Polarized Spectra Analysis

The mean DGD of a link can be measured also by adapting to the round-trip either the wavelength scanning technique [69] or the pseudo-interferometric one [70]. Actually, the mean DGD can be retrieved by analyzing the fluctuation, as a function of frequency, of the power spectrum, $T_{\mathrm{R}}(\omega)$, detected after the back-reflected field is passed through a linear polarizer, aligned with the input SOP. The advantage of this kind of measurement, is that it can be based on a simple set-up like the one shown in Fig. 13 [51]. There are several different analysis that can be used to calculate the mean DGD, $\langle \Delta\tau \rangle$, from the frequency evolution of $T_{\mathrm{R}}(\omega)$.

The first one, the level crossing rate (LCR) analysis, is based on the count of the mean number of times, $n(\upsilon)$, that $T_{\mathrm{R}}(\omega)$ crosses a given level υ in a unit frequency band. Actually, it produces the result [51]:

$$\langle \Delta\tau \rangle = \frac{2n(\upsilon)}{\sqrt{\upsilon}} \,. \tag{16}$$

It is also possible to relate $\langle \Delta\tau \rangle$ to the mean number of extrema, n_{e}, that $T_{\mathrm{R}}(\omega)$ has over an unit frequency band. However, since in this case an explicit expression cannot be found, it is necessary to resort to numerical analyses, which yield approximatively [51]:

$$\langle \Delta\tau \rangle \simeq 1.55\, n_{\mathrm{e}} \,. \tag{17}$$

Note that in [48] it is numerically found that $\langle \Delta\tau \rangle \simeq 1.83\, n_{\mathrm{e}}$, thus confirming the approximate nature of the extrema counting technique.

Finally, a third way of performing the frequency analysis of the signal $T_{\mathrm{R}}(\omega)$ consists in evaluating the variance, σ_T^2, of its power spectral density (PSD). Actually, it has been shown that the relationship between the mean DGD and σ_T^2 reads [51]:

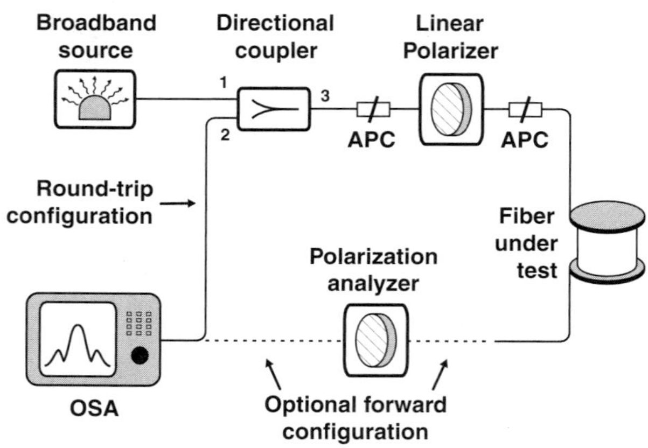

Fig. 13. Schematic of the CW-source experimental set-up for DGD measurement based on polarized spectra analysis. Adapted from [51].

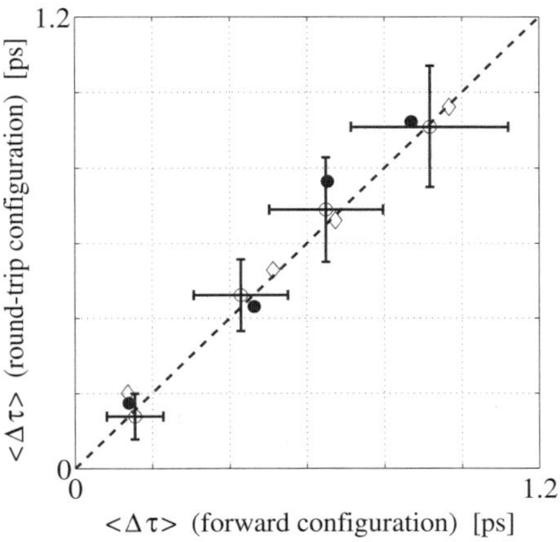

Fig. 14. Values of $\langle \Delta\tau \rangle$ measured in the round-trip configuration vs. $\langle \Delta\tau \rangle$ measured in the forward configuration. Filled dots, empty circles and empty diamonds refer to the PSD analysis, to the LCR analysis, and to the extrema counting, respectively. Error bars refer to the LCR analysis. Measurements were performed over a set of single-mode fibers with length ranging between 3 km and 20 km. Adapted from [51].

$$\langle \Delta\tau \rangle = 4\sqrt{\frac{\pi}{3}}\,\sigma_T. \tag{18}$$

These techniques have been experimentally verified by means of the set-up shown in Fig. 13; experimental data are shown in Fig. 14. The horizontal axis reports the mean DGD measured in the forward configuration, while the vertical axis reports the corresponding value measured in the round-trip configuration. Circles refer to the LCR analysis [i.e., Eq. (16)], diamonds to the extrema counting [Eq. (17)], and filled dots to the PSD analysis [Eq. (18)]. Horizontal and vertical bars represent the uncertainty of the LCR analysis, which may be considered the most reliable.

Also the set-up shown in Fig. 13, can be modified as the one shown in Fig. 11, inserting two switches to perform measurement only on a given and limited portion of the link [48].

6. Fiber Links with PDL

This last section is about reflectometric measurements applied to fiber optic link affected also by polarization dependent loss (PDL). It is shown that the grounding theory is a clean generalization of the theory expounded in previous sections about link not affected by PDL. Nevertheless, when there is distributed PDL in the link a complete local characterization does not seem possible.

When the link is affected by PDL, it is convenient to describe the evolution of the field polarization in term of a bi-dimensional complex vector $\boldsymbol{E}(z,\omega)$, representing the electric field. Let $\boldsymbol{E}_0$ be the input electric field, then the forward propagating field reads $\boldsymbol{E}(z) = \mathsf{F}(z)\boldsymbol{E}_0$, where $\mathsf{F}(z)$ is the Jones matrix of the link. Differentiating this expression respect to z, and neglecting the scalar effects (such as attenuation and chromatic dispersion), the evolution of $\boldsymbol{E}(z,\omega)$ can be described by (the ω-dependence is omitted for brevity)

$$\frac{\mathrm{d}\boldsymbol{E}}{\mathrm{d}z} = \frac{\mathrm{dF}}{\mathrm{d}z}\mathsf{F}^{-1}(z)\boldsymbol{E}(z) = -j\mathsf{H}(z)\boldsymbol{E}(z) = \\ = -j\begin{pmatrix} h_1 & h_2 - jh_3 \\ h_2 + jh_3 & -h_1 \end{pmatrix}\boldsymbol{E}(z)\,.$$

The three complex numbers $h_i(z)$, $i = 1, 2, 3$, can be rearranged to form a 3-dimensional complex vector $\boldsymbol{h} = (h_1,\, h_2,\, h_3)^{\mathrm{T}}$, known as the *propagation vector*. This propagation vector completely describes the local properties of the link, and actually $\boldsymbol{h}(z,\omega) = (\boldsymbol{\beta} + j\boldsymbol{\alpha})/2$, where $\boldsymbol{\beta} = \boldsymbol{\beta}(z,\omega)$ is the local birefringence vector and $\boldsymbol{\alpha} = \boldsymbol{\alpha}(z,\omega)$ is the local dichroism vector, describing the local PDL of the link [71,72].

Similarly, the round-trip electric field is given by $\boldsymbol{E}_{\mathrm{R}}(z) = \mathsf{F}_{\mathrm{R}}(z)\boldsymbol{E}_0$, where $\mathsf{F}_{\mathrm{R}}(z)$ is the Jones matrix representing the round-trip propagation. Owing to reciprocity, $\mathsf{F}_{\mathrm{R}}(z) = \overleftarrow{\mathsf{F}}(z)\overrightarrow{\mathsf{F}}(z) = \mathsf{F}(z)\mathsf{F}^{\mathrm{T}}(z)$; therefore, the evolution of the round-trip field can be described also by means of the *round-trip propagation vector*, $\boldsymbol{h}_{\mathrm{R}}(z)$, which results to be related to $\boldsymbol{h}(z)$ by [37]:

$$\boldsymbol{h}_{\mathrm{R}}(z) = 2\mathsf{M}\mathsf{C}^{\mathrm{T}}(z)\boldsymbol{h}_{\mathrm{L}}(z) = 2\overleftarrow{\mathsf{C}}(z)\boldsymbol{h}_{\mathrm{L}}(z)\,, \tag{19}$$

where $\boldsymbol{h}_{\mathrm{L}}$ is the linear component of $\boldsymbol{h}$, while C is a 3×3 complex matrix. It may be shown that $\mathsf{C}(z)$ is univocally related to $\mathsf{F}(z)$, and that $\overleftarrow{\mathsf{C}}(z) = \mathsf{M}\mathsf{C}^{\mathrm{T}}(z)\mathsf{M}$ is univocally related to $\overleftarrow{\mathsf{F}}(z) = \mathsf{F}^{\mathrm{T}}(z)$. Moreover, $\mathsf{C}(z)$ obeys the following differential equation

$$\frac{\mathrm{dC}}{\mathrm{d}z} = \boldsymbol{h}(z) \times \mathsf{C}(z)\,, \tag{20}$$

and it satisfies the equality $\mathsf{C}^{-1}(z) = \mathsf{C}^{\mathrm{T}}(z)$. All these properties suggest that the matrix $\mathsf{C}(z)$ may be considered as a sort of generalization of the Müller matrix $\mathsf{R}(z)$ to the case of PDL-affected link. Upon comparing Eq. (20) with Eq. (1), and Eq. (19) with Eq. (4), it can be appreciated how simply the theory of reflectometric measurements generalizes to dichroic links.

Likewise $\boldsymbol{\beta}_{\mathrm{R}}(z)$, also $\boldsymbol{h}_{\mathrm{R}}(z)$ does not depend explicitly on the circular component of $\boldsymbol{h}(z)$; information on the circular component might be derived by analyzing the z-evolution of $\boldsymbol{h}_{\mathrm{R}}(z)$, but this task appears to be burdensome and has not been confronted. On the contrary, properties of the linear component of $\boldsymbol{h}(z)$ may be obtained by exploiting the equality $\mathsf{C}^{\mathrm{T}} = \mathsf{C}^{-1}$. Indeed, it is straightforward to show that $\boldsymbol{h}_{\mathrm{R}}(z) \cdot \boldsymbol{h}_{\mathrm{R}}(z) = 4\boldsymbol{h}_{\mathrm{L}}(z) \cdot \boldsymbol{h}_{\mathrm{L}}(z)$, and hence

$$\begin{aligned} \mathrm{Re}[\boldsymbol{h}_{\mathrm{R}} \cdot \boldsymbol{h}_{\mathrm{R}}] &= 4(\beta_{\mathrm{L}}^2 - \alpha_{\mathrm{L}}^2) \\ \mathrm{Im}[\boldsymbol{h}_{\mathrm{R}} \cdot \boldsymbol{h}_{\mathrm{R}}] &= 8\boldsymbol{\beta}_{\mathrm{L}} \cdot \boldsymbol{\alpha}_{\mathrm{L}} = 8\beta_{\mathrm{L}}\alpha_{\mathrm{L}}\cos\theta, \end{aligned} \tag{21}$$

where $\boldsymbol{\beta}_{\mathrm{L}}$ is the linear component of the birefringence vector, $\boldsymbol{\alpha}_{\mathrm{L}}$ is the linear component of the dichroism vector, β_{L} and α_{L} are their respective moduli, and θ is the angle that these two vectors subtend. Unfortunately, the system (21) has three unknowns and only two equations, therefore a unique solution cannot be found. Of course, a unique solution may be found if profitable assumptions about either β_{L}, α_{L} or θ are made.

For example, in the case $\boldsymbol{\alpha}(z)$ and $\boldsymbol{\beta}(z)$ are aligned, $\theta = 0$ and (21) has the unique solution

$$\beta_{\mathrm{L}} = \frac{1}{2\sqrt{2}}\sqrt{|\boldsymbol{h}_{\mathrm{R}} \cdot \boldsymbol{h}_{\mathrm{R}}| + \mathsf{Re}[\boldsymbol{h}_{\mathrm{R}} \cdot \boldsymbol{h}_{\mathrm{R}}]}\,,$$
$$\alpha_{\mathrm{L}} = \frac{1}{2\sqrt{2}}\sqrt{|\boldsymbol{h}_{\mathrm{R}} \cdot \boldsymbol{h}_{\mathrm{R}}| - \mathsf{Re}[\boldsymbol{h}_{\mathrm{R}} \cdot \boldsymbol{h}_{\mathrm{R}}]}\,.$$

Note that, typically, the dichroism affecting a standard telecommunication link is not distributed along the fiber, but is located in some optical devices. Therefore the above theory does not apply to the fibers, rather it may be used to characterize devices, although this requires a very high spatial resolution. However, the theory has one more important application in the following scenario. Suppose that the birefringence properties of an installed fiber section not affected by dichroism has to be analyzed. Suppose also that such section cannot be accessed directly, and that in the fiber link between the access point (where the P-OTDR is connected) and the section under test there is a PDL source. In this case, owing to the PDL, the theory outlined in section 4 does not work, even if the section that should be measured is non dichroic. On the contrary, Eq. (21) can be successfully used because it accounts for PDL. Moreover, since in the fiber section under test α_{L} is null, Eq. (21) yields $\beta_{\mathrm{L}}^2 = \mathsf{Re}[\boldsymbol{h}_{\mathrm{R}} \cdot \boldsymbol{h}_{\mathrm{R}}]/4$ (note, however, that $\boldsymbol{h}_{\mathrm{R}}$ is still a complex vector). Hence, birefringence measurement is still possible.

The whole analysis sketched so far, can be used also to describe the frequency evolution of a field backscattered by a PDL-affected link. In this case, the polarization dependent distortions that act on a forward propagating pulse are represented by the complex vector $\boldsymbol{q}(z,\omega) = (\boldsymbol{\Omega} + j\boldsymbol{\Lambda})/2$ [73], which describes a phase and amplitude distortion of the pulse spectrum. Similarly, the frequency evolution of the round-trip field is governed by the complex vector $\boldsymbol{q}_{\mathrm{R}}$, that is related to $\boldsymbol{q}$ by the expression

$$\boldsymbol{q}_{\mathrm{R}}(\omega) = 2\mathsf{M}\mathsf{C}^{\mathrm{T}}(\omega)\boldsymbol{q}_{\mathrm{L}}(\omega) = 2\overleftarrow{\mathsf{C}}(\omega)\boldsymbol{q}_{\mathrm{L}}(\omega)\,, \tag{22}$$

where symbols have the same meaning as in Eq. (19). Again, upon evaluating the product $\boldsymbol{q}_{\mathrm{R}} \cdot \boldsymbol{q}_{\mathrm{R}}$ it produces the result

$$\begin{aligned} \mathsf{Re}[\boldsymbol{q}_{\mathrm{R}} \cdot \boldsymbol{q}_{\mathrm{R}}] &= 4(\Omega_{\mathrm{L}}^2 - \Lambda_{\mathrm{L}}^2) \\ \mathsf{Im}[\boldsymbol{q}_{\mathrm{R}} \cdot \boldsymbol{q}_{\mathrm{R}}] &= 8\boldsymbol{\Omega}_{\mathrm{L}} \cdot \boldsymbol{\Lambda}_{\mathrm{L}} = 8\Omega_{\mathrm{L}}\Lambda_{\mathrm{L}}\cos\psi \end{aligned} \tag{23}$$

where ψ is the angle subtended by $\boldsymbol{\Omega}_{\mathrm{L}}$ and $\boldsymbol{\Lambda}_{\mathrm{L}}$. Comments made about Eq. (21) apply also to Eq. (23).

6.1. *PDL Measurements*

Usually, the PDL is defined as the ratio of the maximum and minimum output power transmitted through the link while changing the input SOP. Let introduce the *PDL vector* $\boldsymbol{\Gamma}$ [74], that is parallel or anti-parallel to the SOP experiencing the minimum or maximum loss, respectively. Then, according to the definition given above, the PDL ρ expressed in decibels reads

$$\rho = 10\log_{10}\frac{1+\Gamma}{1-\Gamma} = \frac{\gamma}{2}\log\frac{1+\Gamma}{1-\Gamma}\,, \tag{24}$$

where $\gamma = 20/\log 10 \simeq 8.7$, and Γ is the modulus of the PDL vector.

The PDL vector — and hence the PDL — evolves along the link depending on both the local dichroism and on the local birefringence [72]. Owing to the randomness of the birefringence, the vector $\boldsymbol{\Gamma}$ evolves randomly, even if the local dichroism might be deterministic. Consequently, it is possible to prove that ρ is very well described by a Maxwellian random variable, as long as the mean PDL, $\langle\rho\rangle$, is less than 25 dB [75,76].

In analogy with the forward propagation, it is possible to define the *round-trip PDL*, ρ_{R}, as the ratio of the maximum and minimum output power backscattered by the link while changing the input SOP. Similarly, the round-trip PDL vector $\boldsymbol{\Gamma}_{\mathrm{R}}$ can be introduced, so that the round-trip PDL reads

$$\rho_{\mathrm{R}} = \frac{\gamma}{2} \log \frac{1+\Gamma_{\mathrm{R}}}{1-\Gamma_{\mathrm{R}}} .$$

It is also possible to find a relationship between the round-trip PDL and the forward PDL [37]. Actually, the modulus of $\boldsymbol{\Gamma}_{\mathrm{R}}$ can be expressed as

$$\Gamma_{\mathrm{R}} = \frac{2\Gamma_{\mathrm{L}}\sqrt{1-\Gamma_{\mathrm{C}}^2}}{1+\Gamma_{\mathrm{L}}^2-\Gamma_{\mathrm{C}}^2} ,$$

where Γ_{L} and Γ_{C} are the moduli of the linear and circular components of $\boldsymbol{\Gamma}$, respectively. The expression of the round-trip PDL can be rearranged to become

$$\rho_{\mathrm{R}} = \gamma \log \frac{\sqrt{1-\Gamma_{\mathrm{C}}^2}+\Gamma_{\mathrm{L}}}{\sqrt{1-\Gamma_{\mathrm{C}}^2}-\Gamma_{\mathrm{L}}} . \tag{25}$$

Owing to the randomness of the birefringence and to the consequent randomness of $\boldsymbol{\Gamma}$, it is reasonable to assume that $\boldsymbol{\Gamma}$ points uniformly in all directions, and that the modulus and the direction of $\boldsymbol{\Gamma}(z)$ are statistically independent [75]. Furthermore, the analysis can be profitably limited to the case of low PDL, which is the most common in practice, so that Eqs. (25) and (24) yield $\rho_{\mathrm{R}} \simeq 2\rho(\Gamma_{\mathrm{L}}/\Gamma)$. Finally, owing to the statistical assumptions made above, the mean round-trip PDL reads as [37]

$$\langle\rho_{\mathrm{R}}\rangle = \frac{\pi}{2}\langle\rho\rangle . \tag{26}$$

This relationship states that upon measuring the mean round-trip PDL of an optical link, its mean forward PDL can be calculate too. By means of numerical simulations it has been verified that $2\langle\rho_{\mathrm{R}}\rangle/\pi$ differs less than 6% from $\langle\rho\rangle$, even when $\langle\rho\rangle$ is as large as 10 dB [37].

According to these results, the PDL of a link may be measured using a reflectometric technique in the same way as the DGD. Either a pulsed source or a CW source can be used. For instance, the CW set-up shown in Fig. 11 is able to perform PDL measurements, analyzing the field reflected by the fiber far-end face.

7. Conclusions

In this contribution the theory and applications of the P-OTDR have been reviewed. Presently, the most important application of the P-OTDR is the characterization of fiber birefringence, which allows to define mathematical models corresponding, as much as possible, to the reality of the fibers. The definition of such models is not a mere academic issue, but has very important consequences in several practical problems, as for example in the design and manufacturing of low PMD fibers.

P-OTDR finds fruitful applications also in the measurement of PMD statistical properties of an optical link. In this case, besides providing a point-wise characterization of the link, it also simplifies the management the measurement, since both the source of the probe field and the receiver are placed at the same end-face of the fiber link to be tested. Similarly, recent theoretical analyses have shown that also the mean PDL may be measured with a reflectometric technique.

Nevertheless, some specific applications of the P-OTDR — such as spinning profile measurement and local dichroism measurement — still require further investigation, both theoretical and experimental.

8. Acknowledgments

The authors are strongly indebted and grateful to Fabrizio Corsi, Anna Pizzinat, Marco Schiano, Carlo Giacomo Someda and Tiziana Tambosso for the precious help and support they have been providing.

Part of the research activity was carried out under the agreement with Istituto Superiore delle Comunicazioni e delle Tecnologie dell'Informazione (ISCTI, Rome, Italy) and in cooperation with Telecom Italia Lab (Turin, Italy).

A. Derivation of Eqs. (4) and (13)

According to Eq. (3), and recalling Eq. (2), the round-trip birefringence can be calculated as

$$\begin{aligned}\boldsymbol{\beta}_{\mathrm{R}}\times &= \frac{\mathrm{d}\mathsf{R}_{\mathrm{R}}}{\mathrm{d}z}\mathsf{R}_{\mathrm{R}}^{\mathrm{T}} = \\ &= \mathsf{M}\left(\frac{\mathrm{d}\mathsf{R}^{\mathrm{T}}}{\mathrm{d}z}\mathsf{MR} + \mathsf{R}^{\mathrm{T}}\mathsf{M}\frac{\mathrm{d}\mathsf{R}}{\mathrm{d}z}\right)\mathsf{R}^{\mathrm{T}}\mathsf{MRM} = \\ &= \mathsf{M}\frac{\mathrm{d}\mathsf{R}^{\mathrm{T}}}{\mathrm{d}z}\mathsf{RM} + (\mathsf{MR}^{\mathrm{T}})\mathsf{M}(\boldsymbol{\beta}\times)\mathsf{M}(\mathsf{RM}),\end{aligned}$$

where the equality $(\boldsymbol{\beta}\times) = (d\mathsf{R}/dz)\mathsf{R}^{\mathrm{T}}$ has been used. Now, upon multiplying the first term of the last member by $\mathsf{R}^{\mathrm{T}}\mathsf{R} = \mathsf{I}$, and since $(\boldsymbol{\beta}\times)^{\mathrm{T}} = \mathsf{R}(d\mathsf{R}/dz)^{\mathrm{T}} = -(\boldsymbol{\beta}\times)$, one obtain that

$$\begin{aligned}\boldsymbol{\beta}_{\mathrm{R}}\times &= \mathsf{MR}^{\mathrm{T}}[-(\boldsymbol{\beta}\times) + \mathsf{M}(\boldsymbol{\beta}\times)\mathsf{M}]\mathsf{RM} = \\ &= -2\mathsf{MR}^{\mathrm{T}}(\boldsymbol{\beta}_{\mathrm{L}}\times)\mathsf{RM}.\end{aligned}$$

Finally, using the properties of orthogonal matrices (see the appendices of [49]), Eq. (4) is found. Upon substituting the z-derivatives with ω-derivatives, the same calculations lead also to Eq. (13). Similar analyses, detailed in [37], yield Eq. (19).

B. Stochastic Differential Equations

Consider the differential equation

$$\frac{\mathrm{d}\boldsymbol{x}_t}{\mathrm{d}t} = \mathsf{Q}(\boldsymbol{x}_t, t)\boldsymbol{\eta}_t + \boldsymbol{u}(\boldsymbol{x}_t, t), \tag{27}$$

where $\boldsymbol{x}_t$ is a n-dimensional, vector-valued function of $t \in \mathbb{R}$, and $\boldsymbol{\eta}_t$ is a m-dimensional vector of independent white noises, such that

$$\mathrm{E}[\boldsymbol{\eta}_t] = 0 \qquad \mathrm{E}\left[\boldsymbol{\eta}_t\, \boldsymbol{\eta}_{t+\tau}^{\mathrm{T}}\right] = \mathsf{I}_m \delta(\tau)\,,$$

where I_m is the m-dimensional identity matrix. Of course, Q is a $n \times m$ matrix and $\boldsymbol{u}$ is a n-dimensional vector. Equation (27) is said to be a *stochastic differential equation* (SDE) [67].

Owing to the particular properties of $\boldsymbol{\eta}_t$, the theory of standard differential equations does not apply to Eq. (27). However, a complete theory has been developed to face and solve problems involving SDEs. In this appendix only an important result, known as the *Kolmogorov backward equation* (KBE), is reported. A complete and rigorous treatment of SDE theory may be found in [67].

Let $\psi(\boldsymbol{x}_t)$ be a "regular" function of $\boldsymbol{x}_t$, and let

$$\mathrm{E}_t[\psi(\boldsymbol{x}_{t+\tau})] = \mathrm{E}[\psi(\boldsymbol{x}_{t+\tau}) \,|\, \boldsymbol{x}_t]\,, \qquad \tau \geq 0$$

be its mean value in $t+\tau$ conditioned by $\boldsymbol{x}_t$ [62]. The KBE states that for $\tau \geq 0$ the evolution of $\mathrm{E}_t[\psi(\boldsymbol{x}_{t+\tau})]$ is governed by

$$\frac{\partial \mathrm{E}_t[\psi(\boldsymbol{x}_{t+\tau})]}{\partial \tau} = \mathrm{E}_t[(\mathcal{A}\psi)(\boldsymbol{x}_{t+\tau})]\,, \tag{28}$$

with initial condition $\mathrm{E}_t[\psi(\boldsymbol{x}_t)] = \psi(\boldsymbol{x}_t)$. The symbol $\mathcal{A}$ represents a second order differential operator, known as *infinitesimal generator* [67]. As shown in [19], when Eq. (27) refers to a physically realistic model, $\mathcal{A}$ reads as

$$\mathcal{A} = \sum_{j=1}^{n} u_j \frac{\partial}{\partial x_j} + \frac{1}{2} \sum_{j=1}^{n} \sum_{k=1}^{n} \sum_{p=1}^{m} \left(Q_{jp} Q_{kp} \frac{\partial^2}{\partial x_j \partial x_k} + Q_{kp} \frac{\partial Q_{jp}}{\partial x_k} \frac{\partial}{\partial x_j} \right), \tag{29}$$

where x_j and u_j are the jth element of $\boldsymbol{x}_t$ and $\boldsymbol{u}$, respectively, and Q_{jp} is the element of place (j,p) of Q. For completeness, note that Eq. (29) is only one of the two possible expressions of the generator, and it is called the *Stratonovich choice*. The second possibility is the *Îto choice*, which differs from Eq. (29) in lacking the terms involving the first derivative of Q.

By means of the KBE, it possible to determine a differential equation for the ACF of $\psi(\boldsymbol{x}_t)$. Actually, one can write [62]:

$$\begin{aligned} r_\psi(t,\tau) &= \mathrm{E}[\psi(\boldsymbol{x}_t)\, \psi(\boldsymbol{x}_{t+\tau})] = \mathrm{E}[\mathrm{E}[\psi(\boldsymbol{x}_t)\, \psi(\boldsymbol{x}_{t+\tau}) \,|\, \boldsymbol{x}_t]] = \\ &= \mathrm{E}[\psi(\boldsymbol{x}_t)\, \mathrm{E}_t[\psi(\boldsymbol{x}_{t+\tau})]]\,, \end{aligned}$$

so that, taking the derivative with respect to τ, and using (28) one finds

$$\frac{\partial r_\psi(t,\tau)}{\partial \tau} = \mathrm{E}\left[\psi(\boldsymbol{x}_t)\, \frac{\partial \mathrm{E}_t[\psi(\boldsymbol{x}_{t+\tau})]}{\partial \tau}\right] = \mathrm{E}[\psi(\boldsymbol{x}_t)\, \mathrm{E}_t[(\mathcal{A}\psi)(\boldsymbol{x}_{t+\tau})]]\,,$$

which leads to ($\tau \geq 0$):

$$\frac{\partial r_\psi(t,\tau)}{\partial \tau} = \mathrm{E}[\psi(\boldsymbol{x}_t)\,(\mathcal{A}\psi)(\boldsymbol{x}_{t+\tau})] \,. \tag{30}$$

Solving Eq. (30) with initial condition $r_\psi(t,0) = \mathrm{E}\left[\psi^2(\boldsymbol{x}_t)\right]$, the evolution of $r_\psi(t,\tau)$ for every $t \in \mathbb{R}$ and $\tau \geq 0$ can be determined. This provides the evolution also for $\tau < 0$, because of the following property of the ACF:

$$r_\psi(t,\tau) = \mathrm{E}[\psi(\boldsymbol{x}_t)\,\psi(\boldsymbol{x}_{t+\tau})] = r_\psi(t+\tau, -\tau) \,.$$

Likewise, the cross-correlation between two functions $\psi(\boldsymbol{x}_t)$ and $\xi(\boldsymbol{x}_t)$ can be calculated, producing the result ($\tau \geq 0$):

$$\frac{\partial r_{\psi,\xi}(t,\tau)}{\partial \tau} = \mathrm{E}[\psi(\boldsymbol{x}_t)\,(\mathcal{A}\xi)(\boldsymbol{x}_{t+\tau})] \,, \tag{31}$$

with $r_{\psi,\xi}(t,0) = \mathrm{E}[\psi(\boldsymbol{x}_t)\xi(\boldsymbol{x}_t)]$. Again, this provides the evolution of $r_{\psi,\xi}(t,\tau)$ for every $t \in \mathbb{R}$ and $\tau \geq 0$. To analyze the case $\tau < 0$, first $r_{\xi,\psi}(t,\tau)$ has to be determined for $\tau \geq 0$. This may be done using again the KBE, which yields

$$\frac{\partial r_{\xi,\psi}(t,\tau)}{\partial \tau} = \mathrm{E}[\xi(\boldsymbol{x}_t)\,(\mathcal{A}\psi)(\boldsymbol{x}_{t+\tau})] \,, \tag{32}$$

with initial condition $r_{\xi,\psi}(t,0) = r_{\psi,\xi}(t,0)$. Then, the final result is achieved exploiting the relationship $r_{\xi,\psi}(t,\tau) = r_{\psi,\xi}(t+\tau,-\tau)$.

C. Derivation of Eq. (10)

The KBE can be used to calculate the ACF of the round-trip birefringence vector. For simplicity, only $\beta_{\mathrm{R},1}(z)$ — the first component of $\boldsymbol{\beta}_{\mathrm{R}}(z)$ — is considered; the other two components may be analyzed in the same way. Let $\boldsymbol{c}(z) = (c_1,\, c_2,\, c_3)$ be the first column of $\mathsf{R}(z)$, then according to Eq. (4) $\beta_{\mathrm{R},1} = 2\boldsymbol{c}\cdot\boldsymbol{\beta}_{\mathrm{L}} = 2(c_1\beta_1 + c_2\beta_2)$. Moreover, $\boldsymbol{c}(z)$ satisfies the equation [28] $d\boldsymbol{c}/dz = \boldsymbol{\beta}(z)\times\boldsymbol{c}(z)$, with initial condition $\boldsymbol{c}(0) = (1,\, 0,\, 0)$; this equation together with Eq. (6) form the SDE describing the problem under analysis:

$$\frac{\mathrm{d}}{\mathrm{d}z}\begin{pmatrix}\beta_1\\ \beta_2\\ c_1\\ c_2\\ c_3\end{pmatrix} = \begin{pmatrix}\sigma & 0\\ 0 & \sigma\\ 0 & 0\\ 0 & 0\\ 0 & 0\end{pmatrix}\begin{pmatrix}\eta_1\\ \eta_2\end{pmatrix} + \begin{pmatrix}-\rho\beta_1\\ -\rho\beta_2\\ \beta_2 c_3\\ -\beta_1 c_3\\ \beta_1 c_2 - \beta_2 c_1\end{pmatrix} . \tag{33}$$

According to Eq. (29), the infinitesimal generator associated to Eq. (33) reads

$$\begin{aligned}\mathcal{A} = {}& \beta_2 c_3\frac{\partial}{\partial c_1} - \beta_1 c_3\frac{\partial}{\partial c_2} + (\beta_1 c_2 - \beta_2 c_1)\frac{\partial}{\partial c_3} - \\ & \rho\left(\beta_1\frac{\partial}{\partial\beta_1} + \beta_2\frac{\partial}{\partial\beta_2}\right) + \frac{\sigma^2}{2}\left(\frac{\partial^2}{\partial\beta_1^2} + \frac{\partial^2}{\partial\beta_2^2}\right) .\end{aligned} \tag{34}$$

Now the ACF of $\beta_{\mathrm{R},1}(z)$ can be calculated using Eq. (30), and since Eq. (34) yields

$$\mathrm{E_z}[(\mathcal{A}\beta_{\mathrm{R},1})(z+u)] = -\rho \mathrm{E_z}[\beta_{\mathrm{R},1}(z+u)]\,,$$

the ACF $r_{\beta_\mathrm{R}}(z,\, u)$ results to be given by the differential equation

$$\frac{\partial r_{\beta_\mathrm{R}}(z,\, u)}{\partial u} = -\rho r_{\beta_\mathrm{R}}(z,\, u), \tag{35}$$

with initial condition $r_{\beta_\mathrm{R}}(z, 0) = \mathrm{E}\left[\beta^2_{\mathrm{R},1}(z)\right]$. Strictly speaking $\mathrm{E}\left[\beta^2_{\mathrm{R},1}(z)\right]$ is not independent of z, but when $z \gg L_F$ it reaches a steady state for which $\mathrm{E}\left[\beta^2_{\mathrm{R},1}(z)\right] = (8/3)\sigma^2_\beta$ [31]. Now, Eq. (35) can be easily solved for $u \geq 0$ and, recalling that correlations of stationary processes are even functions, Eq. (10) is finally reached.

D. Calculation of the ACF of a Twisted/Spun Fiber

In order to calculate $r_{\beta_\mathrm{R}}(z)$ for a twisted or spun fiber, the theory of SDE should be exploited again. First of all, it is profitable to transform the equations to a rotating frame that compensate for the geometrical rotation caused by the twist/spin. In this new frame the birefringence vector reads $\boldsymbol{\beta}(z) = (\beta_1,\, \beta_2,\, \beta_3)^T$, where β_1 and β_2 are given by Eq. (6), and $\beta_3 = (g-2)\tau'(z)$. Here $\tau'(z)$ is the z-derivative of the twist/spin function $\tau(z)$, while g is a suitable physical constant being equal to 0.14 or 0 for twisted or spun fibers, respectively. In the rotating frame, and using the same formalism as in Aappendix C, the SDE reads

$$\frac{\mathrm{d}}{\mathrm{d}z}\begin{pmatrix}\beta_1\\ \beta_2\\ c_1\\ c_2\\ c_3\end{pmatrix} = \begin{pmatrix}\sigma & 0\\ 0 & \sigma\\ 0 & 0\\ 0 & 0\\ 0 & 0\end{pmatrix}\begin{pmatrix}\eta_1\\ \eta_2\end{pmatrix} + \begin{pmatrix}-\rho\beta_1\\ -\rho\beta_2\\ \beta_2 c_3 - \beta_3 c_2\\ -\beta_1 c_3 + \beta_3 c_1\\ \beta_1 c_2 - \beta_2 c_1\end{pmatrix}, \tag{36}$$

and the relative infinitesimal generator is

$$\begin{aligned}\mathcal{A} = &-\rho\left(\beta_1\frac{\partial}{\partial\beta_1} + \beta_2\frac{\partial}{\partial\beta_2}\right) + \frac{\sigma^2}{2}\left(\frac{\partial^2}{\partial\beta_1^2} + \frac{\partial^2}{\partial\beta_2^2}\right) + \\ &(\beta_2 c_3 - c_2\alpha)\frac{\partial}{\partial c_1} + (c_1\alpha - \beta_1 c_3)\frac{\partial}{\partial c_2} + (\beta_1 c_2 - \beta_2 c_1)\frac{\partial}{\partial c_3}.\end{aligned} \tag{37}$$

It may be proved the expression of $\boldsymbol{\beta}_\mathrm{R}(z)$ in the rotating frame is the same as in the fixed one, consequently $\beta_{\mathrm{R},1}(z)$ is still equal to $2\boldsymbol{c}\cdot\boldsymbol{\beta}_\mathrm{L}$. However, when the infinitesimal generator (37) is applied to $\beta_{\mathrm{R},1}(z)$, a recursive equation involving an infinite series of functions is found [36]. This equation can be rearranged as an infinite-dimensional, first-order differential equation:

$$\frac{\mathrm{d}\boldsymbol{y}}{\mathrm{d}u} = \mathsf{A}\boldsymbol{y}(u), \tag{38}$$

where, $\boldsymbol{y}(u)$ is an infinite-dimensional vector, whose second element is $r_{\beta_\mathrm{R}}(u)$, and the other elements are auxiliary functions. The infinite-dimensional matrix A can be written by blocks as

$$\mathsf{A}(z) = \begin{pmatrix} \mathsf{A}_0 & \mathsf{B}_1 & \varnothing & \dots \\ \mathsf{D} & \mathsf{A}_1 & \mathsf{B}_2 & \varnothing & \dots \\ \varnothing & \mathsf{D} & \mathsf{A}_2 & \mathsf{B}_3 & \varnothing & \dots \\ \dots & \varnothing & \mathsf{D} & \mathsf{A}_3 & \mathsf{B}_4 & \varnothing & \dots \\ & \dots\dots\dots\dots\dots\dots \end{pmatrix},$$

where $\varnothing$ is an empty 3×3 matrix, $\mathsf{D} = 2\rho\mathsf{I}$, and

$$\mathsf{A}_k = \begin{pmatrix} -2k\rho & 0 & -(k+1) \\ 0 & -(2k+1)\rho & \beta_3(z) \\ 0 & -\beta_3(z) & -(2k+1)\rho \end{pmatrix}, \qquad \mathsf{B}_k = 2k\sigma_\beta^2 \begin{pmatrix} 0\;0\;0 \\ 0\;0\;0 \\ 1\;0\;0 \end{pmatrix}.$$

The initial condition of Eq. (38) reads

$$\boldsymbol{y}(0) = \frac{8\sigma_\beta^2}{3} \left(\boldsymbol{y}_0, \boldsymbol{y}_1, \boldsymbol{y}_2, \boldsymbol{y}_3, \dots\right)^T,$$

where $\boldsymbol{y}_k = (0,\, 1/k!,\, 0)$.

This theoretical result holds for any deterministic $\beta_3(z)$. Although an analytical solution of Eq. (38) has not been found, still it is possible to solve the system numerically.

References

1. C. G. Someda, *Electromagnetic waves* (Chapman & Hall, Londra, 1998).
2. D. Marcuse, Bell System Technol. J., **54**, 985–995 (1975).
3. R. Ulrich and A. Simon, Appl. Opt., **18**, 2241–2251 (1979).
4. S. C. Rashleigh and R. Ulrich, Opt. Lett., **2**, 60–62 (1978).
5. H. Kogelnik, R. M. Jopson, and L. E. Nelson, *Polarization-mode dispersion* (in "Optical Fiber Telecommunications, IV B"), I. P. Kaminow and T. Li, editors (Academic Press, San Diego, 2002).
6. E. Collett, *Polarized light, fundamentals and applications* (Dekker, New York, 1993).
7. S. C. Rashleigh, J. Lightwave Technol., **1**, 312–331 (1983).
8. G. J. Foschini and C. D. Poole, J. Lightwave Technol., **9**, 1439–1456 (1991).
9. M. Karlsson and J. Brentel, Opt. Lett., **24**, 939–941 (1999).
10. G. J. Foschini, R. M. Jopson, L. E. Nelson, and H. Kogelnik, J. Lightwave Technol., **17**, 1560–1565 (1999).
11. M. Shtaif, A. Mecozzi, and J. A. Nagel, IEEE Photon. Technol. Lett., **12**, 53–55 (2000).
12. G. J. Foschini, L. E. Nelson, R. M. Jopson, and H. Kogelnik, IEEE Photon. Technol. Lett., **12**, 293–295 (2000).
13. A. Bononi and A. Vannucci, Opt. Lett., **26**, 675–677 (2001).
14. A. J. Barlow, J. J. Ramskov-Hansen, and D. N. Payne, Appl. Opt., **20**, 2962–2968 (1981).
15. M. J. Li and D. A. Nolan, Opt. Lett.,, **23**, 1659–1661 (1998).
16. A. Galtarossa, P. Griggio, A. Pizzinat, and L. Palmieri, Opt. Lett., **27**, 692–694 (2002).

17. A. Pizzinat, B. S. Marks, L. Palmieri, C. R. Menyuk, and A. Galtarossa, IEEE Photon. Technol. Lett., **15**, 819–821 (2003).
18. A. Galtarossa, L. Palmieri, A. Pizzinat, B. S. Marks, and C. R. Menyuk, J. Lightwave Technol., **21**, 1635–1643 (2003).
19. P. K. A. Wai and C. R. Menyuk, J. Lightwave Technol., **14**, 148–157 (1996).
20. P. K. A. Wai, W. L. Kath, C. R. Menyuk, and J. W. Zhang, J. Opt. Soc. Am. B, **14**, 2967–2979 (1997).
21. A. J. Rogers, Appl. Opt., **20**, 1060–1074 (1981).
22. B. Y. Kim and S. S. Choi, Opt. Lett., **6**, 578–580 (1981).
23. J. N. Ross, Appl. Opt., **21**, 3489–3495 (1982).
24. M. Nakazawa, IEEE J. Quantum Electron., **19**, 854–861 (1983).
25. R. E. Schuh, J. G. Ellison, A. S. Siddiqui, and D. H. O. Bebbington, Elett. Lett., **32**, 387–388 (1996).
26. F. Corsi, A. Galtarossa, and L. Palmieri, J. Lightwave Technol., **16**, 1832–1843 (1998).
27. J. G. Ellison and A. S. Siddiqui, IEEE Photon. Technol. Lett., **10**, 246–248 (1998).
28. F. Corsi, A. Galtarossa, and L. Palmieri, J. Lightwave Technol., **17**, 1172–1178 (1999).
29. J. G. Ellison and A. S. Siddiqui, J. Lightwave Technol., **18**, 1226–1232 (2000).
30. A. Galtarossa, L. Palmieri, M. Schiano, and T. Tambosso, OSA Opt. Lett., **25**, 384–386 (2000).
31. A. Galtarossa, L. Palmieri, M. Schiano, and T. Tambosso, OSA Opt. Lett., **25**, 1322–1324 (2000).
32. A. Galtarossa, L. Palmieri, A. Pizzinat, M. Schiano, and T. Tambosso, J. Lightwave Technol., **18**, 1389–1394 (2000).
33. A. Galtarossa, L. Palmieri, M. Schiano, and T. Tambosso, OSA Opt. Lett., **26**, 962–964 (2001).
34. J. G. Ellison and A. S. Siddiqui, IEE Proc. Optoelectron., **148**, 176–182 (2001).
35. M. Wuilpart, P. Megret, M. Blondel, A.J. Rogers, and Y. Defosse, IEEE Photon. Technol. Lett., **13**, 836–838 (2001).
36. A. Galtarossa and L. Palmieri, J. Lightwave Technol., **20**, 1149–1159 (2002).
37. A. Galtarossa and L. Palmieri, J. Lightwave Technol., **21**, 1233–1241 (2003).
38. B. Huttner, J. Reecht, N. Gisin, R. Passy, and J. P. von der Weid, IEEE Photon. Technol. Lett., **10**, 1458–1460 (1998).
39. M. Wegmuller, M. Legré, , P. Oberson, O. Guinnard, C. Vinegoni, and N. Gisin, IEEE Photon. Technol. Lett., **13**, 145–147 (2001).
40. M. Wegmuller, M. Legré, and N. Gisin, J. Lightwave Technol., **20**, 828–835 (2002).
41. N. Zou, M. Yoshida, Y. Namihira, and H. Ito, Electron. Lett., **38**, 115–117 (2002).
42. M. G. Shlyagin, A. V. Khomenko, and D. Tentori, Opt. Lett., **20**, 869–871 (1995).
43. J. Zhang, V. A. Handerek, I. Çokgör, V. Pantelic, and A. J. Rogers, J. Lightwave Technol., **15**, 794–801 (1997).
44. T. Gogolla and K. Krebber, J. Lightwave Technol., **18**, 320–328 (2000).
45. W. W. Hu, K. Inagaki, and Y. Mizuguchi, Opt. Lett., **26**, 193–195 (2001).

46. E. A. Kuzin, J. M. Estudillo Ayala, B. Ibarra Escamilla, and J. W. Haus, Opt. Lett., **26**, 1134–1136 (2001).
47. M. Ferrario, P. Martelli, S. Pietralunga, and M. Martinelli, Tech. Digest OFC (paper WJ3), Atlanta (GA),2003.
48. H. Sunnerud, B. E. Olsson, and P. A. Andrekson, Elett. Lett., **34**, 397–398 (1998).
49. F. Corsi, A. Galtarossa, and L. Palmieri, J. Opt. Soc. Am. A,, **16**, 574–583 (1999).
50. F. Corsi, A. Galtarossa, L. Palmieri, M. Schiano, and T. Tambosso, IEEE Photon. Technol. Lett., **11**, 451–453 (1999).
51. A. Galtarossa, L. Palmieri, M. Schiano, and T. Tambosso, J. Lightwave Technol., **17**, 1835–1842 (1999).
52. B. Huttner, B. Gisin, and N. Gisin, J. Lightwave Technol., **17**, 1843–1847 (1999).
53. H. Sunnerud, B. E. Olsson, M. Karlsson, P. A. Andrekson, and J. Brentel, J. Lightwave Technol., **18**, 897–904 (2000).
54. A. O. Kleivstul and A. S. Sudbø, Electron. Lett., **37**, 621–623 (2001).
55. B. L. Heffner, IEEE Photon. Technol. Lett., **4**, 1066–1069 (1992).
56. R. M. Jopson, L. E. Nelson, and H. Kogelnik, IEEE Photon. Technol. Lett., **11**, 1153–1155 (1999).
57. A. Galtarossa and L. Palmieri, J. Lightwave Technol., **22**, 1103–1115 (2004).
58. J. P. Gordon and H. Kogelnik, Proc. Nat. Acad. Sci., **97**, 4541–4550 (2000).
59. R. C. Jones, J. Opt. Soc. Am., **37**, 107–112 (1947).
60. C. D. Poole, J. H. Winters, and J. A. Nagel, Opt. Lett., **6**, 372–374 (1991).
61. C. D. Poole and J. Nagel, *Polarization effects in lightwave systems* (in "Optical Fiber Telecommunications"), I. P. Kaminow and T. Koch, editors (Academic Press, San Diego, 1997).
62. A. Papoulis, *Probability, random variables and stochastic processes* (McGraw-Hill, New York, 1984).
63. R. E. Schuh, E. S. R. Sikora, N. G. Walker, A. S. Siddiqui, L. M. Gleeson, and D. H. O. Bebbington, Elett. Lett., **28**, 1772–1773 (1995).
64. A. C. Hart, R. G. Huff, and K. L. Walker, U. S. Patent 5 298 047, 1994.
65. L. Arnold, *Stochastic differential equations*, (J. Wiley, New York, 1974).
66. C. W. Gardiner, *Handbook of stochastic methods* (Springer-Verlag, Berlin, 1985).
67. B. Øksendal, *Stochastic Differential Equations* (Springer-Verlag, Berlin, 2000).
68. A. Galtarossa, L. Palmieri, and D. Sarchi, Tech. Digest. OFC (paper FA2), Los Angeles (CA), 2004.
69. C. D. Poole and D. L.Favin, J. Lightwave Technol., **12**, 917–929 (1994).
70. B. L. Heffner, Opt. Lett., **21**, 113–115 (1996).
71. N. J. Frigo, J. Quantum Elect., **22**, 2131–2140 (1986).
72. N. Gisin and B. Huttner, Optics Comm., **142**, 119–125 (1997).
73. Y. Li and A. Yariv, J. Opt. Soc. Am. B, **17**, 1821–1827 (2000).
74. N. Gisin, Optics Comm., **114**, 399–405 (1995).
75. A. Mecozzi and M. Shtaif, IEEE Photon. Technol. Lett., **14**, 313–315 (2002).
76. A. Galtarossa and L. Palmieri, IEEE Photon. Technol. Lett., **15**, 57–59 (2003).

DOI: 10.1007/s10297-004-0013-1
Originally published in J. Opt. Fiber Commun. Rep. 1, 123–140 (2004)

PMD impact on optical systems: Single- and multichannel effects

Magnus Karlsson and Henrik Sunnerud

Photonics Laboratory, Department of Microtechnology and Nanoscience
Chalmers University of Technology, SE-412 96 Gothenburg, Sweden
E-mail: magnus.karlsson@mc2.chalmers.se

Abstract. We discuss the influence of polarization mode dispersion (PMD) on optical communication channels by putting special emphasis on PMD-induced pulse broadening. We present the analytical theory for PMD-induced pulse broadening, first in the deterministic and then for the random case. We then apply this theory to a comparison between a few different simple PMD-compensators. We also discuss the systems implications and outage probabilities that arise for different compensating techniques, including active optical compensation, alternative modulation formats and error correcting codes. Then we consider wavelength division multiplexing (WDM) systems, and present the unique aspects of PMD and PMD compensation that distinguish WDM transmission from the single-channel case. Finally the temporal drift is discussed, and the associated autocorrelation function is presented.

1. Introduction

The title of this contribution to the summer school implies a rather broad discussion of signal impairments due to PMD. As the time and space available necessarily limits the treatment, we decided to emphasize analytic results available for single-channel signal transmission, and in that respect focus on PMD-induced RMS-pulse broadening. This may at first seem to be a rather limited topic, but our hope is to show that quite a lot of information about PMD in general will come as a spin-off of this discussion. In other words, the fundamental physical issues of pulse transmission in randomly birefringent media will be discussed here. It is necessary for the text, since it is of tutorial as well as of fundamental physical nature, to contain a lot of mathematics. However, the difficulties lie mainly in the novel concepts rather than in any lengthy derivations, which any under-graduate student should be able to follow. If then, as a spin-off, some of the joy of playing around with the beautiful mathematics of unitary and orthogonal matrices is conveyed to the readers, the author would be more than delighted.

When it comes to the related and more practical issue of communication system performance and outage probabilities arising from PMD, that is a topic better suited for numerical simulations, and there are published papers that review the results from such simulations, see, e.g., the review [1] and references therein. Yet some useful insight into this topic, both for single-channel systems and for WDM systems will follow from our analytical treatment. We will in the last chapters extend the discussion further to a few particular issues for PMD and PMD compensation in WDM systems together with some drift time issues.

Having said this, let us dive in to the physics, and discuss the aspects of pulse broadening due to PMD!

Polarization mode dispersion is a linear dispersive effect, and although its most important manifestation is broadening of pulses in the time domain, it is, as all linear propagation effects, most easily investigated in the frequency domain. There we can follow each frequency component, and the PMD will then manifest as a polarization- and frequency dependence of the phase delay for each component. In other words, different frequency and polarization components will be delayed with respect to each other and distort the signal in the time domain. The most basic measure of such a distorsion is the root-mean square (RMS) pulse broadening, which is, by use of Parseval's theorem, straightforward to investigate in the frequency domain via the various *moments* of the signal.

2. PMD-induced Pulse Broadening via the Moment Method: Deterministic Case

The nth moment of an electric field distribution in time is defined as

$$W_n = \int_{-\infty}^{\infty} t^n \, \mathrm{E}^\dagger \mathrm{E} \, dt = \left\langle t^n \mathrm{E}^\dagger \mathrm{E} \right\rangle, \tag{1}$$

where $\mathrm{E}(t)$ is the electric field (column) vector of the signal, and the conjugate transpose is denoted by the conventional dagger notation, $\mathrm{E}^\dagger$, so that $\mathrm{E}^\dagger \mathrm{E}$ is a scalar product. The brackets are chosen to denote integration over the whole pulse interval, e.g., from $-\infty$ to $+\infty$ in our case. The zeroth moment, W_0, will thus be the pulse energy. The first moment, W_1, is (in statistic language) the expectation value, or (in mechanic language) the center of mass of the pulse. In our case it simply represents the position of the pulse. The second moment is the variance, or the quadratic deviation. The RMS width σ of an optical pulse can now be expressed using the first three moments as

$$\sigma^2 = \frac{\langle t^2 \mathrm{E}^\dagger \mathrm{E} \rangle}{\langle \mathrm{E}^\dagger \mathrm{E} \rangle} - \frac{\langle t \mathrm{E}^\dagger \mathrm{E} \rangle^2}{\langle \mathrm{E}^\dagger \mathrm{E} \rangle^2} = \frac{W_2}{W_0} - \frac{W_1^2}{W_0^2}, \tag{2}$$

which we might take as the definition of the RMS width. The moments have to be normalized to the pulse energy (unless we consider unity pulse energy), and this is where the denominator comes from.

It is very useful to calculate the moments in the spectral domain. By using Parseval's formula one can derive the first and second moments in the frequency domain as

$$W_1 = \int_{-\infty}^{\infty} t\, \mathrm{E}^\dagger \mathrm{E}\, dt = i \int_{-\infty}^{\infty} \tilde{\mathrm{E}}^\dagger \tilde{\mathrm{E}}_\omega\, d\omega, \tag{3}$$

$$W_2 = \int_{-\infty}^{\infty} t^2\, \mathrm{E}^\dagger \mathrm{E}\, dt = \int_{-\infty}^{\infty} \tilde{\mathrm{E}}_\omega^\dagger \tilde{\mathrm{E}}_\omega\, d\omega. \tag{4}$$

We use the bracket notation, $\langle\ \rangle$, (a bit sloppily though) in the frequency domain as well. Thus the moments can be expressed as

$$W_1 = i \left\langle \tilde{\mathrm{E}}^\dagger \tilde{\mathrm{E}}_\omega \right\rangle = -i \left\langle \tilde{\mathrm{E}}_\omega^\dagger \tilde{\mathrm{E}} \right\rangle, \tag{5}$$

$$W_2 = \left\langle \tilde{\mathrm{E}}_\omega^\dagger \tilde{\mathrm{E}}_\omega \right\rangle = \left\langle |\tilde{\mathrm{E}}_\omega^\dagger|^2 \right\rangle, \tag{6}$$

where ω denotes derivative.

2.1. *Moments in Presence of PMD*

Now, to account for linear dispersive effects in a fiber, the transmission is described by a transfer matrix $\mathbf{U}$ so that the output field after the fiber is $\tilde{\mathrm{E}}_{\mathrm{out}} = \mathbf{U}\tilde{\mathrm{E}}_{\mathrm{in}} = \mathbf{U}a\hat{\mathrm{j}}_{\mathrm{in}}$, where $a(\omega)$ is the scalar amplitude spectrum and $\hat{\mathrm{j}}_{\mathrm{in}}$ is the input Jones vector of the signal. If we neglect losses and polarization dependent losses, the power $\tilde{\mathrm{E}}^\dagger \tilde{\mathrm{E}}$ should be preserved, which means that $\mathbf{U}^\dagger \mathbf{U}$ must equal the unity matrix $\mathbf{I}$, i.e., the transfer matrix $\mathbf{U}$ should be *unitary* [2, 3].

First Moment

According to the definition, the first moment becomes (and from now on we drop index$_{\mathrm{out}}$ on the field vector and use index$_\omega$ to denote differentiation)

$$W_1 = i \left\langle \tilde{\mathrm{E}}^\dagger \tilde{\mathrm{E}}_\omega \right\rangle = i \left\langle (a\mathbf{U}\hat{\mathrm{j}}_{\mathrm{in}})^\dagger (a_\omega \mathbf{U}\hat{\mathrm{j}}_{\mathrm{in}} + a\mathbf{U}_\omega \hat{\mathrm{j}}_{\mathrm{in}}) \right\rangle = \tag{7}$$

$$i \left\langle a^* a_\omega \right\rangle + i \left\langle a^* a \hat{\mathrm{j}}_{\mathrm{in}}^\dagger \mathbf{U}^\dagger \mathbf{U}_\omega \hat{\mathrm{j}}_{\mathrm{in}} \right\rangle.$$

In this expression the first term is the conventional scalar term. The second, polarization dependent, term involves the matrix product $\mathbf{U}^\dagger \mathbf{U}_\omega \equiv \mathbf{A}$. By differentiating the unitarity condition $\mathbf{U}^\dagger \mathbf{U} = \mathbf{I}$ it is easy to show that $\mathbf{A} = -\mathbf{A}^\dagger$. This condition implies that $\mathbf{A}$ can be written as a linear combination of the Pauli matrices (which form the components of the Pauli spin vector œ) and the unity matrix, i.e.,

$$\mathbf{A} = -\frac{i}{2}(2\tau_0 \mathbf{I} + ø_{\mathrm{in}} \cdot œ), \tag{8}$$

where τ_0 and the vector $ø_{\mathrm{in}}$ are real functions of ω. The physical relevance and the notation of these constants will be explained below. The Pauli spin vector is a three-component vector with the Pauli spin matrices as its components

$$œ = \left(\begin{pmatrix} 1 & 0 \\ 0 & -1 \end{pmatrix}, \begin{pmatrix} 0 & 1 \\ 1 & 0 \end{pmatrix}, \begin{pmatrix} 0 & -i \\ i & 0 \end{pmatrix} \right), \tag{9}$$

and it is very useful when going from Jones to Stokes space. For example, a polarization state with the Jones vector $\hat{\mathrm{j}}$ has the corresponding Stokes vector $\hat{\mathrm{j}}^\dagger œ\hat{\mathrm{j}}$. Finally we can insert the expression for **A** in Eq. (8), and we end up with the following expression for the pulse center-of-mass:

$$W_1 = i\langle a^* a_\omega\rangle + \left\langle |a|^2\left(\tau_0 + \frac{1}{2}ø_{\text{in}}\cdot \text{s}\right)\right\rangle, \tag{10}$$

where $\text{s} = \hat{\text{j}}_{\text{in}}^{\dagger} œ\hat{\text{j}}_{\text{in}}$ is the polarization Stokes vector of the signal. We can simplify this expression further by writing the amplitude spectrum in amplitude-phase-angle form, i.e. $a = |a|\exp(-i\varphi)$, so that

$$W_1 = \left\langle |a|^2(\varphi_\omega + \tau_0 + \frac{1}{2}ø_{\text{in}}\cdot \text{s})\right\rangle. \tag{11}$$

Now we can interpret this result as an average transit time, given by the first two terms. The average transit time has two contributions, the first one coming from the signal and the second from the medium. This latter one can alternatively be described using the propagation constant β as $\tau_0 = z d\beta/d\omega = z/v_g$ where z is the propagation distance and $v_g = d\omega/d\beta$ is the group velocity. Obviously, this average propagation time (which is of limited interest in most system applications) together with other effects connected by chromatic dispersion can be included in the spectral phase φ of the signal rather than be taken from the transfer matrix as described above [4]. Thus, from now on we will put $\tau_0 = 0$, and consider it included in φ.

The polarization dependent part of the transit time is proportional to the scalar product between the input Stokes vector of the signal, s and the vector $ø_{\text{in}}$, as originally shown in [5]. The vector $ø_{\text{in}}$ is called the *input PMD vector*, as it is related to the input state of polarization s, and the relation (8) can be taken as the definition of the PMD vector. Similarly, an *output PMD vector* can be defined with respect to the matrix product $\mathbf{U}_\omega\mathbf{U}^\dagger$. The choices of sign in the above definition makes the PMD vector point in the direction of the fastest polarization state, similarly to the convention used in [6], and differing slightly from the notation used in [4]. The polarization directions parallel with the PMD vector is commonly called the *Principal States of Polarization (PSP)*, and besides being the fastest and slowest polarization states, they are also the states that are constant to first order in optical frequency [2, 7]. The length of the PMD vector corresponds to the difference in time delay between the fastest and slowest polarization, and it is hence called the *Differential Group Delay* (DGD) and sometimes denoted $\Delta\tau$.

Second Moment

Turning to the second moment, it can be written as

$$W_2 = \left\langle \tilde{\text{E}}_\omega^\dagger \tilde{\text{E}}_\omega\right\rangle = \left\langle (a_\omega \mathbf{U}\hat{\text{j}}_{\text{in}} + a\mathbf{U}_\omega \hat{\text{j}}_{\text{in}})^\dagger (a_\omega \mathbf{U}\hat{\text{j}}_{\text{in}} + a\mathbf{U}_\omega \hat{\text{j}}_{\text{in}})\right\rangle \tag{12}$$

which results in four terms, of which one is scalar and corresponding to the broadening in absence of polarization effects. Of the remaining three terms, one include the **A** matrix (as defined above), one includes the $\mathbf{A}^\dagger$ matrix, and those terms can then be simplified using Eq. (8). The last term involves the matrix product $\mathbf{U}_\omega^\dagger\mathbf{U}_\omega$, which by multiplication with the unity matrix $\mathbf{U}\mathbf{U}^\dagger$ can be written $\mathbf{U}_\omega^\dagger\mathbf{U}_\omega = \mathbf{U}_\omega^\dagger\mathbf{U}\mathbf{U}^\dagger\mathbf{U}_\omega = \mathbf{A}^\dagger\mathbf{A} = -\mathbf{A}^2$. It can finally be shown that the square of the matrix **A** (neglecting the

τ_0 as motivated above) equals $\mathbf{A}^2 = -\mathbf{I}\o_{\text{in}}^2/4$. Using all this, we can write the second moment as

$$W_2 = \left\langle (|a|_\omega)^2 + \varphi_\omega^2|a|^2 - \varphi_\omega|a|^2\o_{\text{in}} \cdot \mathrm{s} + \frac{|a|^2}{4}|\o_{\text{in}}|^2 \right\rangle \tag{13}$$
$$= \left\langle (|a|_\omega)^2 + |a|^2(\varphi_\omega \mathrm{s} - \frac{1}{2}\o_{\text{in}})^2 \right\rangle .$$

Here we may note that in absence of birefringence, a limit recovered for $\o_{\text{in}} = 0$, we can recover the conventional scalar result, since $\mathrm{s}^2 = 1$. Furthermore, the above formulas for the moments [Eqs. (11),(13)] are rather general, applicable to any birefringence, pulse spectrum and chirp. The only assumptions made is that there are no polarization dependent losses (PDL), and if this condition does not hold true, then the matrix $\mathbf{A}$ does not contain the symmetry $\mathbf{A} = -\mathbf{A}^\dagger$, and Eq. (8) will yield a complex PMD vector [8].

2.2. *General Propagation and Broadening Issues*

The RMS-width can now be expressed by using the moments and Eq. (2). It is instructive to compare with conventional chromatic dispersion-induced broadening. In that case, which is given by the above formulas by putting $\varphi = z\beta(\omega)$ and $\o = 0$ it can be shown that, independently of the dispersive order (i.e., how many terms in the Taylor expansion of $\beta(\omega)$ that are retained), the nth moment will always grow as an nth order polynomial in z [9]. That means that the squared RMS width σ^2 will grow quadratically with z.

A similar result holds true for the generalized treatment including polarization broadening effects given above; i.e., for constant PMD (also called first-order PMD) the PMD vector grows linearly with z [10, 11], and thus the second moment grows quadratically with z. The case is slightly different for random birefringence, as will be discussed in the next section.

Since there is no PDL, the pulse energy W_0 is polarization independent, and to simplify the discussion we assume it to equal unity, so that the RMS width becomes $\sigma^2 = W_2 - W_1^2$. In the case of first order PMD, the expression for the RMS width then simplifies to

$$\sigma^2 = \sigma_0^2 + \frac{\Delta\tau^2}{4}(1 - (\hat{\o}_{\text{in}} \cdot \hat{\mathrm{s}})^2), \tag{14}$$

where σ_0 is the initial RMS width, $\Delta\tau = |\o_{\text{in}}|$ is the DGD and $\hat{\o}_{\text{in}} = \o_{\text{in}}/|\o_{\text{in}}|$ is the input PSP unity vector. From this expression it is evident that the pulse broadening is smallest when the pulse is polarized along the principal states (i.e., when s and $\o_{\text{in}}$ are parallel or anti-parallel). A difference from chromatic dispersion broadening is that there is no dependence on the pulse width in the amount of broadening, and the reason for that is that group velocity dispersion (GVD) is a second-order effect in frequency, and thus dependent on the width of the signal spectrum, whereas first-order PMD is a first-order effect in frequency and not dependent on the signal spectrum. However, in cases where the first-order PMD is compensated away, as, e.g., for transmission systems employing PMD compensation, the broadening will indeed depend on the spectral width.

Finally, it should be emphasized that the combined effect of chromatic dispersion in addition to the PMD will not give rise to any complications such as an increased "synergetic" broadenings larger than any of the two effects considered separately. It is often sloppily emphasized that PMD is "worse" in presence of chirp or dispersion, but this is not true, and the worse performance only stems from the extra chromatic dispersion penalty that can be added to the signal before or after the PMD-induced broadening. Since the dispersion/chirp usually is compensated away separately its inclusion is often not necessary and may actually be more obscuring that illuminating.

2.3. *An Approximate, Useful Formula*

In the following we will use the above formulas with the output vectors $\mathrm{s_{out}}, \varnothing_{\mathrm{out}}$ rather than the input ones used above. This does not matter, as the polarization vectors maintain their relative orientation during propagation through any polarization system that does not contain PDL. An interesting fact is then that for first-order PMD, the RMS width can be written compactly as

$$\sigma^2 = \sigma_0^2 + \frac{1}{4}|\varnothing_{\mathrm{out}} \times \mathrm{s_{out}}|^2, \tag{15}$$

which we could generalize [although incorrectly as we assume $\langle \varnothing_{\mathrm{out}} \cdot \mathrm{s_{out}} \rangle^2 \approx \langle (\varnothing_{\mathrm{out}} \cdot \mathrm{s_{out}})^2 \rangle$ which is only exactly valid for first-order PMD] to

$$\sigma^2 \approx \sigma_0^2 + \frac{1}{4}\left\langle |\varnothing_{\mathrm{out}} \times \mathrm{s_{out}}|^2 \right\rangle = \sigma_0^2 + \frac{1}{4}\left\langle \left| \frac{d\mathrm{s_{out}}}{d\omega} \right|^2 \right\rangle. \tag{16}$$

To get the final step we use the formula $ds/d\omega = \varnothing_{\mathrm{out}} \times \mathrm{s_{out}}$, which the reader may derive from the above definitions as an exercise.This expression is interesting because it can be used to estimate the pulse broadening due to PMD *without knowing the PMD vector explicitly*. This implies that in a measurement situation, it is enough to measure the received polarization state versus wavelength and then use Eq. (16) to get the pulse broadening, without the cumbersome measurement of the PMD vector that usually requires access to both fiber ends and controlled launch of known polarization states, see [12]. This advantage is however obtained under the restriction that first order PMD dominates (i.e., the PMD vector does not significantly change over the pulse spectrum).

3. PMD-induced Pulse Broadening: Random Case

In real transmission fibers the local birefringence is not constant, but the birefringence axes changes randomly along the fiber. This has three significant implications. Firstly, everything (the PMD vector, the pulse broadening, the polarization state etc.) must be statistically treated, and considered in terms of averages and probability distributions. Secondly, since the PMD vector becomes a random function of the propagation distance z it can be shown that the average DGD will grow with $z^{1/2}$ rather than as z for constant birefringence [11, 13, 14]. Thirdly, the PMD vector will become a random function of frequency as well. This last property is of utmost importance when analyzing the pulse broadening.

This randomness of the PMD will give rise to one striking effect. That is that the complete pulse splitting, which is the most common picture used to describe how PMD affects optical signals, will only with low probability arise in a telecom fiber! The reason for this is that pulse splitting is observed only if (i) the PMD is constant (of first order) over the entire pulse spectrum and (ii) the DGD comparable to the pulse width. However, these two conditions are usually not fulfilled simulataneously. A derivation of the autocorrelation function of the PMD vector (which will be discussed below) gives that the PMD vector will change over a spectral width comparable to the inverse average DGD. This implies that if the average DGD is so high to be comparable with the pulse width, then the PMD vector will also change over the pulse spectrum, i.e. not be constant. As a result, large PMD-induced pulse broadenings in telecom fibers will be characterized by a broadening to a continuum rather than a distinct pulse splitting.

The randomness of the PMD vector with frequency is also relevant for the statsitical average of the RMS width. Consider the statistical average of the RMS width in the expression $\sigma^2 = W_2 - W_1^2$. The averaged moments are

$$E[W_1] = \left\langle |a|^2 \varphi_\omega \right\rangle, \tag{17}$$

$$E[W_2] = \left\langle (|a|_\omega)^2 + |a|^2 \varphi_\omega^2 \right\rangle + \frac{1}{4} E[\Delta\tau]^2, \tag{18}$$

since $E[\o] = 0$ and $E[\mathrm{s} \cdot \o] = 0$. However the average of the squared first moment is more cumbersome. We must then move the expectation value inside the two moment integrals, which cannot be separated at this stage. To simplify matters we assume that spectral phase to be constant (i.e., no chirp or dispersion: $\varphi_\omega = 0$), and obtain

$$E[W_1^2] = \frac{1}{4} \int\int |a_1|^2 |a_2|^2 E[(\mathrm{s} \cdot \o_{\mathrm{in},1})(\mathrm{s} \cdot \o_{\mathrm{in},2})] d\omega_1 d\omega_2 \tag{19}$$

where indices 1 and 2 means evaluated at ω_1 and ω_2 respectively. Thus, in order to evaluate this average, we need the autocorrelation function of the PMD vector. This problem was solved in [15], and for the vector components we have $E[\o_j(\omega_1)\o_k(\omega_2)] = \delta_{jk} g(\omega_1 - \omega_2)/3$, where δ_{jk} is Kronecker's delta, the autocorrelation function of the PMD vector is $E[\o(\omega_1) \cdot \o(\omega_2)] = g(\omega_1 - \omega_2)$, and the scalar function g is given by

$$g(x) = 3 \frac{1 - \exp(x^2 E[\Delta\tau^2]/3)}{E[\Delta\tau^2]}. \tag{20}$$

To complicate matters further, we can consider various compensation cases, by assuming certain values of the PMD vector and the signal polarization. We will also assume Gaussian pulses, for which the integrals over the pulse spectrum and the correlation function can be evaluated explicitly.

3.1. Uncompensated Broadening

In the uncompensated case, the input state of polarization is assumed isotropically distributed and uncorrelated with the PMD vector, so that we get

$$E[(\mathrm{s} \cdot \o_{in,1})(\mathrm{s} \cdot \o_{\mathrm{in},2})] = g(\omega_1 - \omega_2)/3,$$

and the average pulse broadening $b_{\rm unc}^2 = E[\sigma^2/\sigma_0^2]$ becomes [15, 16]

$$b_{\rm unc}^2 = 1 + x - \frac{1}{2}\left(\sqrt{1+\frac{4x}{3}} - 1\right), \tag{21}$$

where x is the normalized DGD defined as $x = E[\Delta\tau^2]/(4\sigma_0^2)$. In the limit of small PMD we get $b_{\rm unc}^2 = 1 + 2x/3$ because only 2 of the 3 orthogonal directions in Stokes space causes broadening (the two orthogonal to the PSPs). In the large PMD limit the x-term dominates. It is instructive to discuss the result (21) in some more detail. It can be rewritten in terms of the input pulse width σ_0 and average output pulse width $E[\sigma^2]^{1/2}$, and plotted for different amounts of average DGD in Fig. 1. It can be seen that there exist an optimum input pulse width giving the shortest average output pulse width, and this optimum is given by $\sigma_{0,\rm opt} = E[\Delta\tau]\sqrt{\pi(3\sqrt{2}-4)/8} \approx 0.109 E[\Delta\tau]$, and it is marked with a dashed line in Fig. 1. This plot shows that there is a benefit of using shorter pulse widths (i.e., RZ-like modulation) rather than long pulses (i.e., NRZ-like modulation) modulated data, since the output pulse width on average is shorter. The implication for communications systems is that at a given bitrate (pulse separation) it takes more PMD for short pulses to reach the bit-slot boundary than it does long pulses. Thus RZ performs better than NRZ in PMD-limited systems [17].

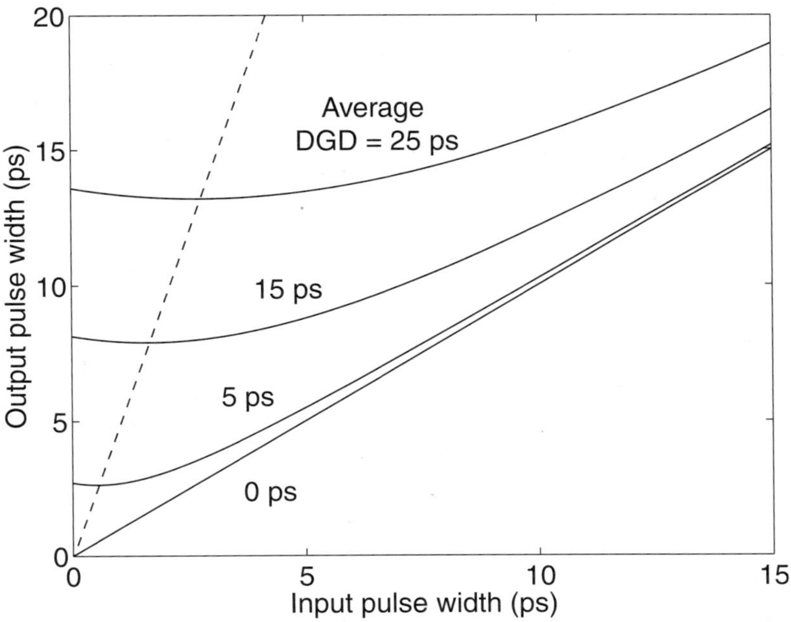

Fig. 1. Plot of output average pulse width versus input pulse width for different amounts of PMD. The dashed line goes through the minima (optimum pulsewidths) of all curves.

3.2. Compensated Broadening

There are various PMD compensation approaches to consider, although not all are analytically tractable. Most cases are discussed in [16, 17]. A useful parameter by which to compare the complexity, cost and the performance of PMD compensators is the number of *degrees of freedom (DOF)* of the compensator, which equals the number of controlled parameters of the compensator. Some non-trivial results concerning DOF are (i) the number of DOF required to transform an arbitrary SOP in to a given, constant SOP (as, e.g., required in PSP-transmission, or post-compensator with constant birefringence axes) is 2, (ii) the number of DOF required to generate an arbitrary first-order PMD vector is 3, (iii) the number of DOF of an arbitrary polarization controller (or equivalently, an arbitary orientation of the Poincaré sphere) is 3.

In the following we have computed the average pulse broadening for some different strategies of PMD compensation.

PSP-transmission (2DOF)

The PSP transmission method is obtained by putting $\mathrm{s} = \langle \o_{\mathrm{in}} \rangle \, / \, \langle |\o_{\mathrm{in}}| \rangle$, so that $E[(\mathrm{s} \cdot \o_{in,1})(\mathrm{s} \cdot \o_{in,2})] = g(\omega_1 - \omega_2)$. The resulting average broadening becomes

$$b_{\mathrm{psp}}^2 = 1 + x - \frac{3}{2}\left(\sqrt{1 + \frac{4x}{3}} - 1\right). \tag{22}$$

which is always lower than the uncompensated, and moreover, in the limit of low PMD has a lower growth rate (x^2 rather than x).

Post Compensation: Cancel $\o(0)$ (3 DOF)

One compensation approach in the receiver is to cancel out the PMD vector at the center frequency with a variable first-order PMD compensator. The input PMD vector of such a system is the vector sum of the PMD vector of the fiber plus that of the compensator, i.e. $\o_{\mathrm{in}} = \o_{\mathrm{fiber}} - \o_{\mathrm{fiber}}(0)$, which yields

$$b_{\mathrm{post},\tau(0)}^2 = 1 + \frac{5x}{3} - \frac{1}{2}\left(\sqrt{1 + \frac{4x}{3}} - 1\right) - 4\left(\sqrt{1 + \frac{2x}{3}} - 1\right). \tag{23}$$

The drawback of this method is that although it cancels out the PMD vector over the central parts of the spectrum, it actually increases the average PMD over other parts of the spectrum. This implies that for sufficiently broad signal spectra, this method will do worse than the uncompensated!

Post Compensation: Cancel $\langle \o \rangle$ (3 DOF)

One can actually do better than the above with three DOF. A better strategy is to use a compensating vector equal to the average of the PMD vector, so that the total input PMD vector becomes $\o_{\mathrm{in}} = \o_{\mathrm{fiber}} - \langle \o_{\mathrm{fiber}} \rangle$. This turns out to give exactly the same performance as the PSP method, i.e.,

$$b_{\mathrm{post},<\tau>}^2 = 1 + x - \frac{3}{2}\left(\sqrt{1 + \frac{4x}{3}} - 1\right). \tag{24}$$

Post Compensation: Optimum 3 DOF

Quite surprisingly it is possible to do even better with a 3 DOF post compensator. In order to do this we replace the PMD vector with the total system PMD vector $\o_{\rm in} = \o_{\rm fiber} + \o_{\rm c}$ in the expression for the pulse RMS width. By then differentiating the expression for the pulse broadening with respect to the compensation vector $\o_c$, and putting this gradient to zero, we get a condition from which we can determine the best compensation vector. It turns out that the optimum $\o_c$ is given by

$$\o_{c,\min} = -\langle \o_{\rm fiber}\rangle - \langle {\rm s}\rangle \frac{\langle {\rm s}\rangle \cdot \langle \o_{\rm fiber}\rangle - \langle {\rm s}\cdot \o_{\rm fiber}\rangle}{1-\langle {\rm s}\rangle^2}. \tag{25}$$

and the broadening corresponding to this choice of compensation vectors is

$$\left(\frac{\sigma}{\sigma_0}\right)^2_{\rm post,opt} = 1 + \frac{1}{4}\left(\langle \o^2_{\rm fiber}\rangle - \langle \o_{\rm fiber}\rangle^2 - \frac{(\langle {\rm s}\rangle \cdot \langle \o_{\rm fiber}\rangle - \langle {\rm s}\cdot \o_{\rm fiber}\rangle)^2}{1-\langle {\rm s}\rangle^2}\right). \tag{26}$$

For this method we have, however, not found a way to compute the average minimum broadening apart from in the limit of small x, when it is $b^2_{\rm post,opt} \approx 1 + x^2/9 + O(x^4)$ to be compared with the PSP method $b^2_{\rm psp} \approx 1 + x^2/3 + O(x^4)$. Since three degrees of freedom are required to generate an arbitrary PMD vector, we can state that this gives the best solution for a 3DOF PMD compensator. Another interesting fact from these equations is that it accounts also for higher orders of PMD, since the factor $\langle {\rm s}\rangle \cdot \langle \o_{\rm fiber}\rangle - \langle {\rm s}\cdot \o_{\rm fiber}\rangle$ is nonzero only for higher orders of PMD. Thus, this optimum accounts for higher order PMD in addition to the first order. The pulse broadening resulting from these various approaches for PMD compensation are plotted, with the uncompensated case, in Fig. 2.

The above approach for computing the average broadening can be generalized further, by using, e.g., the PMD vector and its first derivative at the carrier frequency, or by using the frequency-averaged PMD vector and its first derivative in the above formulas. The result will be slightly more complicated formulas, typically with average broadening formulas depending on x^3 to lowest order, rather than x^2 as in the above results [18].

4. System Performance and Outage Probabilities

In the above chapters we have shown that a certain amount of PMD will cause pulse broadening and hence intersymbol interference (ISI) in optical transmission lines. Generally, the amount of ISI will change randomly with the amount of PMD in the fiber, and then system penalty will also change simultaneously. In system characterization one accounts for this by computing the probability for having (due to PMD) a BER above a certain limit as, e.g., 10^{-12}. This value, often denoted as the *outage probability*, is then a useful measure of how significant PMD limits a transmission system. The pulse broadening formulas derived above give a good general picture of how PMD affects optical transmission. However, to get a good quantitative knowledge, one has to resort to numerical simulations. It then turns out that many other issues than the pulse duration will affects the system performance. The result of a few such quantitative studies on PMD limited systems will be reviewed below.

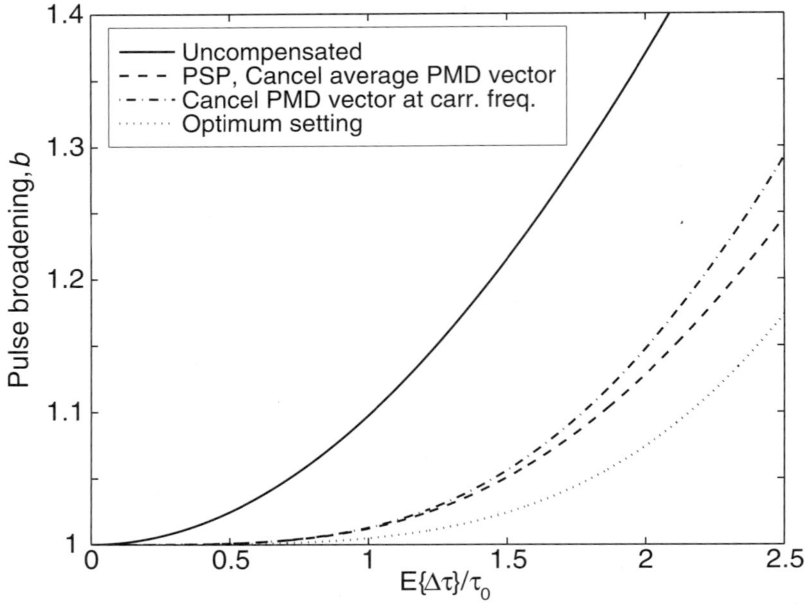

Fig. 2. Average pulse broadening b for the difference cases discussed in the text.

4.1. *PMD Compensation by Use of Robust Modulation Formats*

A straightforward means of reducing the PMD influence is to use modulation formats that have an inherent robustness to PMD. Alternatively, an increased *power margin*, i.e., the power level above which we have a BER of 10^{-12}, can be used, and one may actually trade power margin against PMD tolerance [1, 17]. It is also of interest to consider PMD compensation by use of robust modulation formats, and forward error correcting codes (FEC). Also, as we discussed earlier, RZ is more robust than NRZ to PMD for uncompensated systems, whereas the opposite is true for PMD compensated systems. The reason is that in a PMD compensated system only the lowest orders of PMD are compensated and not the higher orders. It is the higher orders of PMD that are more dominant for larger spectral widths, i.e., for RZ pulses.

Although the fiber nonlinearity prevents the power (and the power margin) to be increased indefinitely, the nonlinearity may actually in carefully designed systems be used to prevent dispersive pulse broadening. Such, nonlinear pulses, so called dispersion-managed solitons, have been shown to give a robustness to PMD comparable to first-order compensated systems [20].

4.2. *Active PMD Compensation*

A typical PMD compensating subsystem consists of a compensating element, and control system that controls the compensating element with respect to an error signal which is to be optimized. Naturally, many different error signals and compensating elements can be considered but remarkably, systems of similar complexity (degrees of freedom) often have similar performance. Common error signals that have been used in compensators are components in the detected RF spectrum and the degree of polarization (DOP). Both perform rather well for low amounts of PMD, although the RF is better and the DOP perform less well for high amounts of PMD.

PMD compensation can also be performed in the electrical domain by using various kind of equalizing circuits, but generally the performance of such are not as good as the optical ones, due the to loss of the optical polarization and phase information that occurs at the quadratic detection of the optical signal. However, these methods have the potential of being significantly less costly than the optical compensators.

A comparison of the different approaches for PMD compensation is summarized in Table 1, which shows the tolerable average PMD in percent of the bitslot for both RZ and NRZ modulation. Here the tolerable PMD is defined as the amount of PMD that gives rise to an outage probability of 10^{-3}. The data for the solitons is a bit different with respect to power margin, as specified in the footnotes. The conventional (i.e. constant dispersion) soliton system is actually worse than the corresponding linear system due to an enhanced interaction. The dispersion-manged solitons, however perform better. The table also covers more complex compensators (with more DOFs), but then problems with suboptima will arise, and the difference between a global optimum and a suboptimum may in some cases be significant.

It is interesting to note from this table that no more than an average DGD of some 30-40% of the bitslot can be tolerated by almost any compensating method, and this indicates the significant transmission obstacle that PMD is.

Table 1. Acceptable average DGD per bit slot, α, for NRZ and RZ when using various PMD mitigation and compensation methods at an outage probability of 10^{-3} and a power margin of 2 dB.

Mitigation/Compensation	NRZ $\alpha(\%)$	RZ $\alpha(\%)$
Uncompensated, 2 dB power margin	17	22
Uncompensated, 7 dB power margin	27	33
FEC, 2 dB power margin	26	28
FEC, 7 dB power margin	35	41
Linear uncompensated[a]	–	16
Conventional solitons[a,b]	–	15
DM solitons[a,b], $S = 4$	–	27
Cancel PMD vector at carrier frequency[c]	36	31
PSP transmission[c]	36	32
Fixed delay[d] (2 DOF)	36	33
Variable delay[d] (3 DOF)	43	35
Two fixed delays[d] (4 DOF) subopt.	45	33
Two fixed delays[d] (4 DOF) glob. opt.	49	43
Fixed delay+polarizer[e] (4 DOF) subopt.	34	30

[a] 1 dB Power margin
[b] $\bar{D} = 0.1$ ps/nm/km
[c] Compensator setting calculated analytically
[d] RF spectrum as feedback
[e] Average power out from the polarizer as feedback

5. WDM Issues

The above discussion has been focused on single-channel effects, which means that it will hold for each channel in a WDM system. Thus one may wonder what particular aspects that will change in a WDM implementation. From a linear transmission point of view the simple answer is: Not very much. However, some particular issues that are of particular relevance to WDM systems limited by PMD will be discussed. These are correlation bandwidth and simulation issues, nonlinear limitations, and PMD compensation in WDM systems.

5.1. Channel Separation and the Correlation Bandwidth

In most cases of relevance, the PMD between adjacent channels is not correlated to any usable significance. This is so because the autocorrelation frequency for PMD is of the order of the inverse average DGD, as can be seen from the autocorrelation function for the PMD in Eq. (20). This means that when the average DGD is so high that it will significantly affect the data on one channel (i.e., on the order of the channel data rate), the autocorrelation bandwidth is typically less than the channel spacing (which must be larger than the inverse data rate). Thus the channels must be treated as independent, from a PMD point of view.

This may cause a problem in simulations of WDM systems affected by PMD. Usually, the random PMD is in simulations emulated by a sequence of randomly oriented, birefringent waveplates. However, if the birefringence of those plates is equal the resulting global Jones matrix will be periodic in frequency, with a period equal to the inverse of the DGD of each plate. Such a periodicity is of course an artificial property of the model and will cause erroneous result in simulations of wide spectra such as systems with many WDM channels. Mathematically, this means that the autocorrelation function will be periodical, with a period equal to the inverse DGD of each element. However, it is very straightforward to avoid this artefact, by using random birefringences in the concatenated birefringent elements.

5.2. Nonlinear WDM Impairments

The onset and influence of fiber nonlinearities will cause WDM transmission to differ from a mere superposition of single channels. The dominating nonlinearites in WDM system transmission are four-wave mixing (FWM, which causes power transfer between the wavelength channels), cross-phase modulation (XPM, which causes frequency shifts of the individual channel bits), and Stimulated Raman scattering (SRS, which causes power transfer from shorter to the longer wavelength channels). Besides these we have self-phase modulation which acts on single channels. Of these FWM is particularly troublesome when upgrading dispersion shifted fiber to WDM, and it will practically limit the use of DSF to very few WDM channels due to the four-wave mixing crosstalk. A common way that has been proposed to relax this problem is to use orthogonal polarization between adjacent WDM channels, since FWM only transfers power between copolarized signals. Thus by carefully adjusting the polarizations of the WDM channels, the FWM crosstalk might be significantly reduced. However, in presence of PMD, the orthogonalization between the adjacent channels will not be maintained. In fact, one may show that on average, the states of polarization at two frequencies ω_1 and ω_2 is correlated as [21, 22]

$$E[\mathrm{s}(\omega_1, z) \cdot \mathrm{s}(\omega_2, z)] = \mathrm{s}(\omega_1, 0) \cdot \mathrm{s}(\omega_2, 0) \exp\left(-\frac{\Delta\omega^2 E[\Delta\tau^2]}{3}\right) \qquad (27)$$

from which we see that the correlation bandwidth is inversely proportional to the average DGD. Since the average DGD squared, $E[\Delta\tau^2]$ grows linearly with z, the polarization correlation will fall off exponentially along the fiber. This means that in a WDM system with PMD, the orthogonality will not be maintained, and as a rule of thumb, the average DGD should be less than $1/(4\Delta f)$ where Δf is the channel separation[22]. Typically in a WDM system with channel separation of 100 GHz, this requires a DGD less than 2.5 ps, which is much more restrictive than the requirement for single-channel PMD robustness discussed above.

5.3. Multichannel PMD Compensators

In principle, PMD compensators should be implemented on each channel in a WDM system. However, a certain amount of PMD compensation will be gained if only one compensating element is used to reduce the PMD influence on several channels simultaneously. There are also some other approaches, specific for WDM systems, that has been proposed. These results will be reviewed in this section.

Since the PMD over the different WDM channels is uncorrelated in most cases of interest, a straightforward approach is to use a few free extra channels for back-up, and simply drop the channel or channels thats suffer too high PMD penalty. This and similar approaches were discussed and quantified by simulations in [23, 24].

To use a single compensator for a multitude of channels, one can use a single-channel device with a large number of degrees of freedom [25], so that it in principle can be tuned to cancel out the PMD vector over a broad wavelength range. There are at least two limitations in the control system that makes this approach problematic. If only one error signal is used, it may be difficult to find an algorithm that ensures low PMD over a broad range when trying to find the optimium. Also, when a large number of DOFs are used, the performance will be limited by the chance of being locked in a local suboptimum. It is here crucial to emphasize the importance of the error signal. A PMD compensator can never perform better than the feedback it gets from its error signal. Unless the error signal reflects the signal quality over the entire bandwidth that should be compensated, the compensator cannot be expected to work satisfactory. One simple step in this direction was suggested in [26], where by using a single detector, the sum of all RF spectra can be used a an error signal. There is one final drawback of broadband compensators that should be mentioned, and that is that the birefringent element used for compensation will increase the average PMD outside the compensating band. This means that for a given compensator, there will be an upper limit of the number of channels compensated. This is illustrated in Fig. 3. The figure shows, schematically, the outage probability versus number of channels. As the number of channels and the used wavelength range increases, sooner or later (at the critical number N_{crit}), a channel will reach outside the compensated range. In that range the average DGD squared is actually *larger* than the uncompensated, being the square product of the

DGDs in the fiber and in the compensator: $E[\Delta\tau^2_{\rm fiber}] + E[\Delta\tau^2_{\rm comp}]$. Obviously this is above the DGD in the fiber, with a corresponding higher outage probability. Then to the question of what exactly $N_{\rm crit}$ is, this will obviously vary from case to case, but we believe it is likely to be around 3–5.

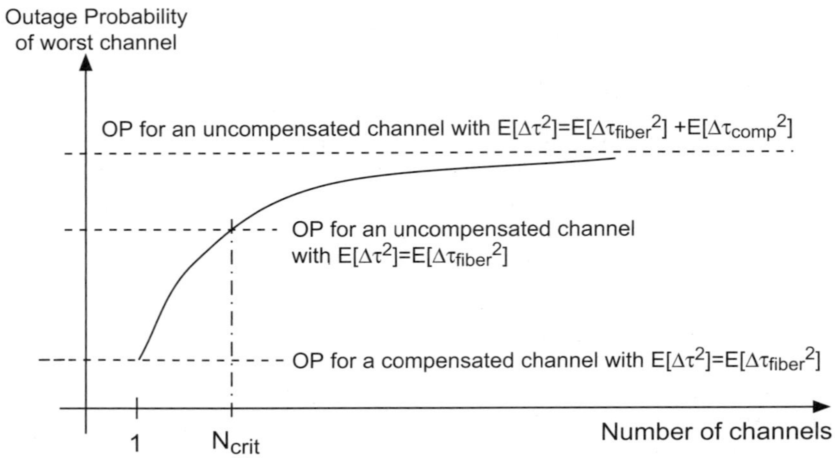

Fig. 3. Schematic plot of outage probability of worst channel in a WDM system utilizing a single compensating element. The compensator adds PMD to the channels outside the compensating bandwidth, so that for sufficiently many channels, above $N_{\rm crit}$, the worst channel will have an OP worse than in the uncompensated system.

A final approach to broadband PMD compensator operation is to distribute the compensators along the transmission line [27, 28]. In that way the PMD are limited before it reaches high values, but on the other hand also here the control system might be a source of problem. An attractive feature of this approache is that compensating elements need not be included—it is enough to just install polarization controllers along the transmission line! This will then avoid the problem illustrated in Fig. 3. However, this approach meet several practical problems: Global optimization might be difficult to ensure, and also instabilities might arise as a number of control loops that affect each other should work together, probably with long delay times for control signals in between due to their physical separation.

6. Temporal Drift Issues

The temporal drift of the PMD is what really makes PMD such a serious obstacle. There are a few things that need to be emphasized in connection with the temporal drift. The first issue is that the typical drift time of a system is a characteristic time scale of every individual transmission link that need to be measured in the field. The basic origin of the drift is mainly mechanic perturbations (giving rise to fast drifts) and temperature drifts (giving rise to much slower drifts), and it need to be pointed out that all polarization changes along the entire link accumulates to give a net global drift time which can be much faster than the local drift time.

Another important issue to understand is that the although the PMD vector and the absolute polarization state drifts, the absolute polarization drifts *faster* than the PMD vector. This is so because here really all changes of the polarization state along the transmission line add up. For the PMD vector, the largest drifts in length are arising from the central parts of the transmission line, and the largest drifts in direction from one of the endpoints. As an example, make the thought-experiment that a polarization controller (PC) is inserted near the input of a transmission line. Then the output polarization state $\mathrm{s}_{\mathrm{out}}$ will be affected by this polarization controller, but not the output PMD vector $\o_{\mathrm{out}}$ The input PMD vector $\o_{\mathrm{in}}$, however, is affected by the PC near the input, as will the input SOP s_{in}.

The difference in drift times between the PMD and the SOP can be put in more mathematical terms by computing autocorrelation functions (ACF) for the different drifts according to [21]

$$E[\mathrm{s}(t_1) \cdot \mathrm{s}(t_2)] = \exp(-|t_1 - t_2|/t_d), \tag{28}$$

$$E[\o(t_1) \cdot \o(t_2)] = E[\Delta\tau^2] \frac{1 - \exp(-|t_1 - t_2|/t_d)}{|t_1 - t_2|/t_d}. \tag{29}$$

If the width at, e.g., half the maximum of these functions are compared, we find that it is nearly a factor of two larger for the PMD ACF than for the SOP ACF, so in this sense one can say that the SOP drifts about a factor of two faster than the PMD vector.

This effect has consequences for the necessary speed of operation of PMD compensators. A PMD compensator based on the PSP-transmission method must respond on the time scale of the PMD drift, whereas a post-compensator with a compensating element near the receiver should respond much faster, on the time scale of the SOP drift.

7. Conclusion

In conclusion, we have derived formulas for the PMD-induced RMS pulse broadening, and discussed how pulses are affected by PMD, both deterministic and random. We have also presented and derived formulas for the average pulse broadening in PMD compensated and uncompensated systems, and reviewed simulation results that are likely to be of help in system design. Typically, the various methods for compensation give a modest factor of approximately 2 of increase in the tolerable average PMD. Finally we ha ve discussed some relevant aspects of PMD with respect to WDM systems and the issues related to the temporal drift of PMD.

Acknowledgments

We wish to acknowledge numerous discussions with and input from Peter Andrekson and Chongjin Xie on PMD and PMD conmpensation. Also the input and previous work of F. Matera on the WDM parts is acknowledged.

References

1. H. Sunnerud, C. Xie, M. Karlsson, and P. A. Andrekson, "Polarization Mode Dispersion in High-Speed Fiber-Optic Transmission Systems", J. Lightwave Technol., **20**:2204–2219 (2002).
2. C. D. Poole and R. E. Wagner, "Phenomenological approach to polarization dispersion in long single-mode fibers", Electron. Lett., **22**:1029–1030 (1986).
3. N. Gisin and J. P. Pellaux, "Polarization mode dispersion: time versus frequency domains", Opt. Commun., **89**:316–323 (1992).
4. M. Karlsson, "Polarization mode dispersion induced pulse broadening in optical fibers", Opt. Lett., **23**:688–690 (1998).
5. L. F. Mollenauer and J. P. Gordon, "Birefringence-mediated timing jitter in soliton transmission", Opt. Lett., **19**:375–377 (1994).
6. J. P. Gordon and H. Kogelnik, "PMD fundamentals: Polarization mode dispersion in optical fibers", Proc. Natl. Acad. Sci., **97**:4541–4550 (2000).
7. D. Andrescani, F. Curti, F. Matera, and B. Daino, "Measurement of the group-delay difference between the principal states of polarization on a low-birefringence terresterial cable", Opt. Lett., **12**:844–846 (1987).
8. N. Gisin and B. Huttner, "Combined effects of polarization mode dispersion and polarization dependent losses in optical fibers", Opt. Commun., **142**:119–125 (1997).
9. D. Anderson and M. Lisak, "Analytic study of pulse broadening in dispersive optical fibers", Phys. Rev. A, **35**:184–187 (1987).
10. C. D. Poole, J. H. Winters, and J. A. Nagel, "Dynamical equation for polarization mode dispersion", Opt. Lett., **16**:372–374 (1991).
11. G. J. Foschini and C. D. Poole, "Statistical theory of polarization mode dispersion in single mode fibers", IEEE J. Lightwave Technol., **9**:1439–1456 (1991).
12. G. Bosco, B. E. Olsson, and D. J. Blumenthal, "Pulsewidth Distortion Monitoring in a 40-Gb/s Optical System Affected by PMD", IEEE Photon. Technol. Lett., **14**:307–309 (2002).
13. N. Gisin, "Solution of the dynamical equation for polarization dispersion", Opt. Commun., **86**:371–373 (1991).
14. M. Karlsson, "Probability density functions of the Differential Group Delay in optical fiber communication systems", J. Lightwave Technol., **19**:324–331 (2001).
15. M. Karlsson and J. Brentel, "The autocorrelation function of the polarization-mode dispersion vector", Opt. Lett., **24**:939–941 (1999).
16. H. Sunnerud, M. Karlsson, and P. A. Andreksson, "Analytical theory for first-order PMD compensation", IEEE Photon. Technol. Lett., **12**:50–52 (2000).
17. H. Sunnerud, M. Karlsson, and P. A. Andrekson, "A Comparison Between NRZ and RZ Data Formats with Respect to PMD-induced System Degradation", IEEE Photon. Technol. Lett., **13**:448–450 (2001).
18. S. Kim, "Analytical calculation of pulse broadening in optical higher order PMD compensation ", J. Lightwave Technol., **20**:1118–1123 (2002).
19. H. Sunnerud, C. Xie, M. Karlsson, R. Samuelsson, and P. A. Andrekson, "A Comparison Between different PMD-compensation techniques", J. Lightwave Technol., **20**:368–378 (2002).

20. C. Xie, M. Karlsson, P. A. Andrekson, H. Sunnerud, and J. Li, "Influences of polarization-mode dispersion on soliton transmission systems", J. Select. Topics Quant. Electron., **8**:575–590 (2002).

21. M. Karlsson, J. Brentel, and A. Andreksson, "Longterm measurement of PMD and polarization drift on optical fibers", J. Lightwave Technol., **18**:941–951 (2000).

22. J. Hansryd, H. Sunnerud, P. A. Andreksson, and M. Karlsson, "Impact of PMD on four-wave mixing crosstalk in WDM systems", IEEE Photon. Technol. Lett., **12**:1261–1263 (2000).

23. S. Särkimukka, J. Djupsjöbacka, A. Gavler, and G. Jacobsen, "Mitigation of Polarization Mode Dispersion in optical multichannel systems", J. Lightwave Technol., **18**:1374–1380 (2000).

24. L. Moller and J. H. Sinskyi, "Analysis of multichannel pmd mitigation strategies based on worst channel equalization", In Proc. of European Conference on Optical Communication, ECOC'01, vol. 2, pages 212–213, 2001.

25. R. Noe, D. Sandel, S. Hinz, M. Yoshida-Dierolf, V. Mirvoda, G. Feise, H. Herrmann, R. Ricken, W. Sohler, F. Wehrmann, C. Glingener, A. Schöpflin, A. Färbert, and G. Fischer, "Integrated $LiNbO_3$ distributed polarization mode dispersion compensator at 20 gb/s transmission system", Electron. Lett., **35**:652–653 (1999).

26. R. Khosravani, S. A. Havstad, Y. W. Song, P. Ebrahimi, and A. E. Willner, "Polarization mode dispersion compensation in WDM systems", IEEE Photon. Technol. Lett., **13**:1370–1372 (2001).

27. M. Karlsson, P. A. Andrekson, and H. Sunnerud, "High-speed otdm and impact of pmd. In Proc. of LEOS 12th annual meeting LEOS,99, vol. 1, pages 343–344, 1999.

28. J. M. Fini and H. A. Haus, "Accumulation of polarization mode dispersion in cascades of compensated optical fibers", IEEE Photon. Technol. Lett., **13**:124–126 (2001).

DOI: 10.1007/s10297-004-0014-0
Originally published in J. Opt. Fiber. Commun. Rep. 1, 201–209 (2004)

Polarization effects and performance of fiber optic recirculating loops

Brian S. Marks[1,2], Gary M. Carter[1,2] and Yu Sun[1]

[1]Department of Computer Science and Electrical Engineering
University of Maryland Baltimore County
1000 Hilltop Circle, Baltimore, MD 21227, USA
[2]Laboratory for Physical Sciences, 8050 Greenmead Drive
College Park, MD 20740, USA
Email: marks@umbc.edu
carter@umbc.edu
yusun@umbc.edu

Abstract. Polarization effects such as PMD and PDL in a straight-line optical fiber communication system are random due to the random birefringence of optical fiber. However, since there is periodicity in a fiber recirculating loop, the polarization effects are quite different than in a straight line. Theoretically and experimentally we analyze the polarization effects in a loop and their effect on performance and the evolution of the state of polarization of a channel. Using this analysis, we are able to better understand the physical impact of the evolution of the state of polarization of a channel on performance. We show that loop-synchronous polarization scrambling is an effective way of breaking this periodicity in an attempt to make the recirculating loop perform more like a straight-line system.

Recirculating loops have been long used as simple testbeds for the evaluation of long-haul, straight-line transmission systems. They provide a relatively inexpensive way to study transmission over vast distances, and especially, the accumulation of nonlinear effects and other optical phenomena that occur over long distances. In fact, transmission distances have been achieved in loops that are unimaginable in any other transmission medium, with transmission distances over 1,000,000 km [1,2].

By their very nature, recirculating loop experiments are periodic in the light path. For this reason, transmission behavior can differ significantly from what one would expect in a typical straight-line system, and can often be improved. In particular, it has been known for some time that polarization effects in recirculating loops are important due to the periodicity. In contrast, of course, straight-line systems exhibit random polarization behavior—birefringence and polarization-dependent losses and gains—as light propagates through the transmission path.

For several years now, our research group at UMBC has been performing experiments on a recirculating loop that has been used to propagate dispersion-managed solitons through distances of over 20,000 km [3]. It is a 107-km loop with a dispersion map comprised of 100 km of dispersion-shifted fiber with normal dispersion, compensated by 7 km of standard anomalous single-mode fiber. Through careful analysis of the physics, we have been able to achieve incredibly close agreement between theory and experimental results that has enabled us to increase performance by optimizing the dispersion map and amplifier location as well as other system parameters. We have been able to demonstrate close comparison between theory and experiment in terms of pulse-width variation in a dispersion map, noise growth in the marks and spaces, timing jitter, and eye diagrams [4, 5]. All models of this loop to date, however, used only a scalar treatment of the propagation equation, neglecting one polarization mode. The reason that this scalar approach works well will become evident in the following discussion.

Recently, we have performed a number of experiments to understand the polarization dependence of the performance of the loop. To optimize performance in the loop experiment, we can vary three polarization controllers: one at the input and two within the loop. As seen in Fig. 1, changing the polarization controllers in the loop has a significant impact on both the bit error rate (BER) and the degree of polarization (DOP) of the signal as a function of distance. The lowest BER corresponds to a setting of the polarization controller that causes a signal to remain mostly polarized for the entire transmission distance. We wanted to better understand this correlation between polarization behavior in the loop and system performance.

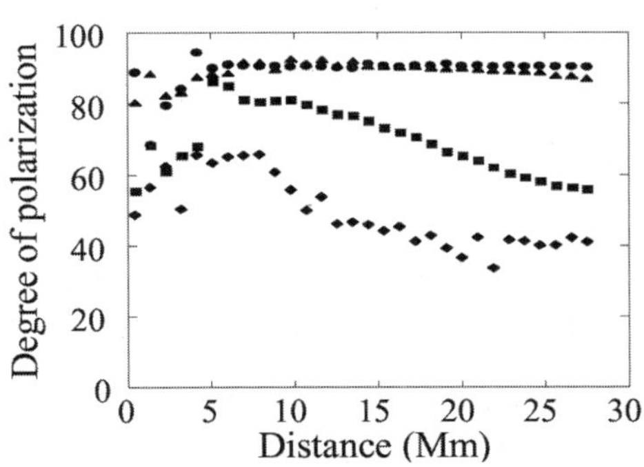

Fig. 1. Degree of polarization as a function of distance for four different settings of one of the in-loop polarization controllers in the recirculating loop yielding BER of 10^{-9} (circles), 10^{-6} (triangles), 10^{-4} (squares), and 10^{-1} (diamonds).

In a straight-line system, one expects the BER or Q to vary randomly as a function of time since environmental conditions randomize the fiber's orientation, and one can obtain a distribution of BER or Q for different fiber realizations. In a loop, we can obtain a similar distribution of Q values as a polarization controller in the loop is varied. We first adjusted the orientations of the three polarization controllers to optimize performance, and then collected statistics of Q values at 5000 km as one of the in-loop polarization controllers was varied randomly. After Q values for 200 settings of the polarization controller were obtained, we determined the probability density function (pdf) given by the solid curve in Fig. 2. The loop's Q factor distribution was then compared to that of a simulated straight-line system with similar parameters as our loop, shown by the dotted curve in Fig. 2, and a simulation of the loop, given by the dashed curve. One can easily see that the pdf for the loop is significantly spread toward higher Q values as compared to that of the straight-line system. Through simulations, we have seen that this Q-distribution is always spread, and often is double-humped, with the lower peak aligned with the single peak of the straight-line Q-distribution. Hence, we discovered that due to the loop's periodicity and the polarization dependence within the loop, performance is artificially enhanced.

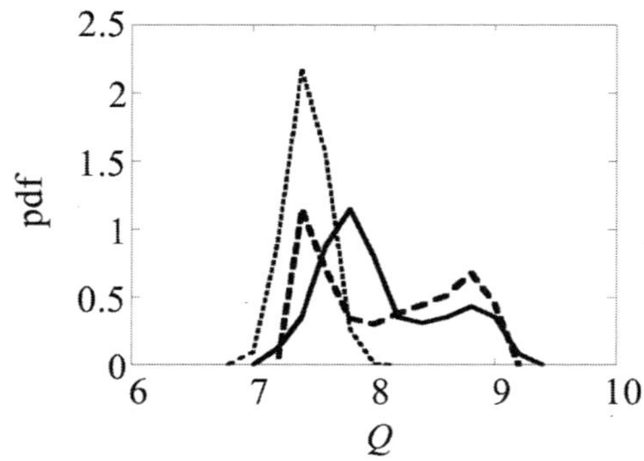

Fig. 2. The Q distribution for different settings of an in-line polarization controller from a loop experiment (solid), from simulation of the loop (dashed), and from a simulation of a straight-line system with similar parameters (dotted). Note that the Q distribution is spread toward higher Q values for the loop results as compared to the simulation of the comparable straight-line system. The round trip PDL was 0.1 dB for all distributions.

In any fiber system, the important polarization effects are polarization-mode dispersion (PMD) in the fiber, polarization-dependent loss (PDL) due to optical isolators and WDM couplers in the amplifiers and other components, and polarization-dependent gain (PDG) in the amplifiers. Because we use relatively new fibers in our loop experiment, the PMD does not lead to significant signal distortion, but does serve to randomly rotate the polarization state of the channel as the signal propagates through the system. Of the polarization effects in the loop, PDL and PDG have the most significant impact on the performance, but they primarily affect performance by changing the signal-to-noise ratio during propagation, and not by distorting the waveform of the signal. Therefore, the gain saturation in amplifiers also plays a role in its interaction with polarization effects.

To perform simulations of our recirculating loop, we use a reduced model that follows only the average Stokes parameters and powers of channels and the noise in the system [6]. Because it does not consider waveform distortion of the signal, it is a significantly faster simulation tool than a full Fourier split-step method, but it considers the effects of PMD, PDL, PDG, gain and loss, and gain saturation in the amplifiers. Using this reduced model not only gives information about the Stokes parameter evolution of a channel, but also provides a ΔQ penalty associated with these polarization effects. In the simulations of our loop, we lump the PDL in one round trip at one point in the loop, just before the loop switch. This is a reasonable reduction for a relatively short loop, and we have measured a round trip aggregate PDL of about 0.1–0.35 dB per round trip, depending on the quality of the isolators we used in the amplifiers.

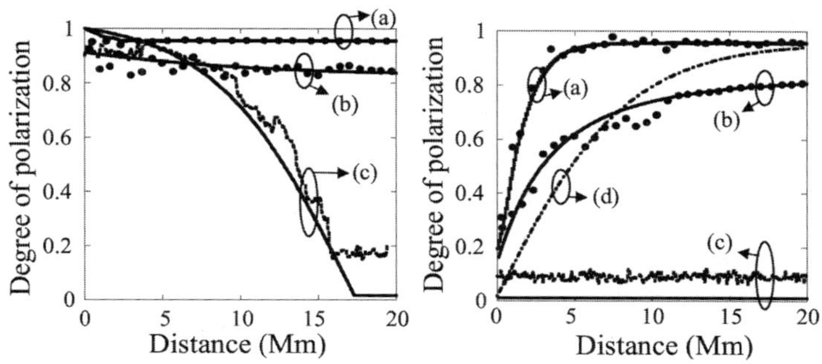

Fig. 3. Experimental (circles and dotted curves) and simulation (solid) results showing the effect of PDL and PDG on the DOP of a signal as it propagates through the recirculating loop. The left plot shows the DOP of the signal and noise together, and the right plot is for noise in the absence of signal for (a) a measured loop PDL of 0.35 dB and simulated PDL of 0.45 dB per round trip, (b) a measured PDL of 0.22 dB and simulated PDL of 0.2 dB per round trip, and (c) measured and simulated PDL of 0.1 dB per round trip. In (a)–(c), the PDG is 0.05 dB per amplifier. Curve (d) shows simulation results for PDL of 0.1 dB per round trip, but PDG = 0 dB.

We propagated a single channel of data in our loop and monitored the DOP of the signal once per round trip as it propagated to about 20,000 km. The experiment and simulation results are plotted in Fig. 3. These results show that if the PDL is measured to be 0.35 dB per round trip, the setting of the polarization controller corresponding to the best performance causes the signal to propagate with a high DOP for the entire propagation distance. We also turned off the signal and allowed only the noise to propagate. In this case, noise, which is initially unpolarized, tends to polarize as it propagates through the system. When the PDL per round trip is sufficiently lowered, the noise stays unpolarized, and the signal becomes more unpolarized as it is allowed to propagate. The cause of this depolarization is the PDG, and curve (d) in Fig. 3 shows that if PDG is turned off, the noise has the opportunity to repolarize, due to the PDL.

In Fig. 1, we showed a relationship between BER and the DOP evolution of a signal for different settings of the polarization controller in the loop. Since the loop was also shown to significantly change the Q-distribution as well, as shown in Fig. 2, we are interested in understanding if the reason for this change was due to the evolution of the polarization state of a channel as it propagates through the system. If changing the polarization controllers in the loop changes the evolution of the polarization state in the loop, we can better understand the reason for the DOP and Q behavior.

In both simulations and experiments, we measured the Stokes parameters of the channel after each round trip of the loop. For trials from the upper end of the Q-distribution shown in Fig. 2, the polarization state on the Poincaré sphere spirals inward to a point and converges quickly, as shown in Fig. 4(a). Trials from the lower end of the distribution exhibit outward spirals, as shown in Fig. 4(c), which can even converge on the opposite side of the Poincaré sphere. Trials from the central portion of the distribution evolve in a nearly circular trajectory on the sphere, as shown in Fig. 4(b), which converges slowly if at all. Each type of behavior will be explained in the analysis presented here.

One can understand the relationship between this spiral behavior and the Q factor in a straightforward way. The PDL in the system has the effect of causing 100% transmission on one side of the Poincaré sphere while causing loss on the other side. In a loop system, the axis of low loss does not change orientation each round trip. Therefore, if the signal continually aligns itself with the polarization state of the low loss each round trip, it will transmit unchanged, while any noise aligned with the opposite side of the sphere—the high-loss axis—suffers loss, increasing the signal-to-noise ratio. In Fig. 4(a), the spiral is on the same hemisphere as the low-loss axis of the PDL, which leads to a high Q. If, on the other hand, the signal continually aligns itself with the high-loss axis of the PDL each round trip, it will suffer loss as compared to the noise, degrading the signal-to-noise ratio. This is shown in Fig. 4(c), where the spiral is on the same hemisphere as the high-loss axis of the PDL. Since the axes of the PDL elements in a straight-line system would be random, this behavior would not be present in a straight line.

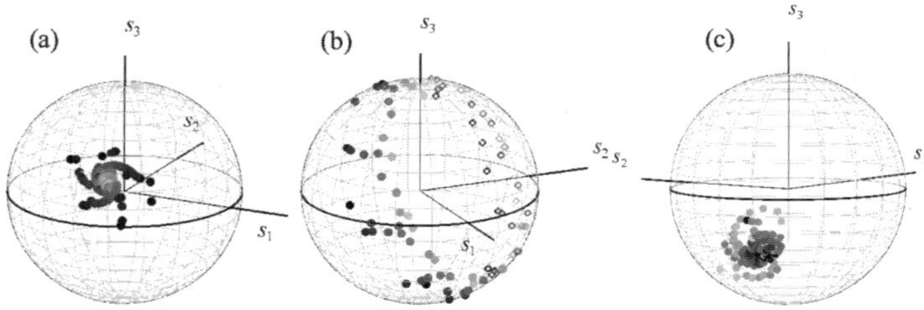

Fig. 4. Experimental data showing the polarization state of a single channel in the recirculating loop on the surface of the Poincaré sphere after each round trip. The grayscale from black to light gray indicates increasing propagation distance, and the unfilled circles indicate points on the far side of the sphere. (a) corresponds to a high Q value and shows a spiral inward. (b) corresponds to a medium Q value and shows a circular trajectory around the sphere. (c) corresponds to a low Q value and shows an outward spiral.

We can understand the evolution of the polarization state on the Poincaré sphere through a simple mathematical analysis. Because a recirculating loop system is inherently periodic, we may write its Jones transfer matrix after n round trips as $\mathbf{T} = \mathbf{M}_{\text{loop}}^n$, where $\mathbf{M}_{\text{loop}}$ is the Jones matrix for one round trip of the loop. This 2×2 Jones matrix has a pair of complex eigenvectors $\mathbf{u}_\pm$ and corresponding eigenvalues. An eigenvector of $\mathbf{M}_{\text{loop}}$ is also an eigenvector of $\mathbf{T}$, and if λ is an eigenvalue of $\mathbf{M}_{\text{loop}}$, then λ^n is an eigenvalue of $\mathbf{T}$. It follows that if the input is an arbitrary field of the form

$$\mathbf{u}_{\text{in}} = c_+\mathbf{u}_+ + c_-\mathbf{u}_-, \tag{1}$$

then the output field vector is

$$\mathbf{u}_{\text{out}} = \mathbf{T}\mathbf{u}_{\text{in}} = c_+\lambda_+^n\mathbf{u}_+ + c_-\lambda_-^n\mathbf{u}_-. \tag{2}$$

If we let $\lambda_\pm = \mid \lambda_\pm \mid \exp(i\phi_\pm)$ and $c_\pm = \mid c_\pm \mid \exp(i\psi_\pm)$, then the corresponding output Stokes vector is $\mathbf{s}_{\text{out}} = (\mathbf{u}_{\text{out}}^\dagger\sigma_3\mathbf{u}_{\text{out}}, \mathbf{u}_{\text{out}}^\dagger\sigma_1\mathbf{u}_{\text{out}}, -\mathbf{u}_{\text{out}}^\dagger\sigma_2\mathbf{u}_{\text{out}})^T$, yielding

$$\mathbf{s}_{\mathrm{out}} = |\, c_+ \,|^2 |\, \lambda_+ \,|^{2n}\, \mathbf{s}_+ + |\, c_- \,|^2 |\, \lambda_- \,|^{2n}\, \mathbf{s}_- + 2\,|\, c_+ \,||\, c_- \,|\,(|\, \lambda_+ \,||\, \lambda_- \,|)^n \times \mathrm{Re}[\mathbf{s}_c e^{-i(n\Delta\phi + \Delta\psi)}], \tag{3}$$

where $\mathbf{s}_\pm$ are the Stokes vectors associated with the Jones eigenvectors $\mathbf{u}_\pm$, $\mathbf{s}_c \equiv (\mathbf{u}_+^\dagger \sigma_3 \mathbf{u}_-, \mathbf{u}_+^\dagger \sigma_1 \mathbf{u}_-, -\mathbf{u}_+^\dagger \sigma_2 \mathbf{u}_-)^T$, $\Delta\phi = \phi_+ - \phi_-$, and $\Delta\psi = \psi_+ - \psi_-$. In the above expressions, the σ_k are the standard Pauli spin matrices. In Eq. (3), if the eigenvalues have different magnitudes, then one obtains spiral motion. The spiral centers are given by one of the polarization eigenstates $\mathbf{s}_\pm$, and the rotation is described by the third term, where the rotation angle of the spiral due to one round trip of the recirculating loop is given by the phase difference of the eigenvalues, $\Delta\phi$. In addition, the ratio of the magnitudes of the eigenvalues gives the relative decay rates of the coefficients, thus yielding the rate of convergence to a spiral center.

In order to understand the spiral behavior in a loop with PDL, we consider a transfer matrix of the form

$$\mathbf{M}_{\mathrm{loop}} = \mathbf{M}_{\mathrm{PDL}} \mathbf{M}_{\mathrm{rot}},$$

where $\mathbf{M}_{\mathrm{rot}}$ is a unitary matrix representing a fixed rotation on the Poincaré sphere giving the random rotation due to the fiber and do to the loop's polarization controller, and $\mathbf{M}_{\mathrm{PDL}}$ represents the effect of PDL in one round trip of the loop. Note that in this formulation, we have neglected the effects of noise, PDG, and amplifier saturation. Our reduced model simulations do consider these effects and show that the predominant effects that alter the polarization state of a signal are the random fiber rotation and the PDL. Since an arbitrary rotation on the Poincaré sphere can be expressed as a rotation through an angle γ about an axis given by a unit vector $\mathbf{s}_{\mathrm{rot}} = (x, y, z)^T$, the matrix $\mathbf{M}_{\mathrm{rot}}$ can be expressed in the form

$$\mathbf{M}_{\mathrm{rot}} = \begin{bmatrix} \cos(\gamma/2) + ix\sin(\gamma/2) & -(z - iy)\sin(\gamma/2) \\ (z + iy)\sin(\gamma/2) & \cos(\gamma/2) - ix\sin(\gamma/2) \end{bmatrix}.$$

The effect of the PDL is modeled by the Jones matrix

$$\mathbf{M}_{\mathrm{PDL}} = \begin{bmatrix} 1 & 0 \\ 0 & 1 - \epsilon \end{bmatrix},$$

where the strength of the PDL is given by a small non-negative parameter ϵ that is less than one. Note that in this notation, the low-loss axis of the PDL is given by $\hat{\mathbf{s}}_{\mathrm{PDL}} = (1, 0, 0)^T$ in Stokes space, but the results are general.

We exploit the fact that ϵ is small by using standard perturbation methods to expand our eigenvector in powers of ϵ, solving for corrections at each order [7]. We seek eigenvalues and eigenvectors that satisfy $(\mathbf{M}_{\mathrm{loop}} - \lambda\mathbf{I})\mathbf{u} = \mathbf{0}$ and expand λ and $\mathbf{u}$ in power series in ϵ as $\lambda_\pm = \lambda_0 + \epsilon\lambda_1 + O(\epsilon^2)$ and $\mathbf{u}_\pm = \mathbf{u}_0 + \epsilon\mathbf{u}_1 + O(\epsilon^2)$. From these expressions, we can compute the Stokes eigenvector in powers of ϵ as well, obtaining $\mathbf{s}_\pm = \mathbf{s}_\pm^{(0)} + \epsilon\mathbf{s}_\pm^{(1)} + O(\epsilon^2)$. Doing this type of expansion yields, to $O(\epsilon)$, the Stokes eigenvectors

$$\mathbf{s}_\pm = \pm\hat{\mathbf{s}}_{\mathrm{rot}} + \frac{1}{2}\epsilon[\hat{\mathbf{s}}_{\mathrm{PDL}} \mp \hat{\mathbf{s}}_{\mathrm{rot}} + (\hat{\mathbf{s}}_{\mathrm{rot}} \times \hat{\mathbf{s}}_{\mathrm{PDL}})\cot(\gamma/2)] + O(\epsilon^2). \tag{4}$$

Note that due to the corrections, the eigenstates in Stokes space are not antiparallel in the presence of PDL, unlike the principal states for PMD [8]. This is illustrated in

Fig. 5. We may use this expansion to compute the output Stokes vector for an arbitrary input state, as well, yielding

$$\begin{aligned}\mathbf{s}_{\text{out}} &= A_+^n \mid c_+ \mid^2 \mathbf{s}_+ + A_-^n \mid c_- \mid^2 \mathbf{s}_- + 2B^n \mid c_+ \mid\mid c_- \mid \\ &\times [\hat{\mathbf{t}}_1 \sin(n\gamma + \Delta\psi) - \hat{\mathbf{t}}_2 \cos(n\gamma + \Delta\psi)] + \cdots, \end{aligned} \tag{5}$$

where $\hat{\mathbf{t}}_1 = (\hat{\mathbf{s}}_{\text{rot}} \times \hat{\mathbf{s}}_{\text{PDL}}) / \mid \hat{\mathbf{s}}_{\text{rot}} \times \hat{\mathbf{s}}_{\text{PDL}} \mid$ and $\hat{\mathbf{t}}_2 = \hat{\mathbf{s}}_{\text{rot}} \times \hat{\mathbf{t}}_1$, and where we are neglecting higher-order contributions. In this expression,

$$A_\pm = 1 - \epsilon(1 \mp \hat{\mathbf{s}}_{\text{rot}} \cdot \hat{\mathbf{s}}_{\text{PDL}}) + O(\epsilon^2)$$

and

$$B = 1 - \epsilon + O(\epsilon^2).$$

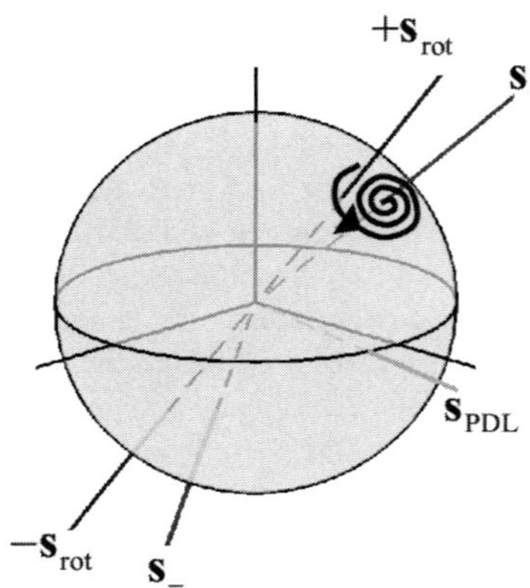

Fig. 5. Pictorial description of the spiral on the Poincaré sphere. Without PDL, the eigenstates are $\pm\mathbf{s}_{\text{rot}}$, but PDL causes the eigenstates $\mathbf{s}_+$ and $\mathbf{s}_-$ to be no longer antiparallel. In this figure, $\mathbf{s}_{\text{rot}}$ is drawn in the same hemisphere as $\mathbf{s}_{\text{PDL}}$, the low-loss axis of the aggregate round-trip PDL, causing the polarization state to spiral toward $\mathbf{s}_+$, the attracting eigenstate.

From Eq. (5), one can see that the B term contains sinusoidal pieces that correspond to the spiral or circular motion on the Poincaré sphere. If $\hat{\mathbf{s}}_{\text{rot}}$ is in the same hemisphere as $\hat{\mathbf{s}}_{\text{PDL}}$ as in Fig. 5, then $A_+ > B > A_-$, and the $\mathbf{s}_+$ term in Eq. (5) dominates as n increases. This means that $\mathbf{s}_+$ is the attracting point, or the center of an inward spiral on the sphere. As n increases, also, there is a rotation by the angle γ around the attractor each round trip, as seen in the B term in Eq. (5). If $c_+ > c_-$ so that the input polarization state is close to the attracting eigenstate $\mathbf{s}_+$, one observes an inward spiral similar to the left-most figure in Fig. 4. In this case, the polarization state of the signal spends most of its time in the same hemisphere as the low-loss axis of the PDL, while half the noise, in the other hemisphere, is attenuated, improving the signal-to-noise ratio. Similarly, if $c_+ < c_-$ so that the input polarization state is on the opposite side of the sphere from the attracting eigenstate, one observes an outward spiral away from the other eigenstate $\mathbf{s}_-$, which is a repeller. This is shown in the right-hand figure of Fig. 4. In this case, the signal's polarization state spends most of its time in the hemisphere of the high-loss axis of the PDL, and therefore suffers degradation, reducing the signal-to-noise ratio. If $\hat{\mathbf{s}}_{\mathrm{rot}}$ is in the opposite hemisphere as $\hat{\mathbf{s}}_{\mathrm{PDL}}$, then $A_+ < B < A_-$ and the $\mathbf{s}_-$ term in Eq. (5) dominates as n increases, and we have similar behavior as before. In the case where $\hat{\mathbf{s}}_{\mathrm{rot}}$ is orthogonal to $\hat{\mathbf{s}}_{\mathrm{PDL}}$, all three terms in Eq. (5) remain equally dominant regardless of the input polarization state. This leads to a circular trajectory on the Poincaré sphere that does not converge to any particular point. In this case, the polarization state of the signal alternates between the hemisphere of low-loss and the hemisphere of high-loss, and has a moderate Q factor because of this.

Clearly, this analysis and simulation show that the periodicity in the loop leads to interesting polarization evolution in the system. The periodicity plus the PDL leads to the spiral trajectories on the Poincaré sphere that, given a particular setting of the polarization controllers, can artificially enhance performance. However, if one is interested in using a recirculating loop to emulate a straight-line system, one wants to break this periodicity. One way in which this has been done is to use loop-synchronous scrambling, in which the polarization state of the channel is rotated on the Poincaré sphere once each round trip. Simulations with the reduced Stokes model have shown that the polarization state remains quite randomized as a function of round-trip number when this scrambling is done. In fact, S. Lee et al. first used this concept in conjunction with an element with high DGD to emulate PMD in a long system [9]. Additionally, our simulations and experiments have shown an improved Q distribution as compared to a straight-line system, when this loop-synchronous scrambling is done using a lithium-niobate based polarization controller.

If one's goal is to eliminate the dependence of polarization effects on performance from the system, this loop-synchronous scrambling may be insufficient. For single-channel systems, PDG causes polarization dependence in a well-polarized signal since it provides excess gain orthogonal to the signal. In a WDM system, PDG is not a problem since PMD causes a randomization of the polarization states of each channel with respect to one another over long distances, and therefore, the DOP decreases. For a single-channel system, however, one can prevent this polarization dependence by scrambling the polarization state of the signal at the input as well as once per round trip. This scrambling causes the DOP to drop thereby reducing the PDG. Simulations with the reduced model show significant narrowing of the Q distribution when this method is implemented. This reduction of polarization dependence in the recirculating loop is a subject of ongoing research in our group.

In conclusion, we have analyzed, experimentally and theoretically, the behavior of recirculating loop systems with polarization effects. We have shown that polarization effects in a loop experiment can complicate their performance and can cause unexpected differences between a loop and the comparable straight-line system. In particular, we understand the performance differences by having characterized the evolution of the polarization state of a channel in a loop system. We also have begun to explore modifications to our loop system to make it behave more like a straight-line system and to make it less polarization sensitive.

References

[1] M. Nakazawa, E. Yamada, H. Kubota, and K. Suzuki, Electron. Lett. **27**, 1270–1272 (1991).

[2] G. Raybon, Y. Su, J. Leuthold, and R.-J. Essiambre, OFC 2002 Technical Digest, Anaheim, CA, paper FD10.

[3] G. M. Carter, R.-M. Mu, V. S. Grigoryan, C. R. Menyuk, P. Sinha, T. F. Carruthers, M. L. Dennis, and I. N. Duling, III, Electron. Lett. **35**, 233–234 (1999).

[4] R.-M. Mu, V. S. Grigoryan, C. R. Menyuk, G. M. Carter, and J. M. Jacob, IEEE J. Select. Topics Quant. Electron. **6**, 248–257 (2000).

[5] R. Holzloehner, H. N. Ereifej, G. M. Carter, and C. R. Menyuk, OFC 2002 Technical Digest, Anaheim, CA, paper ThQ2.

[6] D. Wang and C. R. Menyuk, J. Lightwave Technol. **19**, 487–494 (2001).

[7] J. Kevorkian and J. D. Cole, *Multiple Scale and Singular Perturbation Methods* (Springer-Verlag, New York, 1996).

[8] B. Huttner, C. Geiser, and N. Gisin. J. Select. Topics Quant. Electron. **6**, 317–329 (2000).

[9] S. Lee et al., OFC 2001 Technical Digest, Anaheim, CA, paper WT2.

DOI: 10.1007/s10297-004-0015-z
Originally published in J. Opt. Fiber. Commun. Rep. 1, 283–303 (2004)

PMD compensation techniques

Henning Bülow[1] and Stéphanie Lanne[2]

[1]Alcatel Research & Innovation, Holderaeckerstr. 35
D-70499 Stuttgart, Germany
Email: henning.buelow@alcatel.de
[2]Alcatel Research & Innovation, Route de Nozay
91460 Marcoussis, France
Email: stephanie.lanne@alcatel.fr

Abstract. Mitigating PMD is still a challenge in today's optical communications. After giving the key performance indicators dictated by the statistical nature of PMD, we introduce the classification of PMD compensation schemes into 2 categories: optical and electrical PMD compensators. In a first part, we explain the operation principle of a PMD compensator by taking a detailed look at the basic optical PMD compensator and appropriate feedback signals. More complex multistage structures and a feed-forward adaptation approach are also discussed. This first part is closed by results from a one-year field trial confirming the behavior and performance of a prototype compensator. In the second part of this article, electronic equalization for PMD mitigation is explained. Starting with a discussion on performance and adaptation of linear equalizers suitable for analog electronic signal processing, finally also the Viterbi equalizer basing on digital signal processing is analyzed. A comparative review of mitigation by optical or electronic means concludes the discussion.

PMD in Optical Fiber Transmission Systems

The statistical dimension of PMD, coming from variation in time of the external stresses, makes it a peculiar phenomenon compared to chromatic dispersion. There cannot be guarantee of a maximum penalty due to PMD in a transmission system: whatever penalty one may choose as an upper limit, there is always a chance, called outage probability, to go beyond the limit.

From the transmission matrix of a system exhibiting PMD, one can define mathematically a "first order" (neglecting the frequency dependence of PMD) frame of work: the two orthogonal principal states of polarisation (PSPs). On those PSPs, a pulse is propagated undistorted but the group velocity is different (there is a slow and a fast PSP). The difference in arrival time is called the differential group delay (DGD). This mathematical frame is only valid when the signal spectrum bandwidth is small compared to the frequency dependence of the PSPs and the DGD [1].

The PMD value for a fiber (in ps/$\sqrt{\mathrm{km}}$ or ps) is the mean value over time (or wavelength) of the DGD. It is usually considered that the maximum tolerable PMD is between 10 and 20% of the bit duration [2, 3]. Typical PMD values of installed fibers are greater than 1.5ps/$\sqrt{\mathrm{km}}$, which limits to a few 100 km the transmission distance at 10 Gb/s. For ultra-long-haul systems at 40Gb/s, the limitation of the transmission distance is determined by the PMD of line fiber ($< 0.1\mathrm{ps}/\sqrt{\mathrm{km}}$ for currently manufactured fibers), but also the dispersion compensation modules and the amplifiers. The maximum tolerable PMD value without compensation is rapidly reached. For higher values, compensation is required. The subject of this paper is to review compensation schemes.

General View on PMD Compensation

A slight improvement of the robustness to PMD can be obtained by keeping the receiver threshold in the middle of the eye diagram degraded by ISI. This automatic threshold adjustment leads to the PMD limit of approximately 15% of the bit period T (0.15 T). Modulation formats like return-to-zero (RZ) tolerate a slightly higher bit broadening due to PMD and thus further increase the maximal tolerable PMD (0.18 T). The tolerance to PMD can further be increased if there is room to allocate a power margin (OSNR margin) for PMD of much more than 1 or 2 dB. This margin can be obtained by reducing span loss or by the incorporation of forward error correction (FEC).

However, in general only a limited margin will be reserved for PMD and a higher PMD limit should be overcome. Therefore, several active compensation techniques have been proposed to mitigate the degrading effect of PMD. The majority of these approaches can be classified into optical and into electrical compensation techniques that apply optical signal processing within an optical PMD compensator unit, or post-detection electronic signal processing of the photo diode signal by an electrical equalizer within the receiver (cf., Fig. 1) [4, 5].

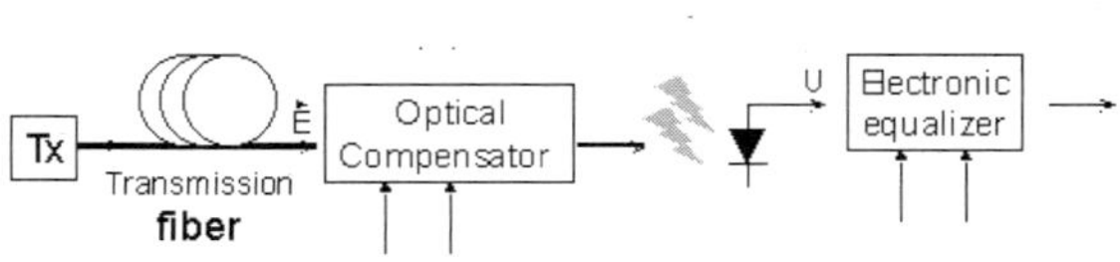

Fig. 1. Optical transmission system with optical and electronic PMD mitigators.

However, neither in the electrical domain nor in the optical domain, the time-varying transfer characteristics of the signal path can be measured without affecting the transmitted signal. Therefore, the inverse transfer function (or matrix) can only be approximated and the amount of the compensation can only be estimated through measurements (used as feedback signals) of the transmitted signal.

Time variations of PMD due to temperature changes are quite slow: a few degrees per second on the Poincaré sphere [2]. However, they can also come from human operation, mechanical vibrations or wind (for aerial fibers), and in some cases they can reach 50 revolutions per second on the sphere [6, 7].

Another way to "compensate" for PMD effect is to align the input state of polarization of the signal to one of the PSPs [8]. However, it requires a feedback signal coming back to the transmitter and hence only works for systems with very long time scales of variation. This means that human disturbance cannot be taken into account for instance.

In the following section, the operation principle of a PMD compensator will be explained by a detailed look at the basic optical PMD compensator and appropriate feedback signals. Then more complex multistage structures and a feed-forward adaptation approach will shortly be discussed. This first part will be closed by results from a one-year field trial. In the second part of this article electronic equalization for PMD mitigation will be explained and a comparative review of mitigation by optical or electronic means will conclude the discussion.

Optical PMD Compensators

Principle of Operation of Basic Optical PMD Compensators

A closer look at PMD mitigation in the optical domain makes it obvious that the principle of operation of a PMD compensator is to reduce the total PMD of line + compensator. Therefore, the compensator is generally composed of highly birefringent elements (polarisation maintaining fibers [9], $LiNbO_3$ delays [10], Bragg gratings [11]) separated by polarisation controllers to mimic the line.

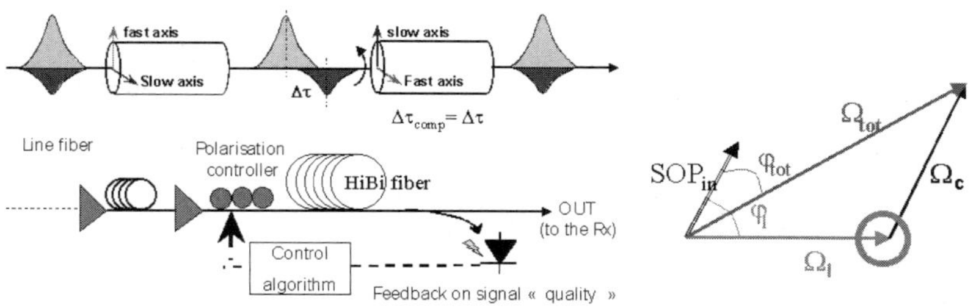

Fig. 2. Basic scheme of a PMD compensator (left) and geometrical representation in the Stokes space (right).

Figure 2 shows the basic scheme of a PMD compensation, also referred to as "first order PMD compensator": one fixed highly birefringent element (e.g., a piece of polarization maintaining fiber), one polarization controller to orientate the axes of the counteractive element, a measure of the signal "quality," used by a control algorithm. The driving algorithm of the counteractive element is a key point. It runs in a tracking mode, optimizing the feedback signal. Since PMD conditions are not dynamically measured the algorithm operates in a "blind" mode: it keeps tracking a local optimum of the feedback signal. Precisely, this is a limiting factor in PMD compensation. As we will see in the following, local, suboptimum operating points can have disastrous impact on the transmission quality [12].

The PMD conditions can be represented to the first order in the Poincaré space by the PMD vector Ω (which orientation is one of the principal state of polarization and which magnitude is the DGD) and the input state of polarization, $\mathrm{SOP_{in}}$ (Fig. 2). The critical parameters of the line are the DGD (i.e., $\mathrm{DGD}_l = \mid \Omega_l \mid$) and the angle ϕ_l between $\mathrm{SOP_{in}}$ and Ω_l. The PMD vector of the compensator Ω_c has a fixed length and its orientation (i.e., two parameters in space) is determined by the polarization controller driven in order to maximize the signal "quality" after the PMD mitigator.

The PMD conditions after the PMD mitigator are represented to the first order by $\Omega_{\mathrm{tot}} (= \Omega_l + \Omega_c)$ and ϕ_{tot}. During operation, the compensator will minimize either (Case A) the total DGD ($\mid \Omega_{\mathrm{tot}} \mid$) or (Case B)$\phi_{\mathrm{tot}}$ [9,13].

In a first step we have looked at the first local maximum of the feedback. This is the way a PMD compensator works, since reaching the absolute maximum of the feedback would require to temporarily go through a minimum of the feedback, yielding a loss of data. Furthermore, it can be shown that an absolute maximum may become a local maximum when PMD conditions randomly vary. In a second step, we have looked at the absolute maximum of the feedback to see the theoretical improvement.

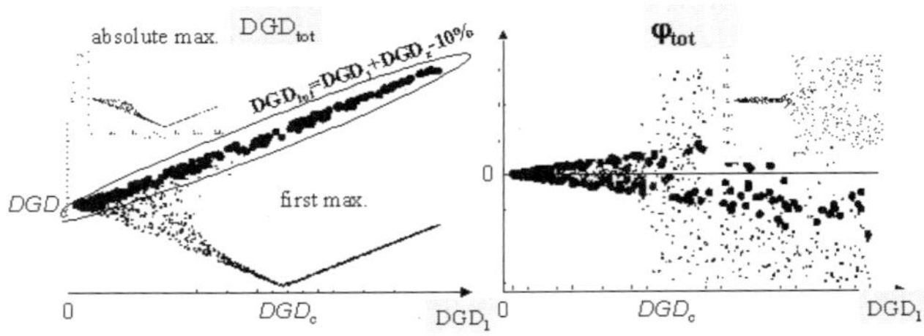

Fig. 3. $\mathrm{DGD_{tot}}$ versus DGD_l (left) and ϕ_{tot} versus DGD_l (right). Main graphs: first maximum search. Inserts: best maximum.

Figure 3 shows $\mathrm{DGD_{tot}}$ versus DGD_l and ϕ_{tot} versus DGD_l respectively. The wide dots are defined by

$$\mathrm{DGD_{tot}} > \mathrm{DGD}_l + \mathrm{DGD}_c - 10\%. \tag{1}$$

$\mathrm{DGD_{tot}}$ always ranges between $\mid \mathrm{DGD}_l - \mathrm{DGD}_c \mid$ and $\mathrm{DGD}_l + \mathrm{DGD}_c$. Figure 3 clearly indicates that for DGD_l lower than $\approx \mathrm{DGD}_c/2$, the compensator will reduce ϕ_{tot} (Case B), while it will almost always minimize or maximize $\mathrm{DGD_{tot}}$ for the largest DGD_l (Case A). However, when $\mathrm{DGD_{tot}}$ is maximized, ϕ_{tot} is reduced (Fig. 3). Such behavior of the compensator has been verified experimentally (Fig. 4). These operating points, corresponding to the lowest values of ϕ_l, can be explained by a minimization of higher-order effects. Note that they exhibit the highest penalties. This means that the worst case of PMD without compensation (largest DGD, equal splitting between PSPs) does not necessarily correspond to the worst case after compensation.

The preceding analysis highlights the necessity to include higher-order PMD in the assessment of the compensators performance. Indeed, since the DGD of the total system is reduced, the PMD value can be increased. Higher orders of PMD scaling with the PMD value, they will no longer be negligible.

Feedback Signals for Optical PMD Compensators

The main feedback signals used in various schemes are

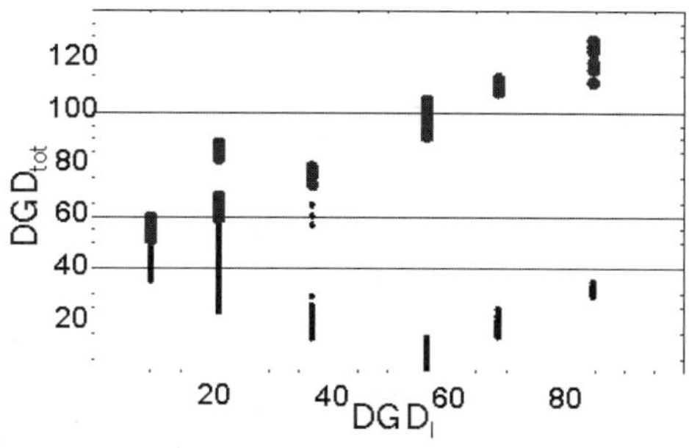

Fig. 4. DGD_{tot} versus $DGD_{|}$, experimental results for $DGD_c = 53$ps. First maximum search.

1. The degree of polarization (DOP) [14]. The pulse splitting induced by PMD results in time varying state of polarization along the pulse and therefore reduces the DOP. It is measured with a polarimeter setup formed by polarizers and four photodiodes. This feedback signal only works with optical compensation but is bit-rate independent.
2. A further feedback signal which has been applied several times is the "spectral line" feedback [10, 15, 16]. It is based on the analysis of the electrical spectrum of the detected signal at compensator output. Since PMD generates a notch in the spectrum at the frequency 0.5/DGD [17], the maximization of spectrum samples, e.g., at 1/4 and 1/2 of the bitrate, indicates a decreasing residual PMD distortion.
3. The analysis of the detected eye diagram at decision time provides a measure with the best correlation to the BER. This can be accomplished by an electronic eye monitor that allows to extract the eye opening or an estimate on the Q factor at PMD compensator output [18, 19]. As shown in Fig. 5, this eye monitor can be realized by an additional monitor decision gate integrated within the receiver electronics and working in parallel to the main decision gate. It analyzes the compensated signal by a sweep of the monitor decision threshold through the eye.

An improvement of the simple PMD compensator was already demonstrated experimentally by applying eye monitor feedback after DOP pre-setting [20]. The eye opening is a commonly used feedback parameter. The opening is the difference of the threshold position at mark and space where the measured BER (pseudo error rate) exhibits a specific target value, e.g., 10^{-2}. The "1"s of the monitor in Fig. 5 are counted by an analog integrator that converts the pseudo error rate into a voltage U_{int}. The level of this pseudo bit-error rate is a trade-off between response time and sensitivity.

The main features of a feedback signal are

- its sensitivity: its variation for a small variation of the PMD conditions;
- its correlation with the BER. The higher the correlation, the better the feedback signal;
- its response time: in order for the PMD compensator to be able to react within the required time, the response time of the feedback signal should be much lower in

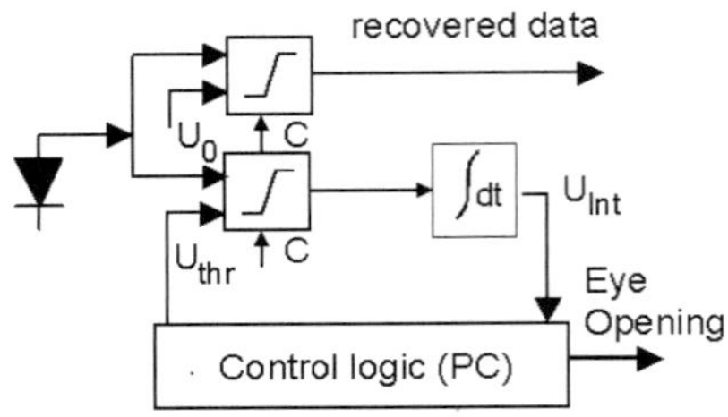

Fig. 5. Receiver (upper decision gate) with eye monitor operating in parallel (lower decision gate)

order for the algorithm to process the information. A compromise has to be made between response time and accuracy.

Improved Concepts for Optical PMD Compensators

Guided by the picture that a real fiber PMD can be modelled by a cascade of many birefringent elements (PMF) and polarization controllers (PC), more efficient PMD compensators have been proposed based on a cascade of two (see Fig. 6) or more [10, 21] of the basic compensator as shown in Fig. 6. The birefringence of each of the polarization maintaining fibers amounts to about 40 % of the bit period [20].

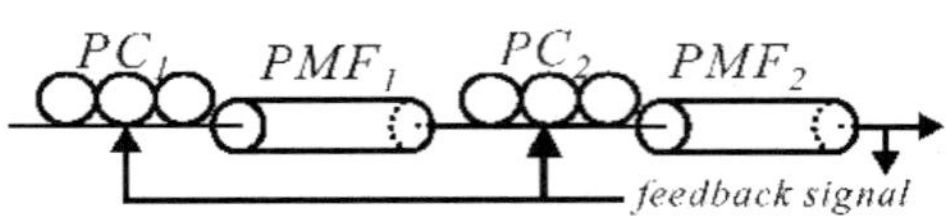

Fig. 6. Two-stage optical PMD compensator.

Detailed numerical and experimental analysis of this two-stage structure at 40 Gbit/s shows that a further improvement compared to a one-stage structure can be achieved [22]. On one hand, this is demonstrated by the measurements drawn in Fig. 7 where for a given PMD generated by a statistical emulator more than 1% (10^{-2} relative occurrence) of PMD draws induce a penalty in excess of 5 dB, which is reduced to 2 dB by a one-stage compensator and further reduced to 1 dB by a two-stage setup [20]. Extrapolation of these experimental results to system relevant low outage probability (10^{-5}) predicts a maximum tolerable PMD (PMD limit) of about 40% of the bit period for the two-stage structure. Note that this performance can only be achieved when for tracking of a drifting PMD distortion a feedback signal is used which is strongly correlated to the BER. For the experiments an eye monitor was incorporated. With DOP as feedback parameter the likelihood increases of being trapped in a nonoptimum local maximum [22].

In theory, the more birefringent elements there are, the more the compensator is able to address all orders of PMD in the link. Though in lab experiments compensation with up to 64 stages have been demonstrated [10, 21], it is questionable whether the increasing number of adaptation parameters can efficiently be controlled in the feedback adaptation scheme in a time scale convenient for real system application. In practice the PMD limit may not always increase with the number of elements.

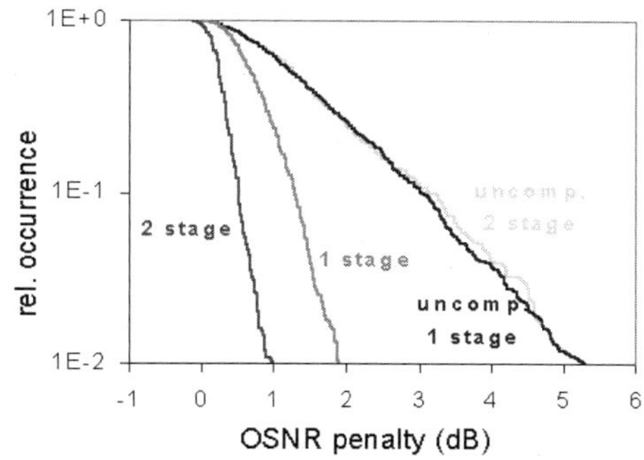

Fig. 7. Occurrence (y-axis) of an OSNR penalty beyond a limit (x-axis) measured without PMD compensator, with a one-stage PMDC, and with a two-stage PMDC (cf., Fig. 6).

Some attention has been paid to PMDC concepts incorporating tunable birefrigent elements. Some technologies, like fiber Bragg gratings, allow to realize a variable DGD in the compensator [11, 23, 24, 25], which has the potential to avoid trapping in local maxima [26]. In a combined feed-forward/feedback adaptation scheme a proper setting of the DGD element was demonstrated thanks to a polarisation scrambler at the input of the line that enables the compensator to extract the actual DGD from the received signal polarization spectrum [23].

Lattice Filter Compensator and Feed-forward Adaptation

From the picture of PMD compensation as to build the exact PMD mirror image of the line, the understanding shifted also to apply signal processing techniques to the signal field [27, 28, 29]. With this shift of the viewpoint the attention also moved away from compensator structures formed by cascaded birefringent elements (Fig. 6) and was focused on optical signal processing structures. Such structures are already well established for group velocity dispersion compensation, namely, lattice filters that have the form of a Mach–Zehnder interferometer (MZI) cascades consisting of phase tunable symmetric MZIs and asymmetric MZIs having a delay t typically less than a bit period. Different to chromatic dispersion compensation, PMD compensation has to be formulated in terms of two filter input ports rather than only one input port [30]. This is illustrated by Fig. 8 showing a representative waveguide structure where the input signal is decomposed into two orthogonally polarized components by a polarization beam splitter (PBS). These two signals are then launched into the two input ports of the lattice filter. Owing to a 90° rotation of one of the two linearly polarized signals of the PBS, the two input signals are able to interfere within the filter. The realization of PMDCs based on different types of lattice filters has been demonstrated with planar lightwave circuits (PLC) with thermooptic phase shifting elements [31, 32, 33].

Besides the potential of a compact realization, a further advantage of integrated optic (IO) lattice filters is the possibility to tap out intermediate feedback signals and to use them for the reduction of the number of adaptation parameters by autonomously operating internal feedback loops [34]. However, with regard to adaptation the biggest

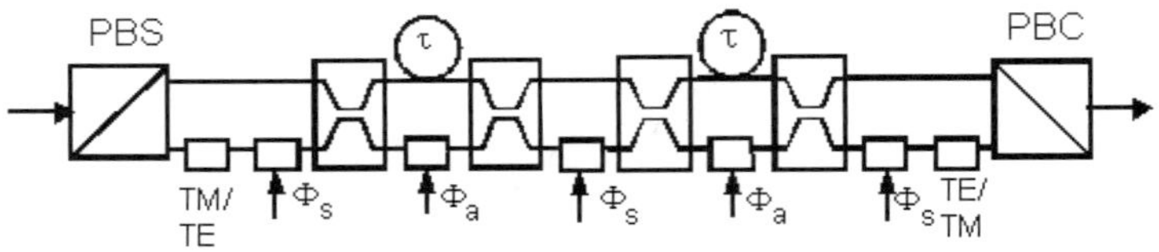

Fig. 8. Lattice filter waveguide structure for PMD compensation.

advantage of IO lattice filters might be that they offer phase stable compensator structure which, in conjunction with a feed-forward adaptation scheme, should allow to overcome the bottleneck of the conventional feedback scheme. As already indicated in the last chapter, in the feedback adaptation a time consuming consecutive dithering of the tuning parameters need to be applied since the scalar feedback signal neither contains information on the tuning direction nor at which parameter tuning is efficient. Moreover, most of the birefringent elements and polarization controllers are not stable and their phase drifts have to be taken into account too.

Different to this, the feed-forward adaptation scheme has the inherent potential to come to a faster and reliable adaptation even of highly efficient multistage structures. Owing to the stable and known transfer function of the lattice filter a measurement of the polarization spectrum of the distorted input signal allows to calculate the appropriate setting of the filter in a control unit, and to command the tuning parameter setting directly in one step. Beside the basic structure of the feed-forward-adapted PMDC sketched out in Fig. 9, an alternative realization is possible where the spectrum is measured at the lattice filter output. A realization of integrated-optic fast scanning narrow band polarimeter was proposed that is based on a thermally tuned repetitively scanning PLC ring resonator [35].

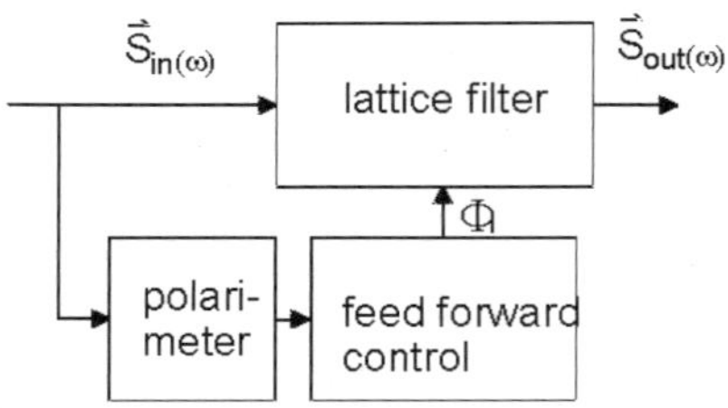

Fig. 9. Feed forward adaptation of a PMD compensator (lattice filter) based on the polarization spectrum measurement of the input field S_{in}.

Field Trials

Several field trials, lasting from a few hours to several months, have proved that PMD compensation is a viable solution [36, 37, 38, 39]. Here we describe in more details a two-month field trial involving a basic single-stage PMD compensator [36].

The objective is to validate the behavior of such an active device in a realistic network environment. The compensator operates in a configuration as close as possible to a real system in terms of polarisation variation speed, DGD evolution over time, long-term measurements. Furthermore, PMD measurements are simultaneously performed in order to interpret the compensator behavior. As we shall see, the results are compliant with our expectations.

Field-trial Setup

The transmission line consists of 3 spans of G652 standard single-mode fibre, totalling 320 km. A dual-stage amplifier including a chromatic dispersion compensation module precedes each span. The launch power is 7 dBm and the dispersion map has been optimized to give the maximum chromatic dispersion tolerance. The residual chromatic dispersion is 300 ps/nm.

The fibers were carefully chosen among France Telecom optical network infrastructure in order to achieve a total PMD value of 41 ps. Such a high value allows the compensator efficiency test in severe conditions.

The signal at the receiver side is split into two arms by a coupler. Only one of them supports the optical PMD compensator. The receiver detects the signal from either one or the other alternatively and can then compare them.

The optical PMD compensator consists of a $LiNbO_3$ polarization controller and a polarization maintaining fiber of 50 ps differential group delay (DGD). A control algorithm maximizing the degree of polarization drives the polarization controller. As will be shown in the following paragraph, the natural behavior of the fiber leads to a good coverage of the PMD conditions. Added to the fact that the transmitter and the receiver are commercial ones, this shows a realistic behavior of the compensator.

Measurement Procedure

Firstly, the periodical measurement cycle includes continuous monitoring of the output state of polarization of the line. The correlation time of the fiber is estimated to be a few hours. With our apparatus, polarization variation speed up to one revolution over the Poincaré sphere per second can be measured.

The BER measurement from both the compensated and uncompensated arm is performed with the same optical power level on the receiver. In order to remain in the same PMD conditions during the whole measurement cycle, and therefore be able to compare the obtained BER, measurement time slots had to be kept compatible with the variation rate of the link.

In addition to BER, the instantaneous DGD on the line is also monitored. During such measurements, requiring the use of a tuneable laser source, the PMD compensator operation is frozen, but the compensator is never reset. Therefore, there is no perturbation of the PMD compensator operation.

The probability density function of the DGD measured on the line over one year is shown in Fig. 10, along with a Maxwellian fit. The agreement is quite good, showing a PMD value of 41 ps.

Results and Discussion

Measurements of the BER are carried out on the naturally evolving fiber over two months with the prototype board. The correlation of the BER with and without PMD compensation is plotted in Fig. 11.

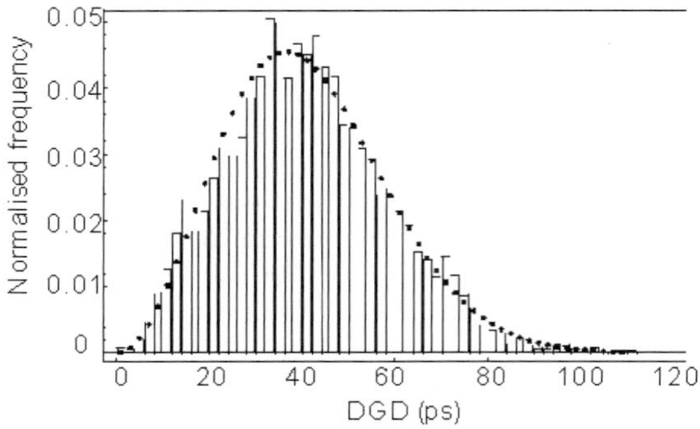

Fig. 10. Histogram: Probability density function of DGD measured over one year on installed SMF fiber. Dotted line: Maxwellian fit with 41 ps of PMD.

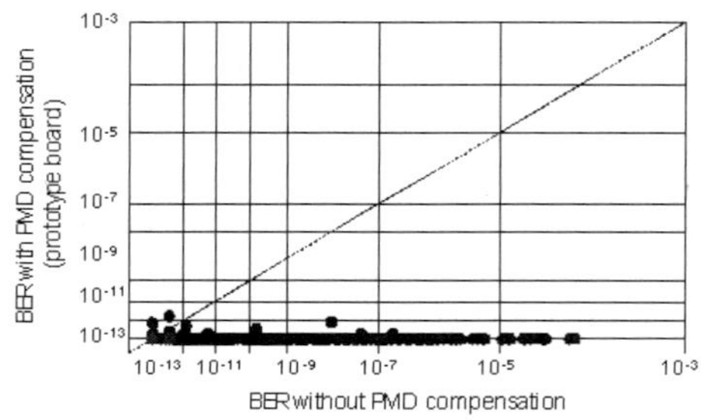

Fig. 11. BER with compensation versus BER without compensation, for two-month data.

First, note that PMD can actually cause very important degradation of the signal: the BER without compensation commonly reaches values above 10^{-6}. For these BER values, the PMD compensator dramatically improves the BER, even in the presence of recorded polarisation variations of more than one revolution over the Poincaré sphere per second; considering a 10° variation of the polarization state over the Poincaré sphere as a significant change of target for the compensator, the tracking speed of the latter must be in the range of tens of milliseconds.

The compensator can degrade slightly the BER, however that occurs only for very low BER (less than 10^{-12}) and BER after compensation remains acceptable. The corresponding data points are above the diagonal in Fig. 11. In order to assess the compensator global performance, some more statistical processing of the data is needed, as exposed in [38]. We shall not detail the results here, but focus on another feature of PMD compensators.

The variations of PMD conditions over time, even if they are slow, can lead the PMD compensator to change its operating point very abruptly. This typical behaviour, due to our PMD compensation algorithm that always tracks the best operating point, is illustrated in Fig. 12. The degree of polarization measured by the PMD compensator, and used as a feedback signal, is shown as a function of time over one day. The optimum operating point found by the compensator at the beginning of the day slowly degrades into a local optimum as PMD conditions evolve. At $t = 19$ h, the optimum operating point tracking leads to an abrupt improvement of the DOP value, and of the BER after compensation (not shown here). In spite of the rapid change in the compensator, no loss of data was observed.

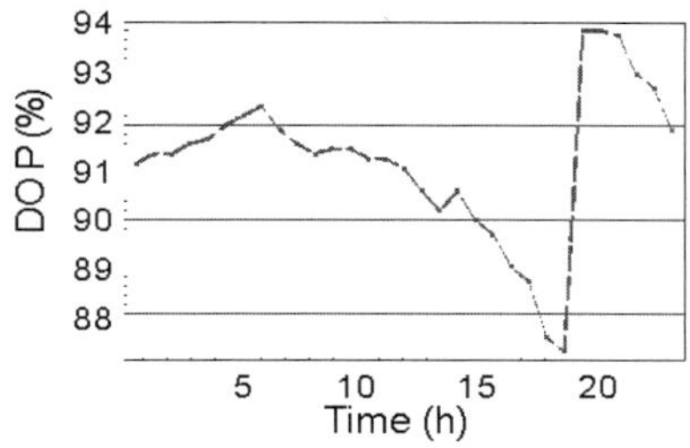

Fig. 12. Degree of polarization (DOP) measured at the output of the PMD compensator over one day.

We have assessed the performance of a simple PMD compensator over installed fiber, keeping all parameters as close as possible to those of a realistic system. Our measurements, taken over two months, clearly show the efficiency of the compensator, which reduces drastically very high BER degradation yielded by PMD. We have also observed typical behavior of active compensation device, on a prototype board. This field trial proved the technical reliability of optical PMD compensators.

Electronic PMD Mitigation

Basic Electronic Equalizers

More than one decade ago the discussion was stimulated to use post-detection electronic signal processing to mitigate penalty induced in the optical domain [40]. Electronic equalization bears the potential of seamless integration in the receiver electronics and thus leading to a cost-effective solution for penalty mitigation. Signal processing schemes well established in lower rate communications have mainly been investigated for application to detected optical signal. They are feed-forward equalizer (FFE), which superimposes differently delayed and weighted input signal replica, decision feedback equalizer (DFE), which subtracts a weighted and delayed part of the already decided bit(s), as well as a concatenation of both (cf., Fig. 13). A DFE realization more suitable for high-speed integration [41] is also shown in Fig. 16. There two decisions with different thresholds are provided simultaneously and one of both is selected by a multiplexer in a digital feedback loop depending on the last decision. Integrated circuits (IC) exclusively designed for 10-Gb/s signal processing and based on SiGe or GaAs technology [42, 43, 44, 45, 46] have been manufactured for lab use and also first product prototypes are already on the market. First ICs suitable for 40-Gbit/s operation might appear in research labs in the next time (2004).

The power penalty for 10-Gbit/s (bit period $T = 100$ ps) operation versus the PMD distortion plane span by the differential group delay DGD) $\Delta\tau$ and the relative power γ in the fast principal state of polarization (PSP) are drawn in Figs. 14(a) and 14(b)

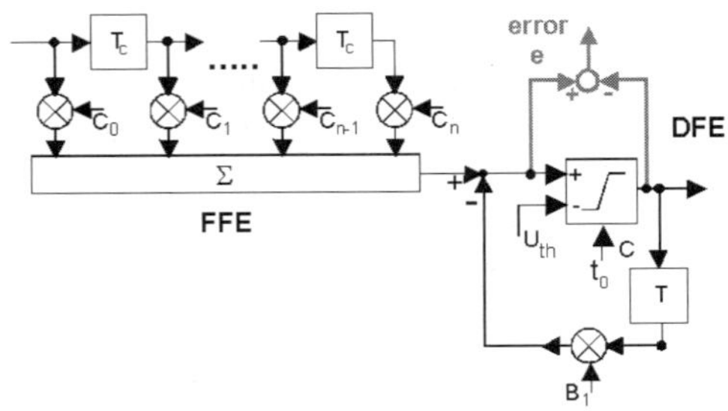

Fig. 13. Electronic equalizer formed by feed-forward equalizer (FFE) and decision feedback equalizer (DFE).

for a standard receiver (Rx) and a receiver incorporating linear equalization by a FFE, respectively. The taps c_i of the FFE are determined for a given PMD by minimizing the mean square error MSE $= \langle [s(t_0) - d(t_0)]^2 \rangle$ at decision time t_0 at FFE output [47]. Signal independent noise has been assumed for the evaluation of the penalty (denoted here as thermal noise, THN). Though the initial penalty of the receiver is reduced by the FFE (21 taps, tap spacing of $T_C = T/2$), the penalty pole at $\Delta\tau = 100$ ps and $\gamma = 0.5$ remains. This penalty pole can only be melted down by a nonlinear equalizer like the DFE that takes into account the inter-symbol interference (ISI) induced by the already decided bits. Fig. 14(c) reveals that the penalty pole at $\gamma = 0.5 (\Delta\tau = 100$ ps) of Fig. 14(b) vanishes and the penalty peak (5.5 dB) shifts to $\gamma = 0.45$.

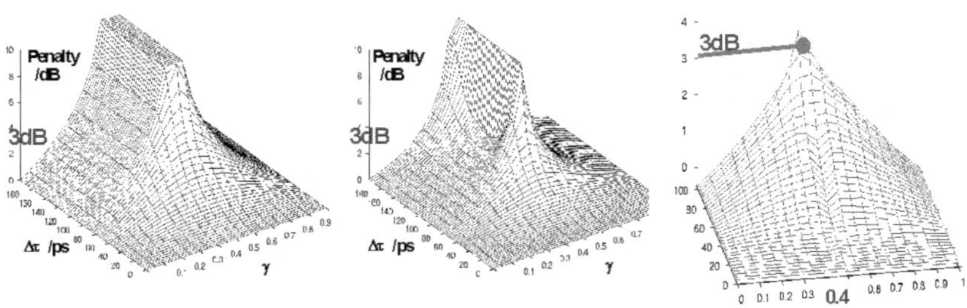

Fig. 14. Residual penalty of an equalizer for a first-order PMD distortion in the presence of stationary (thermal) noise.

Thermal-noise-limited detection exhibits about half of the residual penalty (in dB) compared to optical-noise-limited detection (OSNR limited) [47] that is typical when using optical preamplifiers (EDFA). This difference is underlined by the simulations shown in Fig. 15. The dark dashed and solid lines show the residual penalty of a DFE and a FFE + DFE, respectively, when equalizing a distorted signal for various ratios of the power splitting between fast and slow axis and at a constant DGD of one bit period. In the case of optical noise the penalty peak of the FFE + DFE moves to more than the double dB-value, from 3.5 to approximately 8 dB.

In lab experiments where the equalizers were set to minimum BER the penalty induced by first-order PMD was reduced from 8.1 by 4 dB by an 8 tap FFE and by 4.3 dB using a DFE, and by 6.1 dB with a concatenation of both [46].

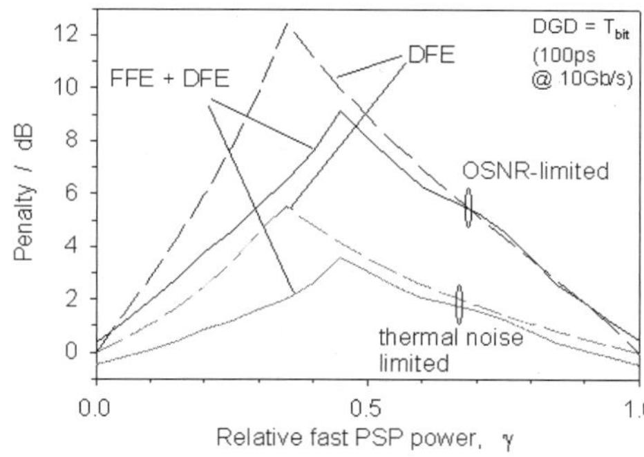

Fig. 15. Residual power penalty calculated for a DFE and a FFE + DFE vs γ assuming a DGD of a bit period.

Adaptation of Electronic Equalizers

For adaptation of the FFE the eye monitor has successfully been used [18, 19]. A consecutive dithering of the FFE taps was applied to maximize an estimate of the Q factor provided by an eye monitor.

The speed limits of this technique could be overcome by the least-mean-square (LMS) adaptation technique, since it provides simultaneously bipolar feedback signals for each tap. In this technique, which is well established at lower rate signal processing, the decision gate output is used as reference signal for the more or less equalized signal at the decision gate input. An error signal e is deduced by subtracting both signals, as shown by Fig. 13. In the LMS algorithm the mean square error $\langle e^2 \rangle$ at decision time t_0 is minimized. For each tap an individual control signal is generated by further correlating the error signal e at t_0 with differently delayed input signals (from the FFE) or with the decided data for DFE adaptation. The correlator (multiplier) output is integrated and fed back to adjust the tap weights. All tap weights of the filters are adjusted in parallel leading to a high-speed adaptation. Simulations exhibited time constants of less than a microsecond for reaction on an abrupt variation of distortion. Nonoptimum setting in the presence of nonstationary detected optical noise can be reduced by the use of an nth-order algorithms which minimizes $\langle e^{2N} \rangle$ [48].

Though the feasibility of the LMS adaptation technique at 10 Gbit/s for FFE and DFE has experimentally been demonstrated, the complex analog electronic processing to generate the tap control signals is still a demanding task at the speed of 10 Gbit/s. For forward-error-correction (FEC) supported transmission systems a concept for FFE and DFE adaptation was proposed [49] that is in some aspects similar to the LMS scheme, but allows a less demanding digital realization of the adaptation logic. The basic scheme is shown in Fig. 16. An digital error signal e is generated by comparing the uncorrected output data an of a DFE and the corrected output signal b_n provided by the FEC decoder. For simplicity, in Fig. 16 the delay introduced by decoding, interleaving, and framing that will affect the processing of a_n and b_n are disregarded. From the comparison of the bits an and b_n the error signal e becomes $+1$, if a “0” was corrected to a “1” and a -1, if a “1” was corrected to a “0”. For a synchronous FFE ($T_C = 1T$) this error information can be digitally correlated (EXOR) with differently delayed bits a_j. Similar to the LMS scheme, a respective tap c_i is adjusted to a maximum decorrelation of the errors e from the detected bits a_i by a digital integration (summation) of the correlator output. Moreover, as also shown by Fig. 16, the thresholds U_0 and U_1 of the DFE can also by adjusted by balancing the number of corrected marks ($e = 1$) and corrected spaces ($e = -1$) after a decided 0 and 1, respectively.

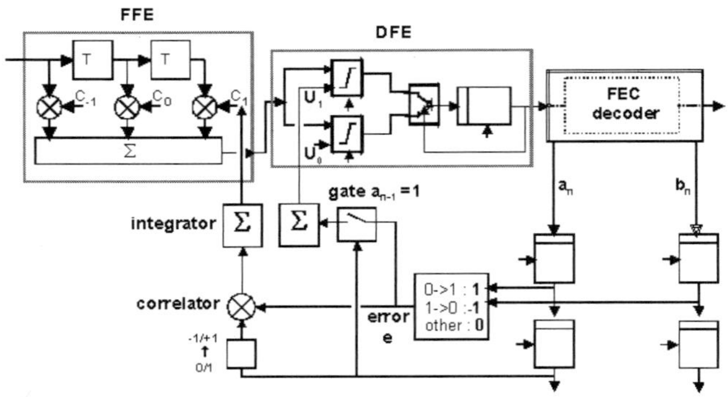

Fig. 16. FEC supported digital adaptation scheme for FFE and DFE. Adaptation of tap weight c_1 and DFE threshold U_1 is detailed.

Viterbi Equalizer

The maximum likelihood signal detection (MLSD) scheme is based on delayed decision after analyzing a whole distorted signal sequence over a few bit slots [40, 42]. MLSD exhibits the ultimate performance of electrical processing schemes. As shown in Fig. 17, the most likely realization might base on digital signal processing (DSP) where the signal is sampled at the bitrate (or faster), then analog-to-digital converted with a 3- or 4-bit converter (ADC) and then the digitized signal is processed in an ASIC.

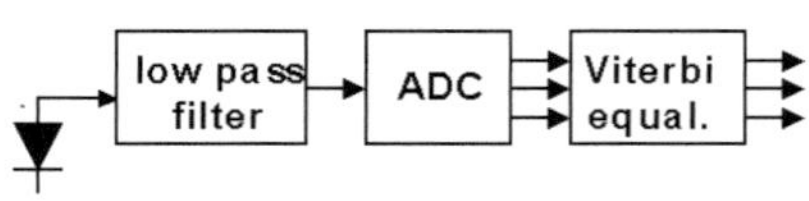

Fig. 17. Basic structure of the Viterbi equalizer (VE) applied for the realization of the maximum likelihood signal detection (MLSD) scheme.

In MLSD the samples s_i of a complete sequence are stored and compared with k internal reference sequences that are generated using an internal model of the signal distortion. This internal channel model contains the expected analog amplitude sample x_i^k for a given transmitted bit combination k as well as assumptions on the probability density function of the deviation of a detected noisy signal sample s_i from these expected values x_i^k. Hence by determining the probabilities $p(x, s)$ that a specific detected signal sample deviates from a channel model value, the probability of a whole reference sequence $P_S^k = \Pi p(x_i^k, s_i)$ can be calculated and by maximimzing the probability P_S^k, or alternative by minimizing the "cost function" $-\log(P_S) = \sum_i -\log(p(x_i^k, s_i))$, the most likely reference sequence x_i is found. The bit sequence

a_i that has generated the analog values x_i together with the channel model delivers the recovered bits.

In order to minimize the high effort for the implementation of the MLSD (e.g., a k-bit input sequence requires a comparison with $2k$ reference sequences of length k), the Viterbi algorithm is used [50, 51]. The effort reduction is achieved by (1) a limited length of the reference sequence (survivor depth) and by (2) keeping only a limited number of reference sequences (survivors). In the Viterbi algorithm the rule which reference sequences are kept without making a non-negligible error are deduced in a state representation of the signal. For example, with a PMD-induced spreading of the transmitted bit over 3 time slots the analog signal level s_i is determined by the ISI of three bits: a_{n-2}, a_{n-1}, a_n. In the state representation the "history" a_{n-2} and a_{n-1} of the actual signal is denoted by a state ($s = 4$ states are possible) whereas the last bit an is denoted as branch (either a "0" or a "1"). With $s = 4$ also 4 survivors are kept and a survivor depth of 16 is implemented in the VE (about 5 times the bit broadening). Typical parameters for a Viterbi equalizer (VE), which is currently under discussion for 10-Gbit/s application, are a 3- or 4-bit ADC, a LPF around 5-GHz bandwidth, the implementation of a four-state Viterbi algorithm. Such a VE is able to take into account a PMD distortion of a bit more than 1-bit period DGD. It exhibits an OSNR penalty of about 4.5 dB at a DGD of 100 ps ($\gamma = 0.5$) and might be able to bring a slight OSNR gain of about 0.3 dB for an undistorted NRZ signal. A modified approach of the VE based on oversampling of the input signal (20 Gsamples/s) exhibits an increased DGD tolerance, up to 1.8 T as demonstrated by the calculated penalty curve in Fig. 18.

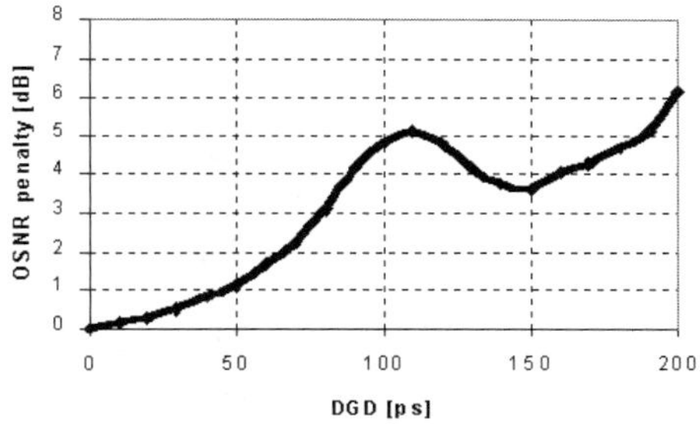

Fig. 18. Calculated penalty of a Viterbi equalizer (4 states) versus DGD for a 10-Gb/s NRZ signal.

Hybrid mitigation schemes based on combined optical and electronic signal processing was analysed for PMD-like diversity detection with 2, 3, or more photo diodes [52, 53] or the concatenation of optical compensator and electronic mitigator to further reduce the remaining distortion [54].

Performance of Compensators

The view at the different PMD mitigator approaches—optical, electronic, single-stage, multistage, etc.—presented in this chapter will be concluded by comparing the performance of all these concepts in terms of maximum PMD. This comparative analysis will help to assess the trade-off between performance/complexity/technology of all the different schemes.

Therefore, for a coarse classification of the different PMD mitigation approaches the "working areas" are sketched out in Fig. 19 showing the residual OSNR-penalty (required margin) versus the maximum tolerable link PMD (PMD-limit) [4]. In the figure the PMD-limit has to be related to the bit period T. The pair penalty/PMD-limit means, that a specific (low) OSNR margin is allocated for the PMDC. Then the outage probability remains below a reasonable low value for the system (often 10^{-5}) if the link PMD does not exceed the PMD-limit. For a PMD beyond this limit even higher allocated margins (what appears as penalty in Fig. 19.) hardly manage to guarantee the low outage.

The most left dark solid curve labeled with "Rx" belongs to an "unprotected" NRZ receiver with automatic threshold adjustment: up to a PMD-limit of 15% of the bit period T a margin of roughly 2 dB has to be allocated.

From numerical assessment of the one-stage compensator (cf., Fig. 2), also often referred to as "simple" PMDC, we know that with a margin of 2 dB a PMD-limit of up to 0.25 T can be tolerated (see shaded curve labeled with "1 stage" in Fig. 19). The second shaded curve labeled with "2 stage" belongs to the two stage setup. It extends the low-penalty operating range up to a PMD-limit of 0.4 T, provided that the a feedback signal is strongly correlated to the BER [39].

Devices with more than two stages realized in fiber-based components or in integrated-optic waveguide structures have the potential to increase the PMD limit and to keep a low penalty if the adaptation can be managed. A possible extension of the working area is indicated by the broad gray line labeled with multistage in Fig. 19.

As we have seen, in electronic PMD equalization— opposite to optical PMD compensation—the loss of information transported in the optical phase and in the polarization due to the photo diode's square law detection process leads to a nonvanishing residual OSNR penalty. On the other hand, electronic equalizers have the potential of being seamlessly and cost-effectively integration in the receiver. Finally, we should not forget that today product prototypes for electronic signal processing exist or are announced only for bitrates of 10 Gbit/s. So far at 40 Gb/s, first circuits have appeared in research labs in 2004.

In Fig. 19 the performance of electronic PMD mitigation is explained by the dark curves. Note that the performance assessment of electronic equalizers is based on first-order simulations only. In order to estimate the PMD-limit we rescaled the DGD value of the x-axis by the PMD-limit value (PMD= $0.31\times$DGD). With this PMD the DGD and with that the penalty is exceeded with a likelihood of 10^{-5} according to the Maxwellian probability density function. According to the pole of the FFE's penalty when moving to an actual DGD of 1 T and beyond [cf., Fig. 14(b)], it achieves a reasonable low penalty only if the DGD does not exceed 0.75 T. Following the aforementioned rule-of-thumb this value corresponds to a PMD-limit of 0.25 T, as it is shown by the dark curve labeled with "FFE".

Attaching a DFE at the output port of the FFE leads not to a significant improvement for actual DGDs below 0.7 T. But beyond 0.7 T the FFE penalty pole is melted down to a finite worst case penalty in the range of 8 dB at a DGD of 1 T [cf., Figs. 14(c) and 15]. This behavior is reflected in the 8-dB penalty plateau at a PMD-limit of 0.3 T and beyond as illustrated by the "FFE + DFE" curve in Fig. 19. However, this curve demonstrates that the electronic equalizer is capable to mitigate strong PMD distortions as well, but a high OSNR margin in the range of these 8 dB has to be allocated if the PMD exceeds about 0.25 T. Incorporating a two or more feedback DFE (DFEx) does

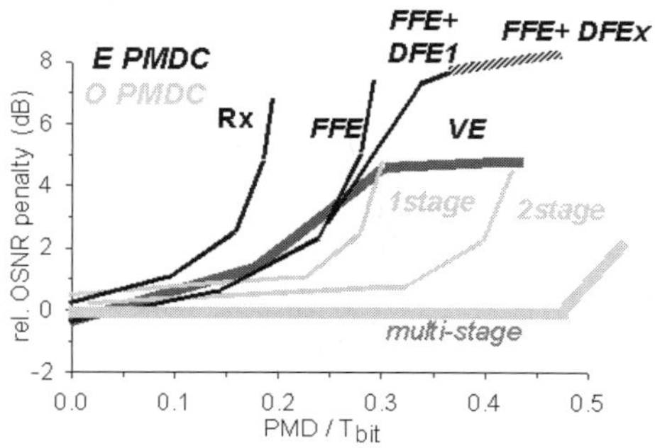

Fig. 19. Residual penalty versus PMD-limit for optical compensators (gray lines, O PMDC) and electronic equalizers (dark lines, E PMDC).

not bring a relieve for DGDs below 1 T [52]. Hence we estimated only a slightly increasing penalty from the FFE + DFE curve beyond 0.3 T when replacing the DFE by a DFE2 or DFE3 (2 or 3 feedback loops).

Since the penalty of the Viterbi equalizer (VE) increases with an increasing DGD, up to 4.5 dB at DGD= $1T$ (100 ps at 10 Gb/s, see Fig. 18), and stays then about constant up to 1.8 T (4 state VE), we estimate the potential working area as indicated by the bright black curve labeled with "VE". At strong distortions which are more likely for high PMD values, the VE outperforms the FFE and DFE, whereas at moderate distortions it exhibits similar performance as FFE + DFE.

References

[1] C. D. Poole and R.E. Wagner, "Phenomenological approach to polarisation dispersion in long single-mode fibres," Electron. Lett., **22** (19), 1029–1030 (1986).

[2] L.M. Gleeson, ESR Sikora, and MJ. O'Mahoney, "Experimental and numerical investigation into the penalties induced by second order polarisation mode dispersion at 10 Gb/s," Proc. ECOC 97, vol.1, 1997, pp. 15–18.

[3] H. Bülow, "Limitation of optical first-order PMD compensation," Proc. OFC/IOOC'99, 1999, WE1.

[4] H. Bülow, and S. Lanne, "Optical and Electronic PMD Compensation," Tutorial Notes, OFC 2003, Atlanta, ThP.

[5] H. Bülow, "Polarization Mode Dispersion Mitigation" in *Encyclopedia of Telecommunications*, J. Proakis (Ed.) (Wiley, 2002).

[6] H. Bülow, W. Baumert, H. Schmuck, F. Mohr, T. Schulz, F. Küppers, and W. Weiershausen, "Measurement of the maximum speed of PMD fluctuation in installed field fiber," Proc. OFC'99, 1999, WE4.

[7] D.S. Waddy, P. Lu; L. Chen, and X. Bao, "Fast state of polarization changes in aerial fiber under different climatic conditions," Photon. Technol. Lett., **13** (9), 1035–1037 (2001).

[8] T. Ono, S. Yamazaki, H. Shimizu, and K. Emura, "Polarization control method for suppressing polarization mode dispersion influence in optical transmission systems," J. Lightwave Technol., **12** (5), 891–898 (1994).

[9] D. Penninckx, F. Roy, S. Lanne, and J-P. Thiéry, "Statistical study of dynamic polarization-mode dispersion compensation based on degree of polarization monitoring," Microwave Opt. Technol. Lett., **26** (1), 41–43 (2000).

[10] R. Noé, D. Sandel, M. Yoshida-Dierolf, S. Hinz, V. Mirvoda, A. Shöpflin, C. Glingener, E. Gottwald, C. Scheerer, G. Fischer, T. Weyrauch, and W. Haase, "Polarization mode dispersion at 10, 20 and 40 Gb/s with various oprical equalizers," J. Lightwave Technol., **17** (9), 1602–1616 (1999).

[11] Z. Pan, Y. Xie, S. Lee, A.E. Willner, V. Grubsky, D.S. Starodubov, and J. Feinberg, "Chirp-free tunable PMD compensation using Hi-Bi nonlinearly-chirped FBGs in a dual-pass configuration," Proc. OFC2000, ThH2, 2000, pp. 113–115.

[12] D. Penninckx and S. Lanne, "Ultimate limits of optical polarization-mode dispersion compensators," Proc. ECOC 2000, P3.8, 2000, pp. 205–206.

[13] M. Karlsson, C. Xie, H. Sunnerud, and P.A. Andrekson, "Higher order polarization mode dispersion compensator with three degrees of freedom," Proc. OFC'01, 2001, MO1.

[14] F. Roy, C. Francia, F. Bruyère, and D. Penninckx, "A simple dynamic polarization mode dispersion compensator," Proc. OFC'99, 1999,TuS4, p. 275.

[15] G. Ishikawa and H. Ooi, "Polarization-mode dispersion sensitivity and monitoring in 40 Gb/s OTDM and 10 Gb/s NRZ transmission experiments," Proc. OFC98, WC5, pp. 117–119, 1998.

[16] F. Heismann, D.A. Fishman, and D.L. Wilson, "Automatic compensation of firstorder polarization mode disperion in a 10Gb/s transmission system," Proc. ECOC'98, WdC11, 1998.

[17] S. Bahsoun, J. Nagel, and C. Poole, "Measurments of temporal variations in fiber transfer characteristics to 20 GHz due to polarization-mode dispersion," Proc. ECOC'90, Sept. 16–20, Amsterdam, 1990, ThG1.5, pp. 1003–1006.

[18] M. Fregolent, S. Herbst, H. Soehnle, and B. Wedding, "Adaptive optical receiver for performance monitoring and electronic mitigation of transmission impairments," Proc. ECOC 2000, Sept. 3–7, Munich, Germany, 2000, 2.1.2.

[19] H. Bülow, F. Buchali, and G. Thielecke, "Adaptation of an Electronic PMD Mitigator by Maximization of the Eye Opening," Proc. ECOC 2000, P3.10, 2000.

[20] F. Buchali, W. Baumert, H. Bülow, J. Poirrier, and S. Lanne, "A 40 Gb/s eye monitor and its application to adaptive PMD compensation," Proc. OFC 2002, 2002, WE6, p. 202.

[21] C. Glingener, A. Schopflin, A. Farbert, G. Fischer, R. Noé, D. Sandel, S. Hinz, M. Yoshida-Derolf, V. Mirvoda, G. Feise, H. Hermann, R. Ricken, W. Sohler, and F. Wehrmann, "Polarization mode dispersion compensation at 20 Gb/s with a compact distributed equalizer in LiNbO3," post deadline paper, OFC/IOOC'99, 1999, PD29.

[22] J. Poirrier, H. Bülow, and F. Buchali, "Optical PMD compensation performance: numerical assessment," Proc. OFC2002, WI3, p. 234.

[23] H. Rosenfeldt, R. Ulrich, E. Brinkmeyer, U. Feiste, C. Schubert, J. Berger, R. Ludwig, H.G. Weber, and A. EhrHardt, "Feed-forward approach for automatic PMD compensation at 80 Gbit/s over 45 km installed single mode fiber," ECOC 2001, postdeadline paper PD.B.1.1.

[24] D. Sobiski, D. Pikula, J. Smith, C. Henning, D. Chowdhury, E. Murphy, E. Kolltveit, and F. Annunziata, "Fast first-order PMD compensation with low insertion loss for 10 Gbit/s system," Electron. Lett., **37** (1), (2001).

[25] L.-S. Yan, C. Yeh, G. Yang, L. Lin, Z. Chen, Y.Q. Shi, and X.S. Yao, "Fast digitally variable differential group delay module using polarization switching," Postdeadline OFC 2002, FA5.

[26] Q. Yu and A.E. Willner, "Performance limits of first-order PMD compensators using fixed and variable DGD elements," Photon. Technol. Lett., **14** (3), 304–306 (2002).

[27] C.K. Madsen, "Optical all-pass filters for polarization mode dispersion compensation," Opt. Lett.**25** (12), 878–880 (2000).

[28] L.Möller, "Broadband PMD Compensation in WDM Systems," Proc. ECOC 2000, Sept 3–7, Munich, Germany, 2000, P1.15.

[29] K. Takiguchi, K. Okamoto, and K. Moriwaki, "Planar Lightwave Circuit Dispersion Equalizer," J. Lightwave Technol., **14** (9), (September 1996).

[30] A. Eyal and A. Yariv, "Design of broad-band PMD compensation filters," IEEE Photon. Technol. Lett., 1088–1090 (Aug. 2002).

[31] T. Saida, K. Takiguchi, S. Kuwahara, Y. Kisaka, Y. Miyamoto, Y. Hashizume, T. Shibata, and K. Okamoto, "Planar Lightwave Circuit Polarization Mode Dispersion Compensator," Proc. ECOC 2001, Amsterdam, NL, Sept. 30–Oct. 4, 2001, Mo.F.2.

[32] M. Bohn, F. Horst, B. J. Offrein, G. L. Bona, W. Rosenkranz, and P. Krummrich; "Simultaneous Adaptive Equalization of Group Velocity and Polarization Mode Dispersion at 40 Gb/s with Integrated Optical FIR-Filters and Electrical Spectrum Monitoring as Feedback," Proc. ECOC 2002, Rimini, Th2.2.4.

[33] C.K. Madsen, E.J. Laskowski, M. Cappuzzo, L. Buhl, S. Chandrasekhar, E. Chen, L. Gomez, A. Griffin, L. Stulz, and A. Wong-Foy, "A Versatile, Integrated Emulator for First- and Higher-Order PMD," Proc. ECOC 2003, Rimin, Th2.2.6.

[34] H. Bülow, "Self-Stabilizing Continuously Tunable Group Delay Line for PMD Compensation," Proc. ECOC2002, Copenhagen, 9.3.5.

[35] C.K. Madsen, P. Oswald, S. Chandrasekhar, L. Buhl, M. Cappuzzo, E.J. Laskowski, E. Chen, L. Gomez, A. Griffin, A. Kasper, L. Stulz, and A. Wong-Foy, "Real-time PMD Measurement Using An Integrated Wavelength-Scanning Polarimeter," Proc. ECOC 2003, Rimini, Th2.2.6.

[36] M.W. Chbat, J-P. Soigné, T. Fuerst, J.T. Anthony, S. Lanne, H. Février, B.M. Desthieux, A.H. Bush, and D. Penninckx, "Long term field demonstration of optical PMD compensation on an installed OC-192 link," Proc. OFC'99, PD12, 1999.

[37] D. A. Watley, K.S. Farley, W.S. Lee, G. Bordogna, B.J. Shaw, and A.P. Hadjifotiou, "Field evaluation of an optical PMD compensator using an installed 10 Gbit/s system," OFC2000, ThB6, pp. 37–39, 2000.

[38] S. Lanne, J-P. Thiéry, D. Penninckx, J-P. Hamaide, J-P. Soigné, B. Desthieux, J. Le Briand, L. Macé, P. Gavignet, "Field optical PMD compensation at 10 Gb/s over installed fibre totalling 35ps of PMD," Proc. ECOC 2000, P3.9, 2000.

[39] T. Ono, Y. Yano, L.D. Garret, J.A. Nagel, M.MJ. Dickerson, and M. Cvijetic, "10 Gb/s PMD compensation field experiment over 452 km using Principal State Transmission method," Proc. OFC 2000, PD44.

[40] J.H. Winters and R.D. Gitlin, "Electrical Signal Procressing Techniques in Long-Haul Fiber-Optic Systems," Trans. Commun., **38** (9), 1439–1453 (1990).

[41] S. Kasturia and J.H. Winters, "Techniques for High-Speed Implementation of Nonlinear Cancellation," J. Select. Areas Commun., **9** (5), 711–717 (1991).

[42] H.F. Haunstein, K. Sticht, A. Dittrich, W. Sauer-Greff, and R. Urbansky, "Design of near optimum electrical equalizer for optical transmission in the presence of PMD," Tech. Dig. OFC 2001, Mar 10–21, Anaheim, USA, 2001, WAA4.

[43] J. Poirrier, A. Gnauck, and J. Winters, "Experimental Nonlinear Cancellation of Polarization-Mode Dispersion," Techn. Dig. OFC'00, Baltimore, Mar. 7–10, ThH4, 2000, pp. 119–121.

[44] L. Möller, A. Thiede, S. Chandrasekhar, W. Benz, M. Lang, T. Jakobus, and M. Schlecthweg, "ISI mitigation using decision feedback loop demonstrated with PMD distorted 10 Gbit/s signals," Electron. Lett., **35** (24), 2092–2093 (1999).

[45] H.Bülow, R.Ballentin, W.Baumert, G.Maisonneuve, G.Thielecke, and T.Wehren, "Adaptive PMD mitigation at 10 Gbit/s using an electronic SiGe equaliser IC," Proc. ECOC'99, Sept. 26–30, Nice, France, 1999, We C3.4, pp. 138–139.

[46] H. Bülow, F. Buchali, W. Baumert, R. Ballentin, and T. Wehren, "PMD mitigation at 10 Gbit/s using linear and nonlinear integrated electronic equaliser circuits," Electron. Lett., **36** (2), 163–164 (2000).

[47] H. Bülow and G. Thielecke, "Electronic PMD mitigation—from linear equalization to maximum-likelihood detection," Techn. Dig. OFC'2001, Mar. 10–21, Anaheim, USA, 2001, WAA3.

[48] U.-V. Koc, K.-Y. Tu, and N. Kaneda, "Adaptive Electronic Equalization Using Higher-Order Statistics for PMD Compensation in Long-Haul Fiber-Optic Systems," Proc. ECOC 2002, Copenhagen, Sept. 2002, 7.1.5.

[49] H. Haunstein, R. Schlenk, K. Sticht, A. Dittrich, W. Sauer-Greff, and R. Urbansky, "Control of Combined Electrical Feed-Forward and Decision Feedback Equalization by Conditional Error Counts from FEC in the Presence of PMD," Tech. Dig. OFC 2003, Atlanta, ThG5.

[50] R.D. Gitlin, J.F. Hayes, and S.B. Weinstein, *Data Communications Principles* (Plenum Press, New York).

[51] J. G. Proakis, *Digital Communications*, 3rd ed., international ed. (McGraw-Hill, 1995).

[52] B.W.Hakki, "Polarization Mode Dispersion Compensation by Phase Diversity Detection," Photon. Technol. Lett., **9** (1), 121–123 (1997).

[53] A. O. Lima, I. T. Lima, Jr., T. Adali, and C. R. Menyuk, "A novel polarization diversity receiver for PMD mitigation," IEEE Photon. Technol. Lett., 2002.

[54] H. Bülow et al., Proc. ECOC 2000, 4.2.4.

[55] H. Bülow, F. Buchali, and G. Thielecke, "Electronically Enhanced Optical PMD Compensation", Proc. ECOC 2000, Sept. 3–7, Munich, Germany, 2000, 4.2.4.

DOI: 10.1007/s10297-004-0016-y
Originally published in J. Opt. Fiber Commun. Rep. 1, 32–62 (2004)

Low-PMD spun fibers

Andrea Galtarossa, Paola Griggio, Luca Palmieri, and Anna Pizzinat

Department of Information Engineering, University of Padova
Via Gradenigo 6/A, 35131 Padova, Italy
E-mail: andrea.galtarossa@unipd.it
griggio@wave.dei.unipd.it
luca.palmieri@dei.unipd.it
anna.pizzinat@unipd.it

Abstract. In the last ten years there has been an increasing interest in the design and realization of low PMD fibers. This goal can be achieved by properly spinning a fiber when it is still in the hot zone. In this chapter, analytical and numerical results on the design of low-PMD fibers by means of spinning techniques are presented.

1. Introduction

It is recognized that polarization mode dispersion (PMD) has become one of the main impairments in high-capacity optical systems. In order to reduce problems due to PMD the research activity is focusing on the feasibility and realization of both PMD compensators and low-PMD fibers. This work is dedicated to the latter case. In particular, it deals with the spinning process of fibers, which proved itself to be an effective way to reduce fiber PMD.

After describing the spinning process and its effect on the local birefringence vector, the reasons why spinning can be effective in reducing the PMD of a fiber are intuitively explained. Then, analytical and numerical results on the PMD of fibers with either unidirectional, or periodic spin in the regimes of constant and random birefringence are presented.

1.1. Main Symbols and Definitions

$\boldsymbol{\beta}(z, \omega)$	local birefringence vector	[rad/m]
β_ω	ω-derivative of the birefringence modulus	[s/m]
$\boldsymbol{\Omega}(z, \omega)$	PMD vector	[s]
$\Delta\tau = \lvert\boldsymbol{\Omega}\rvert$	differential group delay (DGD)	[s]
ω	angular frequency	[rad/s]
$L_B = 2\pi/\langle\beta^2\rangle^{1/2}$	beat length	[m]
L_F	birefringence correlation length	[m]
$A(z)$	spin function	[rad]
$\alpha(z) = dA/dz$	spin rate	[rad/m]
p	spin period	[m]
A_0	spin amplitude	[rad]

2. The Spin Process

A spun fiber is obtained by applying a torque during the drawing process while the fiber is still in the viscous state at a temperature close to the silica melting point. In the earliest experiments [1], [2], this was accomplished by spinning the preform during the drawing. Actually, such spun fibers were designed with the aim of realizing sensors [2]–[4], although already in 1981, Barlow et al. recognized that spun fibers "possess negligible polarization dispersion, an attribute that may assume importance in future telecommunication systems" [1]. Later, when spinning really began to be used as a tool to reduce PMD, the preform rotation became incompatible with the production drawing speed. In [5] it is reported that, for a drawing speed of 10 m/s, the preform would have to be spun at 6000 rpm in order to achieve a suitable spin pitch, but this is not practical in commercial fiber production. As a consequence, for most spun fibers produced nowadays [5]– [10] the torque is applied directly to the fiber right after the melting point, by means of a device called spinner as outlined in Fig. 1.

In any of the two cases the spin is applied in the molten phase, thus it is commonly assumed that it does not induce torsional stresses, and that its unique effect is to rotate the orientation of fiber birefringence, without affecting its modulus [1], [5], [11], [12]. Indeed, preliminary experimental results supporting this assumption have been recently reported [13].

On the contrary, when a fiber is twisted, the torque is applied on the cold fiber with the silica completely solidified. Consequently, a torsional stress exists, and it gives rise to circular birefringence owing to the photo-elastic effect [14], [15].

In order to understand the effects of both spin and twist on the birefringence and on the PMD of a fiber, it is necessary to recall first of all the local birefringence vector that, in the Stokes space, is

$$\boldsymbol{\beta}(z, \omega) = (\beta_1,\ \beta_2,\ \beta_3)^{\mathsf{T}}\ .$$

For standard telecommunication fibers $\boldsymbol{\beta}(z, \omega)$ is a random function of z, whose properties may be described by appropriate statistical models [16], [17]. Moreover, it is commonly assumed [14], [15], [17], that in telecommunication fibers the circular birefringence is negligible, i.e., $\beta_3 = 0$, and that only the modulus of birefringence depends on ω, so that $\boldsymbol{\beta}(z,\omega) = \boldsymbol{\beta}_l(z,\omega) = \beta_l(z,\omega)\hat{\boldsymbol{\beta}}(z)$, where the index l indicates linear birefringence, $\beta_l(z,\omega)$ is the modulus and $\hat{\boldsymbol{\beta}}(z)$ the direction. Thus, the ω-derivative of β_l is $\beta_\omega(z) = \partial\beta_l(z,\omega)/\partial\omega = \beta_l(z,\omega)/\omega$. Additionally, the general definition of beat length is $L_B = 2\pi/\langle\beta_l^2\rangle^{1/2}$.

When the spin is applied, the local birefringence vector undergoes a rotation around the longitudinal axis of the fiber that corresponds to the vector $(0,\ 0,\ 1)^{\mathsf{T}}$ in Stokes space. If the angle of the rotation imposed on the fiber is the spin function $A(z)$, then the birefringence vector of the spun fiber is

$$\mathbf{R}_3[2A(z)]\boldsymbol{\beta}_l(z)\,,$$

where

$$\mathbf{R}_3(\phi) = \begin{pmatrix} \cos\phi & -\sin\phi & 0 \\ \sin\phi & \cos\phi & 0 \\ 0 & 0 & 1 \end{pmatrix}. \tag{1}$$

Incidentally, if both a spin $A(z)$ and a twist $\gamma(z)$ are applied, then the birefringence vector of the spun and twisted fiber is

$$\mathbf{R}_3[2A(z) + 2\gamma(z)]\boldsymbol{\beta}_l(z) + (0,\ 0,\ g\gamma(z))^{\mathsf{T}}\,,$$

where $g \simeq 0.146 \div 0.18$ is the stress-optic or rotation coefficient [14], [15], [18].

It is useful to introduce also the spin rate function that is the derivative with respect to distance of the spin and is measured in rad/m: $\alpha(z) = dA(z)/dz$.

To date several spin functions have been proposed in the literature. In particular it is possible to distinguish two main classes:

– Unidirectional spin. The fiber is rotated always in the same direction and with a constant rate [9], [19] –[21]:

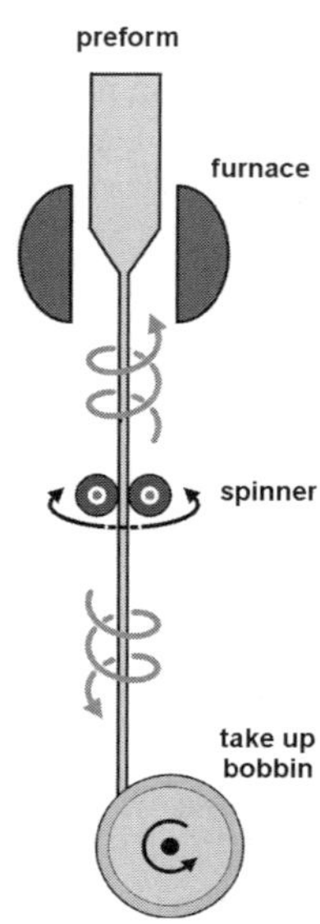

Fig. 1. Drawing process of spun fibers.

$$A(z) = \alpha_0 z = \frac{2\pi}{p} z\,, \tag{2}$$

α_0 is the constant rate and p is the spin period that is the length of fiber required to perform a complete rotation of the fiber around its longitudinal axis.

– Alternated spin. The torque is applied alternately in clockwise and counterclockwise direction. The spatial frequency of the changes of direction may be not constant as suggested in [5] and [6], or constant. The latter case corresponds to a periodic spin that may be sinusoidal, triangular, trapezoidal, frequency modulated, or in any case expanded in Fourier series [9], [22]:

$$A(z) = A_0 q(z)\,, \quad |q(z)| \leqslant 1\,, \quad q(z+p) = q(z)\,.$$

The choice among so many spin functions is principally determined by technological issues, and by the effectiveness in reducing the PMD of a fiber typically measured in terms of its differential group delay (DGD). For this purpose it is useful to define a spin induced reduction factor (SIRF) as the ratio between the DGD of a spun fiber $\langle \Delta\tau(z) \rangle$, and the DGD the same fiber would have if it were not spun $\langle \Delta\tau_{\text{un}}(z) \rangle$,

$$\text{SIRF} = \frac{\langle \Delta\tau(z) \rangle}{\langle \Delta\tau_{\text{un}}(z) \rangle}\,. \tag{3}$$

In the next sections the effects of unidirectional and periodic spin functions are analyzed in the cases of fibers with constant birefringence, i.e., in the "short length regime" and of fibers with random birefringence, i.e., in the "long length regime". In any case the main purposes are to calculate the DGD of the spun fiber, and to find spin functions or spin parameters that minimize the SIRF.

Actually, both these issues involve the analysis of the PMD dynamic equation [23]:

$$\frac{\partial \boldsymbol{\Omega}_f(z,\omega)}{\partial z} = \boldsymbol{\beta}_f(z,\omega) \times \boldsymbol{\Omega}_f(z,\omega) + \frac{\partial \boldsymbol{\beta}_f(z,\omega)}{\partial \omega}\,, \tag{4}$$

where the subscript f refers to the fixed reference frame and $\boldsymbol{\Omega}$ is the PMD vector. Additionally, the direction of $\boldsymbol{\Omega}_f$ corresponds to the principal states of polarization, whereas its modulus is the DGD, $|\boldsymbol{\Omega}_f| = \Delta\tau$.

2.1. *Rotating Reference Frame*

The analytical treatment of spun fibers can be simplified by transforming to a rotating reference frame that compensates for the rotation induced by the spin and/or by the intrinsic rotation of $\boldsymbol{\beta}_l(z)$. Note that the modulus of a vector does not change when the vector undergoes a rotation, so the DGD is independent of the reference and it results $\Delta\tau = |\boldsymbol{\Omega}| = |\boldsymbol{\Omega}_f|$.

The change of coordinate system is implemented by $\boldsymbol{\Omega} = \mathbf{R}_3^{\mathsf{T}}(\phi)\boldsymbol{\Omega}_f$, and $\boldsymbol{\beta} = \mathbf{R}_3^{\mathsf{T}}(\phi)\boldsymbol{\beta}_f$, where ϕ is the rotation angle and $\mathbf{R}_3$ has been defined in Eq. (1). Then, the dynamic equation in the new reference frame is

$$\begin{aligned} \frac{\partial \boldsymbol{\Omega}(z,\omega)}{\partial z} &= \left(\mathbf{R}_3^{\mathsf{T}} \boldsymbol{\beta}_f - \mathbf{R}_3^{\mathsf{T}} \frac{\partial \mathbf{R}_3}{\partial z} \right) \times \boldsymbol{\Omega} + \mathbf{R}_3^{\mathsf{T}} \frac{\partial \boldsymbol{\beta}_f}{\partial \omega} \\ &= \left[\mathbf{R}_3^{\mathsf{T}} \boldsymbol{\beta}_f - \begin{pmatrix} 0 \\ 0 \\ d\phi/dz \end{pmatrix} \right] \times \boldsymbol{\Omega} + \mathbf{R}_3^{\mathsf{T}} \frac{\partial \boldsymbol{\beta}_f}{\partial \omega}\,. \end{aligned} \tag{5}$$

Note that if $\phi = 2A(z)$, then $d\phi/dz = 2\alpha(z)$, that is the spin rate. This corresponds to an apparent circular birefringence that is present in the reference frame rotating at twice the spin function.

2.2. *Why Should the Spin Reduce PMD?*

Before going into the details of the analytical treatment, it is worthwhile to sketch intuitively the physical mechanisms by which spin may reduce a fiber PMD. Roughly speaking, it is possible to distinguish two different mechanisms. In the first case, consider a randomly birefringent fiber; if the fiber is sufficiently long, its mean DGD reads [24], [17]

$$\langle \Delta\tau_{\mathrm{un}}(z)\rangle = \sqrt{\frac{\pi}{3}}\frac{8}{\omega L_B}\sqrt{L L_F}\,, \tag{6}$$

where ω is the angular frequency, L is the fiber length, L_B is the beat length, and L_F is the birefringence correlation length. This last quantity describes how fast the random birefringence evolves along the fiber; the shorter L_F, the faster the evolution of birefringence.

By spinning the fiber in a suitable way, the fiber birefringence should be forced to evolve faster, so that L_F, and hence the mean DGD are reduced. However, this mechanism is not very effective, because $\langle \Delta\tau_{\mathrm{un}}(z)\rangle$ scales with the square root of the correlation length.

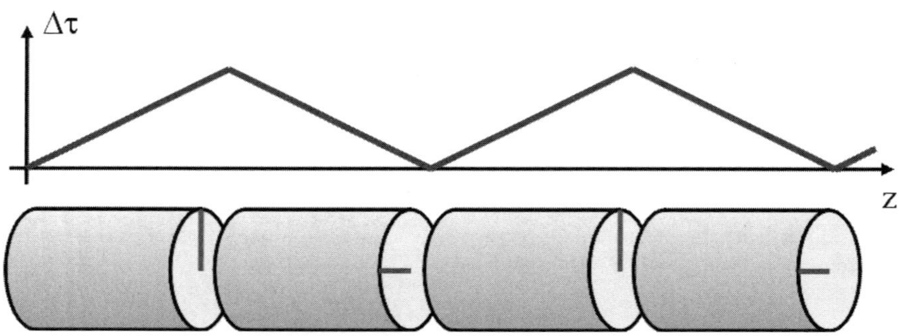

Fig. 2. DGD compensation by periodic spin.

To explain the second mechanism by which spin may reduce PMD, consider a polarization maintaining fiber (whose birefringence is constant and deterministic in modulus and orientation), and suppose that the fiber is sliced in equal sections. These sections are birefringent plates that can be periodically rotated to align the fast optical axis of a section with the slow axis of the preceding one. In this way, all the DGD accumulated in a section, is completely compensated by the following one, resulting in a periodic evolution of the DGD as shown in Fig. 2. Upon recalling that the DGD of a polarization maintaining fiber grows linearly with distance, the second mechanism may lead to a very strong DGD reduction. Unfortunately, the spin process now described requires sudden rotations, which are not possible in practice. However, in the following sections it is shown that even a slow periodic spin can be conveniently calibrated so that a periodic compensation of the DGD is still obtained.

In the case of random birefringence, the fiber cannot be sliced in sections with equal DGD, hence a perfectly periodic evolution cannot be achieved. Actually, the DGD of a randomly birefringent spun fiber inevitably grows with the square root of fiber length. Nevertheless, well calibrated spin may still reduce the rate of growth by almost two orders of magnitude.

3. Fibers with Constant Birefringence

3.1. *Unidirectional Spin*

When an unidirectional spin function (2) is applied to a fiber with constant birefringence, the PMD dynamic equation (4) can be analytically solved [20], and the DGD is

$$\Delta\tau(z) = \frac{1}{\omega\sqrt{1+4L_B^2/p^2}}\sqrt{\left(\frac{2\pi}{L_B}z\right)^2 + \frac{2L_B/p}{\sqrt{1+4L_B^2/p^2}}\sin\left(\sqrt{1+4\frac{L_B^2}{p^2}}\frac{2\pi}{L_B}z\right)}$$
$$\simeq \frac{2\pi z}{\omega L_B\sqrt{1+4L_B^2/p^2}}\,. \tag{7}$$

The last approximation holds because after a short transient the amplitude of the oscillating term becomes negligible with respect to the one that determines the linear growth of the DGD. Recalling that the DGD of the corresponding unspun fiber is $\Delta\tau_{\mathsf{un}} = 2\pi z/(\omega L_B)$, the SIRF is well approximated by

$$\mathrm{SIRF} \simeq \frac{1}{\sqrt{1+4L_B^2/p^2}}\,. \tag{8}$$

Figure 3 (upper part) shows the SIRF as a function of the distance z for the unspun fiber with $L_B = 10$ m, and for 3 constantly spun fibers with the same L_B and $p = 10$, 3 and 1 m, respectively. It can be noted that after a very short oscillating transient the SIRF has a constant value independent of the fiber length.

In Fig. 3 (lower part) the SIRF is represented as a function of the constant spin rate for three values of the beat length: 1, 5 and 20 m. As it can be observed also in Eq. (8), the SIRF depends on the ratio L_B/p, thus the unidirectional spin is effective in reducing the DGD only if the spin period is shorter than the beat length.

3.2. *Periodic Spin*

In the case of periodically spun fibers two parameters come into play: they are the spin amplitude A_0 and the spin period p. The analysis of the PMD properties of these fibers can be simplified in a reference frame that rotates around the vector $(0,\,0,\,1)^{\mathsf{T}}$ with an angle $\phi = 2A(z)$. Following the procedure presented in Section 2.1, the dynamic equation becomes

$$\frac{\partial \boldsymbol{\Omega}}{\partial z} = \begin{pmatrix} 0 & 2\alpha(z) & 0 \\ -2\alpha(z) & 0 & -\beta_l \\ 0 & \beta_l & 0 \end{pmatrix}\boldsymbol{\Omega} + \begin{pmatrix} \beta_\omega \\ 0 \\ 0 \end{pmatrix}. \tag{9}$$

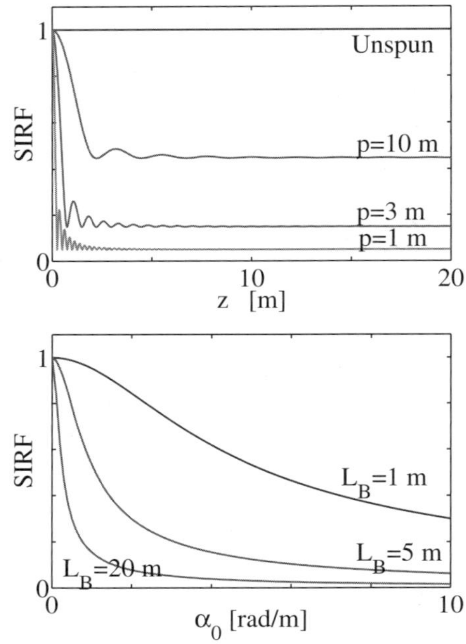

Fig. 3. Upper: evolution of the SIRF as a function of distance for a fiber with $L_B = 10$ m. Lower: evolution of the SIRF as a function of the constant spin rate, for different values of L_B.

The solution of Eq. (9) with a generic spin rate cannot be found, though the well established theory of differential equations with periodic coefficients (see for example [25]), is helpful. However, it is possible to express $\Delta\tau$ as a function of only the first component of the PMD vector. In fact, z-deriving the quantity $\Delta\tau^2 = \boldsymbol{\Omega} \cdot \boldsymbol{\Omega}$, and exploiting Eq. (9) yields

$$\Delta\tau(z) = \sqrt{2\beta_\omega \int_0^z \Omega_1(l)dl}\,. \tag{10}$$

A possible solution to the problem of reducing the fiber DGD is suggested by Eq. (10). It is clear that if $\Omega_1(z)$ is a periodic function, then the DGD will grow with the square root of distance, or even it will be itself periodic, when the average value of Ω_1 over a period is zero.

When the spin rate is a periodic function, Eq. (9) represents an inhomogeneous linear system with periodic coefficients. It may be shown [22], [25] that the behavior of Eq. (9) depends on that of the associated homogeneous system

$$\frac{d\boldsymbol{y}}{dz} = \begin{pmatrix} \beta_l \\ 0 \\ -2\alpha(z) \end{pmatrix} \times \boldsymbol{y} = \mathbf{X}(\boldsymbol{v}(z))\boldsymbol{y}(0)\,, \tag{11}$$

where $\mathbf{X}(\boldsymbol{v}(z))$ represents a rotation around a suitable vector $\boldsymbol{v}(z)$ through an angle equal to $|\boldsymbol{v}(z)|$. This result can be intuitively proved recalling that Eq. (11) is formally the same equation that describes the polarization evolution on the Poincaré sphere. Furthermore, $\mathbf{X}$ is the fundamental matrix of the system (see Appendix A).

According to the two lemmas reported in Appendix A, the second step is to understand if the homogeneous system (11) has periodic solutions. This task can be achieved by calculating the eigenvalues of the matrix $\mathbf{X}(\boldsymbol{v}(p))$, in fact Eq. (11) has a periodic solution of period p, if–and only if–the fundamental matrix evaluated for $z = p$ has a unit eigenvalue [25]. For rotation matrices as $\mathbf{X}$, this is indeed the case because the eigenvalues of $\mathbf{X}(\boldsymbol{v}(p))$ are 1 and $\exp(\pm j|\boldsymbol{v}(p)|)$.

Because the homogeneous system always has at least one periodic solution, it is possible to exploit Lemma 1 reported in Appendix A. Thus, the inhomogeneous system (9) has periodic solutions of period p if—and only if—the following solution is satisfied

$$\frac{1}{p}\int_0^p y_1(z)dz = 0, \tag{12}$$

where $y_1(z)$ is the first component of the periodic solution of Eq. (11). Moreover, it may be shown that if Eq. (12) is satisfied, then the periodic solution of Eq. (9) has zero mean value and, according to Eq. (10), the DGD is itself periodic of period p. Thus, the target is to look for a periodic spin rate such that the corresponding periodic solution of system (11) satisfies condition (12). It is important to note that Eq. (12) holds for any periodic spin function, and makes it possible to implement a numerical method for the choice of the optimum spin parameters:

1. Choose a periodic spin profile and its period p.

2. Choose a value of the spin amplitude A_0, and calculate the fundamental matrix $\mathbf{X}(z)$ of (11) for z in $[0,\ p]$; this coincides with calculating the 3D Mueller matrix of the corresponding constantly birefringent spun fiber.

3. Determine the eigenvector $\hat{\boldsymbol{v}}$ of $\mathbf{X}(p)$ that corresponds to the unit eigenvalue, $\mathbf{X}(p)\hat{\boldsymbol{v}} = \hat{\boldsymbol{v}}$.

4. Calculate the solution $\boldsymbol{y}$ of Eq. (11) with $\hat{\boldsymbol{v}}$ as initial solution; recalling the form of Eq. (11), this is equivalent to calculating the evolution of the state of polarization (SOP) when $\hat{\boldsymbol{v}}$ is the input SOP.

5. Evaluate the average over one period of the first component of $\boldsymbol{y}$.

6. If the average is 0 then the DGD of the fiber is periodic; otherwise go back to step No. 2.

As an example, the numerical method reported above has been applied to search the optimum spin amplitude for a trapezoidal spin as the one shown in fig. 4 with a period $p = 5$ m. The solution has $A_0 = 1.58$ for a fiber with $L_B = 20$ m. The correctness of this result has been verified by resorting to simulations. The corresponding behavior of the first component of the PMD vector and of the DGD are shown in Fig. 4.

Actually, the above discussion presents a drawback. In fact the application of the optimization routine requires the knowledge of the fiber parameters, in particular of the beat length L_B, whereas it is hardly difficult to know this parameter before the drawing process. Moreover, condition (12) is quite difficult to be understood and its dependence on the spin parameters is not at all clear. These two issues can be overcome

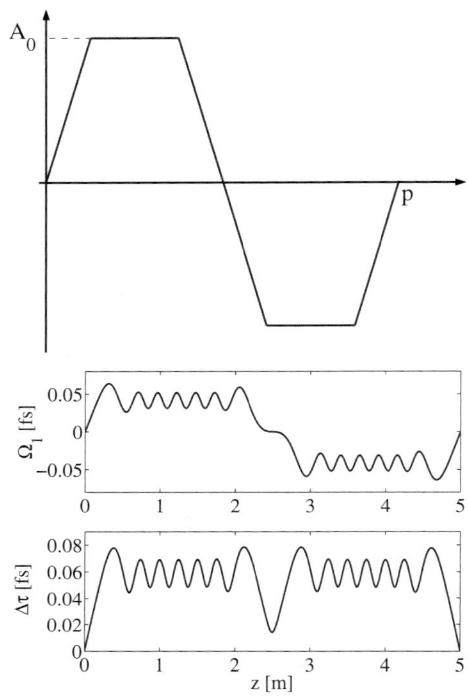

Fig. 4. Upper: A trapezoidal spin function. Lower: The periodic evolution of the first component of the PMD vector and of the DGD as a function of the distance are reported, for a trapezoidal spin function with optimized parameters.

by introducing the assumption that the square of the spin period is smaller than the square of the beat length, i.e., $p^2 \ll L_B^2$. Usually, for telecommunication fibers, L_B is in the order of a few tens of meters [26]; on the other hand, a spin period of a few meters can be practically implemented [9], [8]. Hence, the assumption $p^2 \ll L_B^2$— also called "short period assumption" (SPA)— does not represent a significant constraint on the validity of the results.

Under the SPA, and upon expanding the solution of Eq. (11) in Fourier series, which can be done without loss of generality because all quantities involved in the problem are periodic, it can be shown that the solution of the problem does no more depend on the beat length value [22]. Furthermore, by adding the assumption of a spin profile $A(z)$ with only odd harmonics, the condition for a periodic DGD simplifies to

$$\frac{1}{p}\int_0^p \cos A(z) dz = 0 \,. \tag{13}$$

For some spin functions the analytical treatment is even simpler and the set of spin parameters that correspond to a periodic DGD can be determined in closed form. As an example, consider a sinusoidal spin function $A(z) = A_0 \sin(2\pi z/p)$. In this case, condition (13) reads

$$\frac{1}{p}\int_0^p \cos A(z) dz = J_0(2A_0) = 0 \,, \tag{14}$$

where J_0 is the first-kind Bessel function of order 0. This result has been verified by means of numerical simulations with the waveplate model. Figure 5 shows the evolution of the DGD for a fiber with $L_B = 20$ m. The upper curve corresponds to the unspun fiber, the other two curves correspond to the same fiber with a sinusoidal spin with $p = 2$ m. The amplitude $A_0 = 2.76$ m is such that $2A_0$ is the second zero of J_0 and the DGD is periodic.

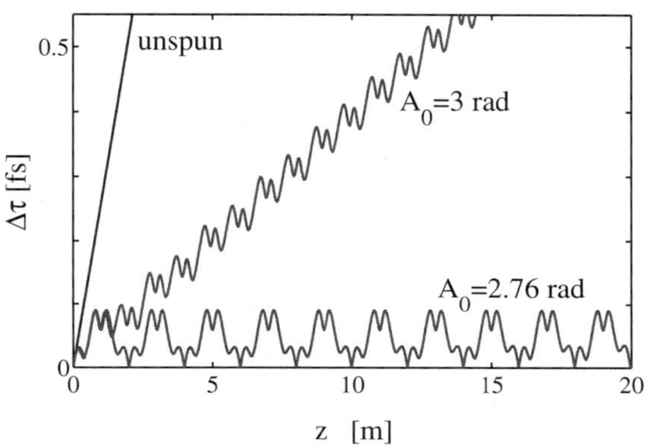

Fig. 5. Evolution of DGD as a function of distance for a fiber with $L_B = 20$ m and sinusoidal spin.

Figure 6 reports the SIRF as a function of the sinusoidal spin amplitude for the same fiber with $L_B = 20$ m and $p = 2$ m. It can be noted that the SIRF is well approximated by $|J_0(2A_0)|$.

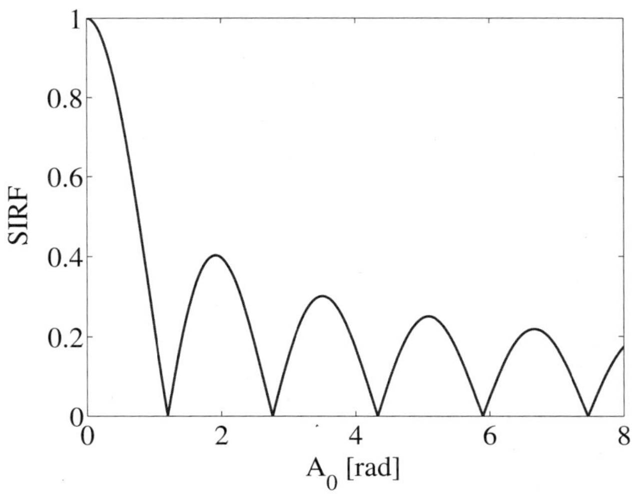

Fig. 6. Evolution of SIRF as a function of the sinusoidal spin amplitude.

Similar results can be achieved also by means of a perturbative analysis [11], [27], [28] under the condition $p \ll 2A_0 L_B$. Actually, in the literature two different kinds of perturbative analysis have been proposed to date. In [11] the perturbative analysis is performed by expanding the birefringence vector in Taylor series. Only the average of the birefringence components over a spin period is kept and this new vector is called "suppressed linear birefringence". Another possibility is to apply a straightforward perturbation expansion to the dynamic equation (9), in the rotating reference frame. Under the condition $p \ll 2A_0 L_B$, at a first order approximation, Eq. (9) becomes

$$\begin{aligned} \frac{d\Omega_1}{dz} &= 2\alpha(z)\Omega_2(z) + \beta_\omega\,, \\ \frac{d\Omega_2}{dz} &= -2\alpha(z)\Omega_1(z)\,, \\ \frac{d\Omega_3}{dz} &= 0\,. \end{aligned} \tag{15}$$

This system can be solved and the solution for Ω_1 integrated according to Eq. (10), obtaining the expression for the DGD

$$\Delta\tau(z) \simeq \Delta\tau_{\text{un}} \sqrt{\left(\int_0^p \cos[2A(z)]dz\right)^2 + \left(\int_0^p \sin[2A(z)]dz\right)^2}\,. \tag{16}$$

The corresponding SIRF is

$$\text{SIRF} \simeq \sqrt{\left(\int_0^p \cos[2A(z)]dz\right)^2 + \left(\int_0^p \sin[2A(z)]dz\right)^2}\,. \tag{17}$$

It can be noted that Eq. (17) is independent of the distance z, moreover there may be choices of $A(z)$ such that SIRF $= 0$. Actually, the spin functions with SIRF $= 0$ are the same that produce a periodic DGD according to the previous theory based on linear differential equations with periodic coefficients. For instance, if $A(z)$ has only odd harmonics, then the second integral in Eq. (17) is always equal to zero, thus the condition for a periodic DGD [Eq. (13)] coincides with the condition SIRF $= 0$. It is important to note that also the result obtained in Eq. (17) with the perturbative approach under the condition $p \ll 2A_0 L_B$ is independent of the beat length. However, a general condition for the periodic DGD can be found only by resorting to the theory of differential equations with periodic coefficients as done previously.

4. Fibers with Random Birefringence

The perturbations acting on telecommunication fibers and causing the PMD, have a random nature because they depend on the cabling conditions, and on the environmental temperature. Under a theoretical point of view it becomes important to identify a suitable mathematical model for the random local birefringence model.

4.1. Choosing the Birefringence Model

In the literature, four different models of birefringence have been proposed:

– the model with random modulus and random orientation of birefringence [17],

- the model with random orientation and constant modulus of birefringence [17],
- the polarization maintaining model that considers the birefringence as the sum of a constant term and of a white Gaussian noise process [24],
- the waveplate model [16], [30].

There are two main reasons for the definition of so many models. The first is that only with the recent development of polarization optical reflectometry [31], it has been possible to obtain an experimental validation of the models. In particular, the random modulus model is the only one fiber model proposed to date that is consistent with experimental results [32], [33]. The second reason for the existence of four models of birefringence is that, though only one model can represent the behavior of real fibers, the others are still useful because they allow theoretical analyses that otherways would not be possible. The general validity of the results obtained with different models is due to the fact that whereas $\boldsymbol{\beta}$ is a local property, the other quantities, like $\boldsymbol{\Omega}$, synthetize the global effects of the birefringence. As a consequence, when the fiber is "sufficiently" randomly perturbed, then the global effects of the birefringence are independent of the model that has been used.

This argument may not hold when a fiber is spun. In fact the spin acts on a local scale and interacts strongly with the birefringence. This intuition is supported by the verification (reported in the next sections) that the use of different birefringence models yields different DGD values. In the following, short descriptions of the random modulus model (RMM), and of the fixed modulud model (FMM) are reported. Since it comes out that the analysis with the RMM is very complex, it is convenient to carry out the theoretical treatment by means of the FMM and then to validate the results by means of numerical simulations with the RMM.

Incidentally, it is important to note that the RMM and the FMM describe the random evolution of the intrinsic birefringence of a fiber without spin. When the spin is applied, the local birefringence vector undergoes the rotation described in Section 2.

The Random Modulus Model (RMM)

According to the RMM, the two components of the linear birefringence vector $\boldsymbol{\beta}_l = (\beta_1, \beta_2, 0)^\mathsf{T}$, are independent Langevin processes [17]:

$$\frac{d\beta_i}{dz} = -\rho\beta_i(z) + \sigma\eta_i(z)\,, \qquad i = 1, 2\,, \tag{18}$$

where $\eta_1(z)$ and $\eta_2(z)$ are independent white-noise sources. As a consequence, the modulus $\beta_l(z, \omega)$ is a Rayleigh distributed random variable. Parameters ρ and σ define the statistical properties of the birefringence. In particular, the birefringence correlation length is $L_F = 1/\rho$, and the beat length reads $L_B = 2\pi\sqrt{\rho}/\sigma = \sqrt{2}\pi/\sigma_\beta = 2\pi/\langle\beta_l^2\rangle^{1/2}$. Note that the parameter σ_β is related to the statistical properties of the modulus of birefringence by means of $\langle\beta_l^{2h}\rangle = h!(2\sigma_\beta^{2h})$.

The Fixed Modulus Model (FMM)

According to the FMM, the local birefringence vector is

$$\boldsymbol{\beta}_l(z,\omega) = \beta_l(\omega)\left(\cos 2\theta(z),\, \sin 2\theta(z),\, 0\right)^{\mathsf{T}}, \tag{19}$$

where $\theta(z)$ is a Wiener process, i.e., it obeys the equation $d\theta/dz = \sigma\eta(z)$, and β_l is the birefringence modulus. In this expression, σ is a constant parameter, and $\eta(z)$ is a white-noise source [17]. Then, the birefringence correlation length and the beat length are $L_F = 1/(2\sigma^2)$ and $L_B = 2\pi/\beta_l$, respectively.

4.2. *General Analysis*

The following analysis is independent of the spin function and holds both for unidirectional and periodic profiles.

The RMM

The analysis of spun fibers using the RMM starts by transforming the reference frame in order to compensate for the intrinsic rotation of the birefringence and for the rotation induced by the spin [34]. This task can be achieved by using the matrix $\mathbf{T}(z) = \mathbf{R}_3^{\mathsf{T}}[2(A(z)+\theta(z))]$. The angle $\theta(z)$ in the RMM is defined in a similar way as in the FMM, i.e., considering $\beta_1(z) = \beta_l(z)\cos 2\theta(z)$ and $\beta_2 = \beta_l(z)\sin 2\theta(z)$ for the unspun fiber. In this case however, $\beta_l(z) = (\beta_1^2+\beta_2^2)^{1/2}$ is a random process. The dynamic equation in the rotating reference frame can be obtained after calculating the matrix $d\mathbf{T}/dz$, therefore, the z-derivative of $\theta(z)$ has to be calculated:

$$\frac{d\theta}{dz} = \frac{\sigma}{2\beta_l^2}\left(\eta_2\beta_1 - \eta_1\beta_2\right).$$

Finally, the dynamic equation in the new reference frame is

$$\frac{\partial \boldsymbol{\Omega}}{\partial z} = \begin{pmatrix} \beta_l \\ 0 \\ -2\alpha - \sigma(\eta_2\beta_1 - \eta_1\beta_2)/\beta_l^2 \end{pmatrix} \times \boldsymbol{\Omega} + \begin{pmatrix} \beta_l/\omega \\ 0 \\ 0 \end{pmatrix}, \tag{20}$$

where $\alpha(z)$ is the z-derivative of $A(z)$. Equations (20) and (18) form a system of stochastic differential equations that can be rewritten as

$$\frac{d}{dz}\begin{pmatrix} \Omega_1 \\ \Omega_2 \\ \Omega_3 \\ \beta_1 \\ \beta_2 \end{pmatrix} = \frac{\sigma}{\beta_l^2}\begin{pmatrix} -\beta_2\Omega_2 & \beta_1\Omega_2 \\ \beta_2\Omega_1 & -\beta_1\Omega_1 \\ 0 & 0 \\ \beta_l^2 & 0 \\ 0 & \beta_l^2 \end{pmatrix}\begin{pmatrix} \eta_1 \\ \eta_2 \end{pmatrix} + \begin{pmatrix} 2\alpha\Omega_2 + \beta_l/\omega \\ -\beta_l\Omega_3 - 2\alpha\Omega_1 \\ \beta_l\Omega_2 \\ -\rho\beta_1 \\ -\rho\beta_2 \end{pmatrix}. \tag{21}$$

By means of the theory of stochastic differential equations [35], it is possible to calculate the infinitesimal generator associated to Eq. (21), and using Dynkin's formula, an equation for the evolution of the mean square DGD can be obtained:

$$\frac{d\langle\Delta\tau^2\rangle}{dz} = \frac{2}{\omega}\langle\beta_l\Omega_1\rangle. \tag{22}$$

Using Dynkin's formula again, it can be found

$$\begin{aligned}
\frac{d\langle \beta_l \Omega_1 \rangle}{dz} &= \frac{\langle \beta_l^2 \rangle}{\omega} + 2\alpha \langle \beta_l \Omega_2 \rangle - \rho \langle \beta_l \Omega_1 \rangle \,, \\
\frac{d\langle \beta_l \Omega_2 \rangle}{dz} &= -2\alpha \langle \beta_l \Omega_1 \rangle - \rho \langle \beta_l \Omega_2 \rangle - \langle \beta_l^2 \Omega_3 \rangle \,, \\
\frac{d\langle \beta_l^2 \Omega_3 \rangle}{dz} &= -2\rho \langle \beta_l^2 \Omega_3 \rangle + 2\sigma^2 \langle \Omega_3 \rangle + \langle \beta_l^3 \Omega_2 \rangle \,.
\end{aligned}$$

Actually, it may be shown that the iterative application of the generator to any new term to be found, leads to an infinite sequence of relationships that can be expressed recursively as follows [34]:

$$\begin{aligned}
\frac{d\langle \beta_l^{2h} \Omega_3 \rangle}{dz} =& 2h^2\sigma^2 \langle \beta_l^{2(h-1)} \Omega_3 \rangle - 2\rho h \langle \beta_l^{2h} \Omega_3 \rangle + \langle \beta_l^{2h+1} \Omega_2 \rangle \,, \\
\frac{d\langle \beta_l^{2h+1} \Omega_1 \rangle}{dz} =& \frac{\langle \beta_l^{2(h+1)} \rangle}{\omega} + 2\sigma^2 h(h+1) \langle \beta_l^{2(h-1)+1} \Omega_1 \rangle \\
& - (2h+1)\rho \langle \beta_l^{2h+1} \Omega_1 \rangle + 2\alpha \langle \beta_l^{2h+1} \Omega_2 \rangle \,, \\
\frac{d\langle \beta_l^{2h+1} \Omega_2 \rangle}{dz} =& 2h(h+1)\sigma^2 \langle \beta_l^{2(h-1)+1} \Omega_2 \rangle - 2\alpha \langle \beta_l^{2h+1} \Omega_1 \rangle \\
& - \rho(2h+1) \langle \beta_l^{2h+1} \Omega_2 \rangle - \langle \beta_l^{2(h+1)} \Omega_3 \rangle \,,
\end{aligned} \tag{23}$$

with initial conditions $\langle \beta_l^{2h} \Omega_3 \rangle = \langle \beta_l^{2h+1} \Omega_1 \rangle = \langle \beta_l^{2h+1} \Omega_2 \rangle = 0$ at $z = 0$ for all h.

The set of equations (23) is unlikely to be solved explicitly due to its evident complexity. To obtain a numerical solution, it should be possible to truncate the infinite dimensional system. Unfortunately, the sequence reported in Eq. (23) does not always converge depending on the birefringence and spin parameters. In general the solution of the truncated system is much faster than performing simulations, but when L_F approaches, or is bigger than L_B, truncating the system gives incorrect results [34]. The negative consequence of this discussion is that it has not been possible to carry out useful analytical results with the RMM. Thus it is preferable to resort to the FMM and then to verify the validity of the achievements by means of simulations with the RMM.

The FMM

According to the FMM, and with respect to a reference frame that compensates for the rotation of the intrinsic birefringence $\theta(z)$ and of the spin, the local birefringence vector reads $\boldsymbol{\beta}(z,\omega) = (\beta_l,\ 0,\ -2[\alpha_0 + \sigma\eta(z)])^{\mathsf{T}}$.

By means of the theory of stochastic differential equations (SDE), and starting from the dynamic equation, an equation for the evolution of the mean square DGD can be obtained. In fact, applying the Dynkin's formula [35], it comes out

$$\frac{\partial \langle \Delta\tau^2 \rangle}{\partial z} = \langle \mathcal{A} \Delta\tau^2(z) \rangle = 2b_\omega \langle \Omega_1(z) \rangle \,, \tag{24}$$

where the infinitesimal generator is

$$\begin{aligned}\mathcal{A} =&(-2\sigma^2\Omega_1 + 2\alpha\Omega_2 + b_\omega)\frac{\partial}{\partial\Omega_1} + (-2\alpha\Omega_1 - 2\sigma^2\Omega_2 - \beta_l\Omega_3)\frac{\partial}{\partial\Omega_2} + \\ &\beta_l\Omega_2\frac{\partial}{\partial\Omega_3} + 2\sigma^2\left(\Omega_2^2\frac{\partial^2}{\partial\Omega_1^2} - 2\Omega_1\Omega_2\frac{\partial^2}{\partial\Omega_1\partial\Omega_2} + \Omega_1^2\frac{\partial^2}{\partial\Omega_2^2}\right).\end{aligned} \tag{25}$$

Using Dynkin's formula again, the following differential equation is also found:

$$\begin{aligned}\frac{\partial\langle\boldsymbol{\Omega}\rangle}{\partial z} &= \begin{pmatrix} -2\sigma^2 & 2\alpha(z) & 0 \\ -2\alpha(z) & -2\sigma^2 & -\beta_l \\ 0 & \beta_l & 0 \end{pmatrix}\langle\boldsymbol{\Omega}\rangle + \begin{pmatrix} b_\omega \\ 0 \\ 0 \end{pmatrix} \\ &= \mathbf{Q}(z)\langle\boldsymbol{\Omega}\rangle + \boldsymbol{u},\end{aligned} \tag{26}$$

where, the birefringence strength is $\beta_l = 2\pi/L_B$, and the square diffusion coefficient is $\sigma^2 = 1/(2L_F)$. Equation (26) must be solved in order to determine the evolution of $\langle\Delta\tau^2(z)\rangle$ and it can be noted that system (26) differs from Eq. (9) obtained in the case of deterministic birefringence only for the terms on the diagonal.

4.3. *Unidirectional Spin*

In the case of unidirectional spin with constant rate, Eq. (26) is a system of differential equations with constant coefficients. As long as $\sigma^2 \neq 0$, $\mathbf{Q}$ is nonsingular and the solution of Eq. (26) for zero initial condition reads

$$\langle\boldsymbol{\Omega}\rangle = \exp(\mathbf{Q}z)\int_0^z \exp(-\mathbf{Q}z')\cdot\boldsymbol{u}dz' = \mathbf{Q}^{-1}[\exp(\mathbf{Q}z) - \mathbf{I}]\boldsymbol{u}\,, \tag{27}$$

where $\mathbf{I}$ is the identity matrix. Then, the evolution of the mean square DGD is obtained by integrating the first component of Eq. (27), and produces the result

$$\langle\Delta\tau^2\rangle = 2\beta_\omega\boldsymbol{e}_1\cdot(\mathbf{Q}^{-1})^2\left[\exp(\mathbf{Q}z) - \mathbf{Q}z - \mathbf{I}\right]\boldsymbol{u}\,, \tag{28}$$

where $\boldsymbol{e}_1 = (1,\ 0,\ 0)$. The characteristic polynomial of $\mathbf{Q}$ is

$$p(s) = s^3 + 4\sigma^2 s^2 + (4\alpha_0^2 + 4\sigma^4 + \beta_l^2)s + 2\beta_l^2\sigma^2\,,$$

and, upon applying the Routh-Hurwitz theorem [36], it is possible to prove that all the roots of $p(s)$ have strictly negative real part. Therefore, the quantity $\exp(\mathbf{Q}z)$ decays to 0 as $z \to +\infty$, and, upon performing explicit calculations, Eq. (28) may be approximated as

$$\langle\Delta\tau^2\rangle \simeq \frac{8\pi^2 L_F}{\omega^2 L_B^2}z - \frac{8\pi^2 L_F^2}{\omega^2 L_B^2}\left(1 + \frac{4L_B^2}{p^2}\right). \tag{29}$$

Surprisingly, this result shows that asymptotically the rate at which $\langle\Delta\tau^2\rangle$ grows as a function of z, does not depend on the spin rate α_0. The positive effects of the spin are all due to the constant negative term in Eq. (29), and increase with the spin rate. This means that as the fiber length increases, the benefit on the DGD reduction due to the spin decreases. Such a behavior differs substantially from that of periodically spun fibers [37], where, as it will be explained in the next section, the spin affects also the

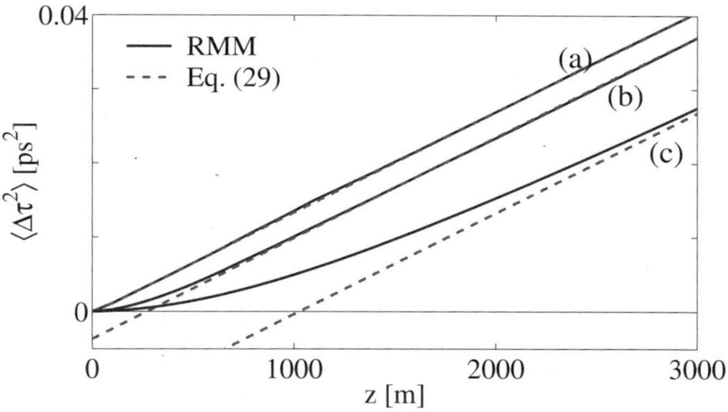

Fig. 7. Evolution of $\langle \Delta\tau^2(z) \rangle$ for a fiber with $L_B = 8.86$ m and $L_F = 20$ m. Curve (a) corresponds to the unspun fiber; curves (b) and (c) refer to constantly spun fibers with $p = 5$ and 2.5 m, respectively. Adapted from [21].

rate of growth of $\langle \Delta\tau^2 \rangle$, and the effectiveness of the spin does not depend on fiber length.

In order to validate the theoretical result presented in Eq. (29), Monte-Carlo numerical simulations have been performed with the FMM and the RMM on a set of 100,000 fiber realizations, for different values of the spin period $p = 2\pi/\alpha_0$ and birefringence parameters. As an example, Fig. 7 shows the evolution of the mean square DGD as a function of distance for a fiber with $L_B = 8.86$ m and $L_F = 20$ m. Curve (a) corresponds to the unspun fiber, whereas curves (b) and (c) correspond to spun fibers with $p = 5$ and 2.5 m respectively. Only the results obtained with the RMM have been reported in Fig. 7, because the corresponding ones obtained with the FMM are superimposed. Note that the good agreement with theoretical results (dashed curves) extends the practical validity of Eq. (29) to the RMM and confirms that in the case of constant spin rate, the DGD reduction is all achieved in the first part of the propagation, which actually represents the so called "short length regime". It can also be observed that the shorter is the spin period, the longer is the distance at which the DGD reaches its asymptotic behavior. Actually, Fig. 7 suggests that the value of z for which the asymptote (29) crosses the z-axis may be defined as a "transient length", and from Eq. (29) it can be found:

$$L'_T = L_F \left(1 + 4\frac{L_B^2}{p^2}\right) . \tag{30}$$

Such length should be understood as an inferior bound to the distance at which the mean square DGD converges to its asymptotic trend. Note that in the case of unspun fiber, $L'_T = L_F$, supporting the intuitive meaning of the transient length, but when $p < L_B$, L'_T may be much longer than L_F, so that the DGD converges to its asymptotic value only after several kilometers.

It is interesting to consider also the statistical properties of the PMD components. To achieve this task, in general it is not correct to refer to the rotating frame. The only exception is represented by Ω_3, which is not affected by the rotation, therefore from Eq. (28), and recalling that $\exp(\mathbf{Q}z)$ tends to zero as $z \to \infty$, it can be found:

$$\langle \Omega_3 \rangle \simeq -\boldsymbol{e}_3 \mathbf{Q}^{-1} \boldsymbol{u} = -\frac{\alpha_0}{\sigma^2 \omega} , \tag{31}$$

where $\boldsymbol{e}_3 = (0,\, 0,\, 1)$.

The calculation of $\langle \Omega_{f1} \rangle$ and $\langle \Omega_{f2} \rangle$ is somewhat more involved because it must be performed in the fixed frame. The starting point is that

$$\Omega_{f1}(z) = \Omega_1(z) \cos[2\alpha_0 z + 2\theta(z)] + \Omega_2(z) \sin[2\alpha_0 z + 2\theta(z)] ,$$
$$\Omega_{f2}(z) = \Omega_1(z) \sin[2\alpha_0 z + 2\theta(z)] - \Omega_2(z) \cos[2\alpha_0 z + 2\theta(z)] .$$

Then, the mean values of $\Omega_i(z) \cos[2\alpha_0 z + 2\theta(z)]$ and $\Omega_i(z) \sin[2\alpha_0 z + 2\theta(z)]$, for $i = 1,\, 2$ have to be evaluated. Upon applying the methods of stochastic differential equations, a system of 8 linear differential equations with constant coefficients as reported in [38] is found. Then, by means of the Routh-Hurwitz theorem, it can be verified that all the eigenvalues of the system have real parts smaller than $-\sigma^2$; hence, $\langle \Omega_{f1} \rangle$ and $\langle \Omega_{f2} \rangle$ converge to zero on a length scale not longer than L_F. Thus, as $z \to \infty$, the mean value of the PMD vector, $\boldsymbol{\Omega}_f(z)$, reads

$$\langle \boldsymbol{\Omega}_f \rangle = \begin{pmatrix} 0 \\ 0 \\ -\alpha_0/(\sigma^2 \omega) \end{pmatrix} . \tag{32}$$

Furthermore, it is possible to calculate the statistical powers of the PMD vector components. Also in this case, the calculation of the mean square of the third component can be performed in the rotating frame, whereas for the other two components it must be done in the fixed frame [38]:

$$\langle \Omega_{f1}^2(z) \rangle = \langle \Omega_{f2}^2(z) \rangle \sim \frac{\beta_l^2}{3\omega^2 \sigma^2} z - \frac{32\alpha^2 + 5\beta_l^2 - 4\sigma^4}{36\omega^2 \sigma^4} , \tag{33}$$

$$\langle \Omega_3^2(z) \rangle \sim \frac{\beta_l^2}{3\omega^2 \sigma^2} z - \frac{2(\alpha^2 + \beta_l^2 + \sigma^4)}{9\omega^2 \sigma^4} . \tag{34}$$

From the last three equations (32), (33) and (34) it can be observed that the statistical properties of $\Omega_3(z)$ differ from those of the other two components of $\boldsymbol{\Omega}(z)$; in particular, note that $\langle \Omega_3 \rangle \neq 0$, whereas in the case of unspun fibers the three components of the PMD vector are Gaussian random variables with zero mean value. As a consequence, the probability density function (pdf) of the DGD of constantly spun fibers might no longer be Maxwellian. To verify this assertion the pdf of the DGD has been estimated performing numerical simulations with the FMM and the RMM in the same conditions of Fig. 7, curve (b). The results are shown in Fig. 8 for different fiber lengths as explained in the legend; only the results of the RMM are reported since those of the FMM are indistinguishable. Solid curves refer to numerical data, whereas the dashed curves represent the Maxwellian distributions corresponding to the mean values of the numerical ensembles.

The dashed curves in Fig. 9 are Gaussian distributions with mean and variance given by theoretical expressions calculated previously; they show a good agreement with the numerical results.

It can be noted that for a fiber length of 3 km the DGD poorly fits with a Maxwellian distribution, notwithstanding that $\langle \Delta\tau^2 \rangle$ has already reached its asymptotic regime (see Fig. 7). In fact, for short fiber lengths the squared non-zero asymptotic average of Ω_3

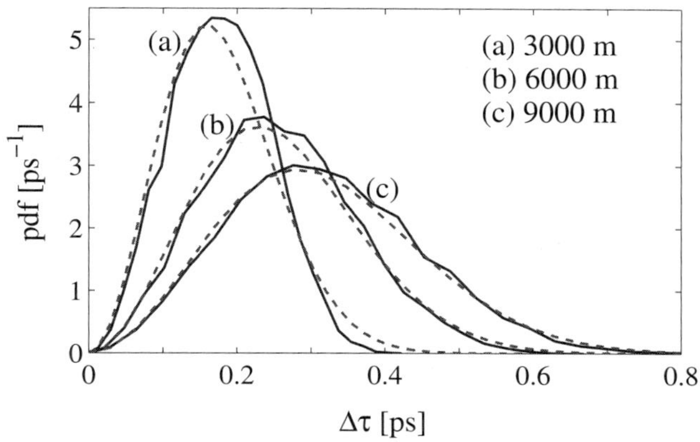

Fig. 8. Solid curves represent the pdf of $\Delta\tau$ for 100000 fiber realizations with $L_B = 10$ m, $L_F = 20$ m and $p = 5$ m. Dashed curves represent the Maxwellian distribution corresponding to the mean values of the ensembles. Adapted from [21].

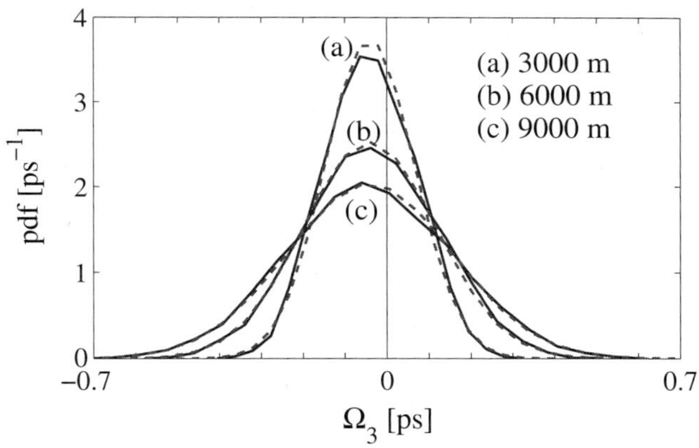

Fig. 9. Probability density function of Ω_3 in the same conditions of Fig. 8. The solid curves are obtained with the RMM, and the dashed curves correspond to Gaussian distributions with first and second moments calculated as in the text. Adapted from [21].

is comparable with $\langle \Omega_3^2(z) \rangle$. However, as the fiber length increases, $\langle \Omega_3(z) \rangle$ and the constant terms of $\langle \Omega_i^2(z) \rangle$ tend to be negligible compared with the z-dependent terms of $\langle \Omega_i^2(z) \rangle$ ($i = 1, 2, 3$), so that the components of $\boldsymbol{\Omega}(z)$ tend to have zero mean and same standard deviation, and the Maxwellian PDF is well approximated again.

The same analysis performed for the PMD vector, can be applied also to the second-order PMD (SOPMD) vector and the results are qualitatively similar. In particular it turns out that the spin-rate does not affect the asymptotic growth of the SOPMD, whereas it lengthens the duration of the short-length regime [38]. The asymptotic expression of the SOPMD modulus is

$$\langle|\boldsymbol{\Omega}_\omega|^2\rangle \sim \frac{\beta_l^4}{3\omega^4\sigma^4}z^2 - \frac{\beta_l^2(28\alpha^2+7\beta_l^2+\sigma^4)}{9\sigma^6\omega^4}z + \frac{34(4\alpha^2+\beta_l^2)^2+(116\alpha^2+5\beta_l^2)\sigma^4+4\sigma^8}{54\sigma^8\omega^4}. \tag{35}$$

The leading term does not depend on the spin rate, so that $\langle|\boldsymbol{\Omega}_\omega|^2\rangle$ of constantly spun fibers grows asymptotically with z^2 at the same speed of an unspun fiber, with the same birefringence parameters. Nevertheless, the spin affects the evolution of the SOPMD modulus in the short length regime, as is clearly shown in Fig. 10, where $\langle|\boldsymbol{\Omega}_\omega(z)|^2\rangle$ is plotted for different spin periods. Continuous curves have been obtained by means of numerical simulations, dashed curves refer to Eq. (35). The parameters used in the simulations are the same as for Fig. 7, in particular $L_B = 8.86$ m, $L_F = 8$ m, and the curves labeled with (a) correspond to the unspun fiber, those marked (b) and (c) correspond to $p = 5$ m and $p = 2.5$ m, respectively.

The analogy with first-order PMD can be pushed further considering the transient length of the SOPMD. Actually, recalling that the transient length represents an inferior bound to the distance at which the SOPMD reaches its asymptotic trend, we can choose as a possible candidate the position of the minimum of the convex and strictly positive parabola described by Eq. (35). Of course, this is a rather arbitrary definition, yet it matches the concept of transient length. Performing the explicit calculation, the position of that minimum is

$$L_T'' \simeq \frac{7L_F}{3}\left(1+\frac{4L_B^2}{p^2}\right), \tag{36}$$

which is an approximate equation that holds when $L_B^2 \ll 112\pi^2 L_F^2 \simeq 1100 L_F^2$ [38]). For typical telecommunications fibers L_B and L_F have the same order of magnitude [26], [33], therefore the above approximation is always well verified. Note that L_T'' depends on the spin in the same way as L_T', hence also the transient length of the SOPMD may get very long as the spin period decreases.

4.4. *Periodic Spin*

In the case of a periodic spin function, Eq. (26) is a system of differential equations with periodic coefficients that should be solved in order to determine the mean square DGD. Unfortunately, an explicit solution for Eq. (26) has not been written yet, instead it is possible to simplify it.

As a first step, it can be shown that the homogeneous system associated to Eq. (26) has no periodic solutions, and consequently, according to Lemma 2 in Appendix A, Eq. (26) has a unique periodic solution.

As a second step, by means of the Lyapunov's second method [25], it is possible to prove that for any initial condition the solution of Eq. (26) converges to a periodic orbit of period p, that is its unique periodic solution. In fact, recalling that the solution of any inhomogeneous system is given by Eq. (50), and using the Floquet-Lyapunov theorem (52), it may be shown that

$$\boldsymbol{\Omega}(z+p) = \boldsymbol{\Omega}(z) + \mathbf{X}(z)\left\{\left[\exp(\mathbf{K}p)-\mathbf{I}\right]\boldsymbol{\Omega}(0) + \int_{-p}^{0}\exp(-\mathbf{K}t)\,\mathbf{F}^{-1}(t)\,\boldsymbol{v}\,dt\right\}, \tag{37}$$

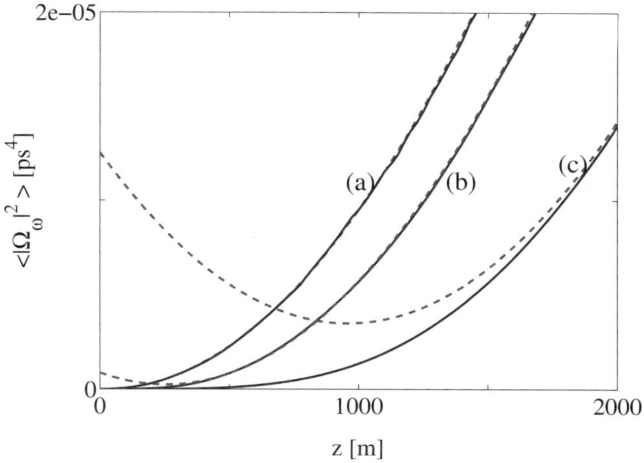

Fig. 10. Evolution of the mean square modulus of SOPMD as a function of distance. The dashed curves correspond to Eq. (35), whereas the solid curves are numerical results obtained with the RMM in the same conditions of Fig. 7. Adapted from [38].

where $\mathbf{X}(z) = \mathbf{F}(z)\exp(\mathbf{K}z)$ is the fundamental matrix. Note that the term between curly brackets is bounded and independent of z, so that the asymptotic behavior of $\boldsymbol{\Omega}(z+p)$ is governed by $\mathbf{X}(z)$. Furthermore, by means of Lyapunov's second method [25], [39] it can be shown that $\mathbf{X}(z)$ is asymptotically stable, in other words $\mathbf{X}(z)$ tends to zero as $z \to \infty$. As a consequence, from Eq. (37), it follows that $\boldsymbol{\Omega}(z+p) \to \boldsymbol{\Omega}(z)$, thus $\boldsymbol{\Omega}$ converges asymptotically to the unique periodic orbit for any initial condition.

The previous discussion implies that the asymptotic evolution of $\langle \Delta\tau^2(z)\rangle$ is determined by the periodic solution of Eq. (26). In fact, owing to Eq. (24), the mean square DGD may be written

$$\langle \Delta\tau^2(z)\rangle = 2b_\omega \int_0^z \langle \Omega_1(t)\rangle dt \simeq 2b_\omega z\,\frac{1}{p}\int_0^p \langle \Omega_1(t)\rangle dt\,, \tag{38}$$

where it has been assumed, without loss of generality, that $\langle \Delta\tau^2(0)\rangle = 0$.

Within the Short Period Assumption

The next step is to find the unique periodic solution of Eq. (38), and since each term involved in Eq. (26) is periodic, the equation can be expanded in a Fourier series and the short period assumption $p^2 \ll L_B^2$ can be introduced as done in the case of fibers with constant birefringence [22]. Under this condition Eq. (26) simplifies to

$$\frac{d\boldsymbol{y}}{dz} = \begin{pmatrix} -2\sigma^2 & 2\alpha(z) \\ -2\alpha(z) & -2\sigma^2 \end{pmatrix} \boldsymbol{y}(z) + b_\omega \begin{pmatrix} 1 \\ C^{-1}S \end{pmatrix}, \tag{39}$$

where $\boldsymbol{y} = (\langle \Omega_1\rangle, \langle \Omega_2\rangle)^\mathsf{T}$, and C and S are defined as

$$C = \frac{\gamma}{p^2} \int_0^p \int_0^p \exp(-2\sigma^2 u) \cos 2[A(t) - A(t-u)] dt\, du \,,$$
$$S = \frac{\gamma}{p^2} \int_0^p \int_0^p \exp(-2\sigma^2 u) \sin 2[A(t) - A(t-u)] dt\, du \,, \tag{40}$$

with $2\sigma^2 = 1/L_F$ and $\gamma = 2\sigma^2 p/[1 - \exp(-2\sigma^2 p)]$.

The unique periodic solution of Eq. (39) can be explicitly written, and it reads [25]:

$$\boldsymbol{y}(z) = b_\omega [\mathbf{I} - \mathbf{R}(z,0)]^{-1} \int_0^p \mathbf{R}(z, p-t) \boldsymbol{f}\, dt \,, \tag{41}$$

where $\boldsymbol{f} = (1,\, C^{-1}S)^\mathsf{T}$, $\mathbf{I}$ is the identity matrix, $\mathbf{R}(z,t) = \mathbf{Y}(z+p)\mathbf{Y}^{-1}(z+t)$ and the fundamental matrix of (39), $\mathbf{Y}(z)$, is

$$\mathbf{Y}(z) = \exp(-2\sigma^2 z) \begin{pmatrix} \cos 2A(z) & \sin 2A(z) \\ -\sin 2A(z) & \cos 2A(z) \end{pmatrix} .$$

Now that an explicit expression for $\langle \Omega_1(z) \rangle$ has been obtained, also $\langle \Delta\tau^2(z) \rangle$ might be calculated and, recalling that in long-length regime $\langle \Delta\tau^2 \rangle = 3\pi \langle \Delta\tau \rangle^2/8$, it comes out:

$$\langle \Delta\tau \rangle = \sqrt{\frac{C^2 + S^2}{C}}\, \langle \Delta\tau_{\mathsf{un}} \rangle \,, \tag{42}$$

where $\langle \Delta\tau_{\mathsf{un}} \rangle$ is the mean DGD the same fiber would have if it were not spun, and C and S are defined in Eq. (40).

Note that Eq. (42) holds in the long length regime $z \gg L_F$, for any periodic spin function $A(z)$, under the hypothesis $p^2 \ll L_B^2$. As already stated this constraint is not very restrictive. Since the beat length of telecommunication fibers is usually in the order of tens of meters [26], the short period assumption is well obeyed even with periods of a few meters, which are commonly used [9].

Furthermore, from Eq. (42), it is straightforward to obtain

$$\mathrm{SIRF} = \sqrt{\frac{C^2 + S^2}{C}} \,. \tag{43}$$

One can object that the derivation of Eq. (42) is based on a somewhat unrealistic model of birefringence.

To verify that Eq. (42) holds also for the more realistic RMM, several numerical simulations, for different spin functions and different values of birefringence parameters, have been performed. In all cases the theoretical DGD given by Eq. (42) agreed very well with the numerical estimate obtained from the RMM. As an example, Fig. 11 shows the numerical estimates of the SIRF (circles and triangles) compared with Eq. (43) (continuous line) for a sinusoidal spin profile. The RMM and the FMM have been implemented in the dynamic equation and Monte Carlo simulations have been performed over an ensemble of 15,000 fibers 10 km long emulated by 10^6 plates. The very good agreement with theoretical predictions extends the practical validity of Eq. (42) to the RMM. Again the agreement is very good, indicating that the simplifications made to derive Eq. (42) were reasonable.

One of the main features of Eq. (42) is that it simplifies the calculation of the DGD of spun fibers. In fact, the SIRF summarizes all the spin effects on fiber DGD. This factor depends on $A(z)$ and on birefringence correlation length, while it is independent

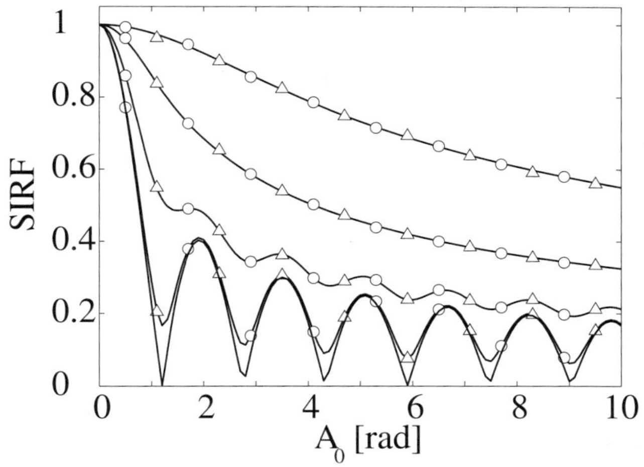

Fig. 11. Evolution of the SIRF as a function of spin amplitude, for a sinusoidal spin function with $p = 4$ m and $p^2 \ll L_B^2$. The curves are obtained from Eq. (43). The solid curves, from the upper to the lower, correspond to $L_F = 0.1,\ 0.3,\ 0.8$, and 3 m, respectively. The lowest solid curve represents the case $L_F \to +\infty$. Triangles and circles are numerical estimates of the SIRF obtained from the FMM and the RMM, respectively for $L_B = 17.7$ m. Adapted from [39].

of L_B because of the short period assumption. To evaluate the SIRF, the somewhat complicated integrals given in Eq. (40) have to be calculated; yet, these integrals can be easily evaluated with a proper numerical quadrature routine [40]. As an example, Fig. 12 refers to the following spin profile:

$$A(z) = A_0 \left(\sin \frac{2\pi z}{p} + \cos \frac{4\pi z}{p} \right) \tag{44}$$

and shows the evolution of the SIRF as a function of the amplitude A_0, for $p = 4$ m. From the upper to the lower, continuous lines correspond respectively to the following values of L_F: 0.1, 0.3, 0.8 and 2 meters. Dashed line refers to the case $L_F \to +\infty$.

It can be noted that the SIRF is always less than 1, meaning that the spin reduces the DGD in any case. This result was confirmed for every spin function that has been numerically checked, and, indeed, it seems physically reasonable.

Figure 12 also shows that for a given spin function, the shorter is L_F, the smaller is the DGD reduction induced by the spin. To understand this behavior, it is useful to note that the birefringence vector evolves both because of intrinsic random fluctuations of birefringence and because of spin. If L_F is very short, then the intrinsic fluctuations are very fast and tend to hide those induced by the spin. As a result, spin is not effective. For the same reason, for a given L_F, as the spin amplitude increases (i.e., spin evolution becomes faster), the SIRF decreases, even if not monotonically.

As the birefringence correlation length increases, the curves tend to the case $L_F \to +\infty$, as is the case for polarization maintaining fibers. Note that for $L_F = 2$ m, the SIRF is already quite close to this limit. The expression of the SIRF in the case $L_F \to +\infty$ may be simplified. In fact, if $L_F \gg p$, then in both integrals of Eq. (40) $\exp(-2\sigma^2 u) \simeq 1$, and $\gamma \simeq 1$. As a consequence, one finds:

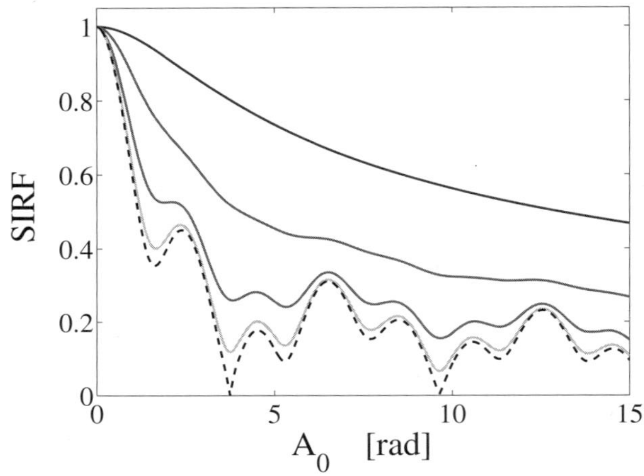

Fig. 12. Evolution of SIRF as a function of spin amplitude, for the spin function given in Eq. (44) with $p = 4$ m. Continuous curves, from the upper to the lower, refers to $L_F = 0.1, 0.3, 0.8$ and 2 m, respectively; the dashed curve represents the case $L_F \to +\infty$.

$$S \simeq \frac{1}{p^2} \int_0^p \int_0^p \sin 2[A(t) - A(t-u)] dt\, du = 0\,,$$

which holds because the innermost integral is a p-periodic, odd function of u. Therefore, SIRF simplifies to $\sqrt{C}$, where C can be rearranged as

$$C \simeq \left[\frac{1}{p}\int_0^p \cos 2A(u)\, du\right]^2 + \left[\frac{1}{p}\int_0^p \sin 2A(u)\, du\right]^2 . \tag{45}$$

According to Figs. 11 and 12 , when $L_F \to +\infty$ there exist values of A_0 for which the reduction factor vanishes. These amplitudes correspond to those spins that make the DGD of a deterministic fiber a periodic function (see Section 3). This result explains why, as numerically observed in[22], spin functions optimized for a polarization maintaining fiber allow quasi-optimal DGD reduction even when the fiber birefringence evolves randomly.

For example, the mean DGD of a polarization maintaining fiber spun with the function $A(z) = A_0 \sin(2\pi z/p)$ may be calculated. According to what is stated above, $\langle \Delta\tau \rangle = \sqrt{C}\langle \Delta\tau_{\text{un}} \rangle$, and using (45), $C = J_0^2(2A_0)$, where $J_0(x)$ is the zero-order Bessel function of first kind. This result enables to readily identify the optimal values of the spin amplitude.

Finally, it may be shown that there are two classes of spin functions for which Eq. (43) simplifies. The first of them is the class of spin functions whose Fourier series contains only odd harmonics for which, one finds $A(z) = -A(z + p/2)$ [22]. The second class consists of those functions which satisfy $A(z + q) = A(-z + q)$, where q is an arbitrary quantity. As an example, the spin defined in Eq. (44) belongs to this class for $q = p/4$. If the spin function belongs to one of these two classes, then, exploiting the symmetry properties of $A(z)$, it may be proved that $S = 0$, and hence $\text{SIRF} = \sqrt{C}$, notwithstanding that in this case C is still given by Eq. (40).

Without the Short Period Assumption

To our knowledge, it has not been found yet a closed form solution for the mean DGD when the short period assumption does not hold. In order to investigate the system's behavior in that case, we performed numerical simulations for both the RMM and the FMM fiber models, and calculated the SIRF as the ratio betwen the average DGD of the spun and corresponding unspun fiber. We estimated the average DGD over an ensemble of 15,000 fibers, considering a fiber of length $z \gg 100L_F$, so that the transient behavior has completely died out [41].

As a first case, it has been fixed $p = 4$ m and $L_B = 4.4$ m. The curves in Figs. 13(a) and 13(b), correspond to the FMM and the RMM respectively, and from the upper to the lower, are obtained for $L_F = 0.5$, 1, 2, 3, 16, and 50 m.

They show that there is a fairly good agreement between the two models only for the shortest values of L_F, when the spin is less effective. On the contrary, as the correlation length increases, the difference between the two models becomes evident. In particular, in this regime the SIRF curves obtained with the FMM in Fig. 13(a) have marked local minima, whose values tend to zero as L_F increases. Conversely, the SIRF curves obtained with the RMM in Fig. 13(b) have local minima that do not tend to zero, whose positions do not correspond to those obtained with the FMM.

As a second case, consider $p = 18$ m and $L_B = 4.4$ m. The solid curves, from the upper to the lower, in Figs. 14(a) and 14(b) correspond to the SIRF function for the following values of the correlation length: $L_F = 1$, 3, 5, 10, 18, 30, and 50 m. From Figs. 14(a) and 14(b), which refer to the FMM and the RMM respectively, it can be seen that in this case the spin is not effective in reducing the mean DGD. Moreover, the two models only agree for the shortest values of L_F. Also in this regime the main difference between the FMM and the RMM is that, as soon as $L_F \geqslant L_B$, the SIRF has marked minima in the FMM that do not appear when the RMM is used.

In summary, the simulations show that the two models only consistently agree in the short period limit. This result confirms that the formula obtained for the mean DGD of a periodically spun fiber holds well for both birefringence models when the spin period is shorter than the beat length. Conversely, when the spin period is of the same order or bigger than the beat length, the simpler analysis that can be performed by means of the FMM leads to a mean DGD that differs significantly from the mean DGD that is obtained with the RMM. Therefore, the behavior of randomly-birefringent, periodically-spun fibers differs significantly from the case of unspun fibers, where the mean DGD is independent of the model used for the random birefringence [17]. Additionally, in Figs. 11–14 the mean DGD of a spun fiber decreases as L_F increases, a result that also differs from the case of unspun fibers.

Explanation of the Difference

The key to understanding the difference between plots (a) and (b) in Figs. 13 and 14 is that in the FMM the modulus of the birefringence is fixed, while in the RMM it varies randomly along each fiber realization [42]. It is easiest to understand the impact of the varying birefringence in the limit $L_F \gg p$, while keeping the fiber length $z \gg L_F$. In this limit, each fiber realization can be viewed as a concatenation of long fiber sections, in which the length is of order L_F and the birefringence is constant. Since an ergodic theorem holds for an ensemble of optical fibers, so that the ensemble

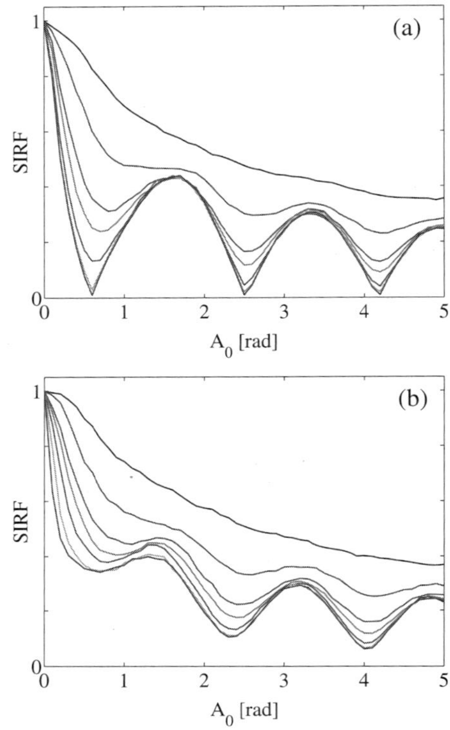

Fig. 13. Numerical estimates for the variation of the SIRF as a function of spin amplitude, for a sinusoidal spin function with $p = 4$ m and $L_B = 4.4$ m. Plots (a) and (b) correspond respectively to the FMM and the RMM. The solid curves, from the upper to the lower, refer to $L_F = 0.5,\ 1,\ 2,\ 3,\ 6,\ 16,$ and 50 m, respectively. Adapted from [34].

average is equivalent to a long spatial average over a single realization [17], [43], it follows that the SIRF in the RMM must equal the sum of the SIRFs for each different birefringence modulus in the FMM, weighted by its probability of occurring. This probability is Rayleigh-distributed, according to

$$f(\beta_l) = \frac{\beta_l}{\sigma_\beta^2} \exp\left(-\frac{\beta_l^2}{2\sigma_\beta^2}\right) . \tag{46}$$

The result is to smooth out the SIRF and to eliminate the zeros. Figure 15 shows the SIRF for the FMM and for the RMM with $p = 4$ m and $L_B = 4.4$ m for $L_F = 50$ m. The smoothing is clearly visible. When L_F is small, a similar smoothing is expected to occur. However, there is no longer a simple relationship between the FMM and the RMM.

Why then is there no difference between the FMM and the RMM in the short period limit? The explanation is that the SIRF is independent of the fiber beat length in the short period limit as shown in [22] and [37]. To better understand this point the SIRF has been calculated for the first model of birefringence for $L_F = 50$ m as a

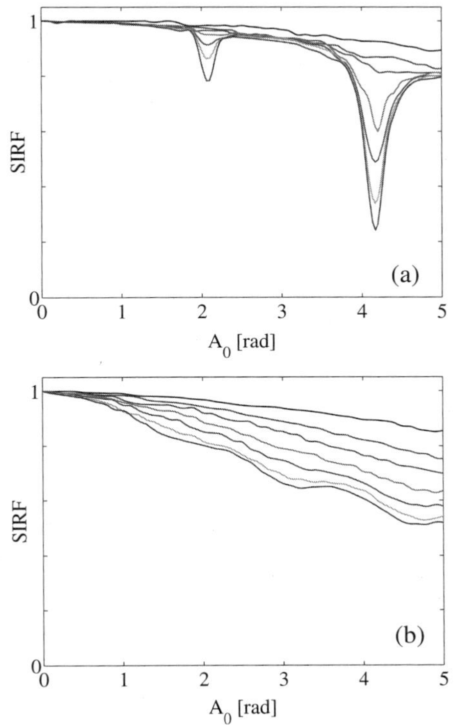

Fig. 14. Numerical estimates for the variation of the SIRF as a function of spin amplitude, for a sinusoidal spin function with $p = 18$ m and $L_B = 4.4$ m. Plots (a) and (b) correspond respectively to the FMM and the RMM. The solid curves, from the upper to the lower, refer to $L_F = 1, 3, 5, 10, 18,$ 30 and 50 m, respectively. Adapted from [34].

function of the spin amplitude A_0 and of the ratio p/L_B. Figure 16 shows the result, the gray scale corresponds to different SIRF values, so that black refers to SIRF$= 0$. Figure 16 corresponds to the sinusoidal spin function $A(z) = A_0 \sin(2\pi z/p)$ with $p = 4$ m, whereas L_B varies from 0.3 m up to approximately 300 m. The horizontal axis shows the ratio p/L_B on a logarithmic scale. One sees that when $p < L_B$ the zones corresponding to any given value of the SIRF are straight lines parallel to the horizontal axis, confirming their independence of L_B.

Thus, since the SIRF in the FMM is independent of L_B, when one averages over the different values of L_B in the FMM to obtain the RMM, one obtains exactly the same curve that he started out with. By contrast, if $p \gtrsim L_B$, then the SIRF zeros or any other fixed SIRF values are no longer parallel to the horizontal axis, indicating that the SIRF in the FMM now depends on L_B. Thus, the result is the smoothing effect that is visible in Fig. 15.

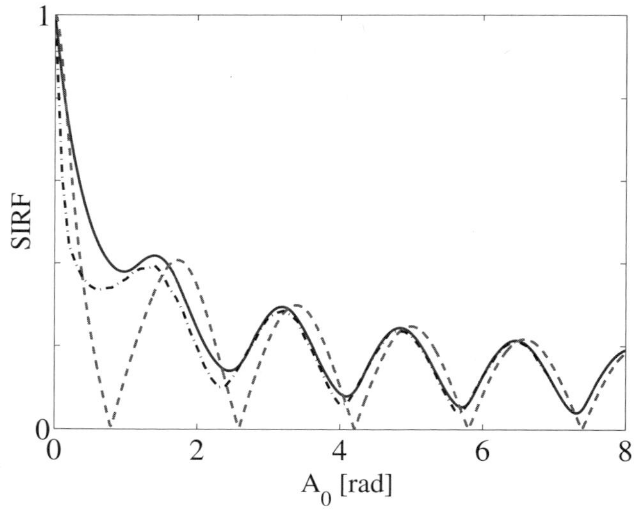

Fig. 15. Evolution of the SIRF as a function of the spin amplitude A_0 for $L_B = 4.4$ m and $L_F = 50$ m. The dashed curve is obtained with the FMM. The solid curve corresponds to the RMM and is obtained by weighting the FMM data according to the Rayleigh distribution of the local birefringence. The dashed-dotted curve is the same reported in Fig. 13 (b) for $L_F = 50$ m. Adapted from [42].

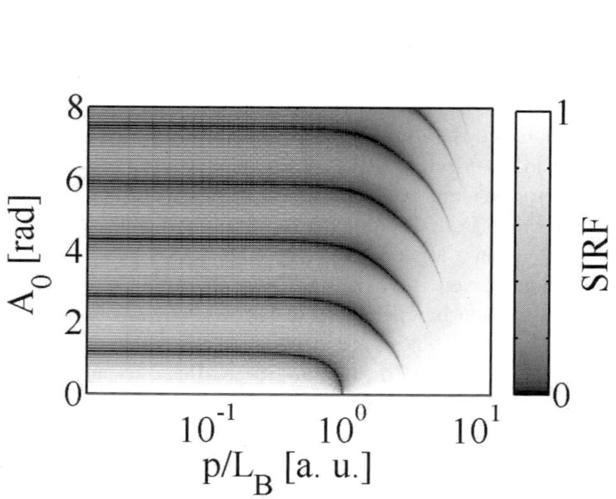

Fig. 16. Evolution of the SIRF as a function of the spin amplitude A_0 and of the ratio p/L_B in the case $L_F = 50$ m for the FMM. Adapted from [42].

5. Conclusions

In this chapter, the PMD behavior of spun fibers has been analyzed in detail, and techniques for choosing the spin parameters that minimize the DGD of a fiber have been put in evidence.

In particular fibers with a constant or a periodic spin function have been considered in the short length and long length regimes.

As a result it comes out that the PMD behavior of constantly spun fibers differs substantially from that of periodically spun fibers.

In the short length regime and unidirectional spin the DGD reduction depends on the ratio between the spin period and the fiber beat length. This means that the only way to reduce PMD is to spin the fiber as fast as possible. Conversely, in the short length regime and periodic spin, a method for the optimization of the spin parameters has been derived; it allows to obtain a DGD that does not increase as a function of distance but is periodic. Moreover, if the conditions $p^2 \ll L_B^2$ or $p \ll 2A_0L_B$ are satisfied, the optimal spin parameters are independent of the fiber beat length.

In the long length regime and unidirectional spin it has been found unexpectedly that the DGD of the spun fiber is asymptotically the same of the corresponding unspun fiber. Actually, the distance at which the mean DGD of a constantly spun fiber reaches the asymptotic regime, may be much longer compared with unspun fibers. An effective DGD reduction may be achieved by using constantly spun fibers with $p^2 \ll L_B^2$—that corresponds again to spin as fast as possible—and of length such that they are still in the transient regime. Conversely, in the long length regime and periodic spin, the same spin parameters that allow a periodic DGD in the short length regime, provide the best DGD reduction in the long length regime. Also in this case the short period assumption ($p^2 \ll L_B^2$) plays a key role both in the theoretical analysis and in the practical implementation of optimization issues.

In the long length regime, the choice of a proper mathematical model for the birefringence plays a fundamental role because when the short period assumption does not hold, the use of different models leads to different DGD values.

In spite of the number of achieved results, the study of low PMD spun fibers is quite new, and the effects on higher-order PMD and nonlinear interactions have still to be analyzed in detail.

Acknowledgments

We are indebted with Carlo Giacomo Someda for useful suggestions and continous encouragement, with Curtis R. Menyuk, Brian S. Marks and Davide Sarchi for fruitful collaboration.

Part of the research was supported by Pirelli Cavi e Sistemi Telecom, and it was performed under the agreement with Istituto Superiore delle Comunicazioni e Tecnologie dell'Informazione of Rome.

A. Linear Differential Equations with Periodic Ccoefficients

Consider a first-order linear differential equation

$$\frac{d\boldsymbol{x}}{dt} = \mathbf{A}(t)\boldsymbol{x}(t) + \boldsymbol{f}(t), \tag{47}$$

where $\boldsymbol{y}(t)$ and $\boldsymbol{f}(t)$ are an n-dimensional vector functions, and $\mathbf{A}(t)$ is a $n \times n$ matrix function. Its corresponding homogeneous system reads

$$\frac{d\boldsymbol{y}}{dt} = \mathbf{A}(t)\boldsymbol{y}(t). \tag{48}$$

and its adjoint system is defined as

$$\frac{d\boldsymbol{z}}{dt} = -\mathbf{A}^*(t)\boldsymbol{z}(t), \tag{49}$$

where $\mathbf{A}^*(t)$ is the transpose conjugate of $\mathbf{A}(t)$. The solution of Eq. (47) with initial condition $\boldsymbol{x}(0)$ is [25]

$$\boldsymbol{x}(t) = \mathbf{X}(t)\left\{\boldsymbol{x}(0) + \int_0^t \mathbf{X}^{-1}(t')\boldsymbol{f}(t)dt'\right\}, \tag{50}$$

where $\mathbf{X}(t)$ is the *fundamental matrix* associated with Eq. (47), which is a $n \times n$ matrix function obeying the homogeneous equation

$$\frac{d\mathbf{X}}{dt} = \mathbf{A}(t)\mathbf{X}(t), \tag{51}$$

with initial condition $\mathbf{X}(0) = \mathbf{I}$, the identity matrix, and every solution of the homogeneous system can be written as $\boldsymbol{y}(t) = \mathbf{X}(t)\boldsymbol{y}(0)$.

When both $\mathbf{A}(t)$ and $\boldsymbol{f}(t)$ are periodic with period p, Eq. (47) belongs to the class of Hill equations that are linear differential equations with periodic coefficients [25]. By the Floquet-Lyapounov theorem [25], the fundamental matrix of a system of differential equations with periodic coefficients may be expressed as

$$\mathbf{X}(t) = \mathbf{F}(t)\exp(\mathbf{K}t) \tag{52}$$

where $\mathbf{F}(t)$ is a periodic nonsingular matrix with period p, such that $\mathbf{F}(0) = \mathbf{I}$ and $\mathbf{K}$ is a constant matrix. The eigenvalues of $\mathbf{K}$ are known as the *characteristic exponents* of Eq. (47). We are interested in finding a periodic solution of period p for Eq. (47). It may be shown that the behavior of the inhomogeneous system depends on that of the homogenous one and on the properties of the fundamental matrix. In particular it is possible to distinguish two cases, which lead to the following results [25].

Lemma 1 *If the homogeneous system* (48) *has d linearly indipendent p-periodic solutions $\boldsymbol{y}_1(t), \ldots, \boldsymbol{y}_d(t)$. Then*

1. *The adjoint system* (49) *also has d linearly indipendent p-periodic solutions $\boldsymbol{z}_1(t), \ldots, \boldsymbol{z}_d(t)$.*

2. *The inhomogeneous system* (47) *has p-periodic solutions* **if and only if** *:*

$$\frac{1}{p}\int_0^p \boldsymbol{f}(t) \cdot \boldsymbol{z}_k(t)dt = 0 \qquad (k = 1, \ldots, d). \tag{53}$$

3. *If conditions* (53) *are satisfied, any* p*-periodic solution of the inhomogeneous system is given by*

$$\boldsymbol{x}(t) = \boldsymbol{x}_0(t) + \sum_{k=1}^{d} c_k \boldsymbol{y}_k(t)\,, \tag{54}$$

where $\boldsymbol{x}_0(t)$ *is some particular* p*-periodic solution of* (47) *and* $c_1, \ldots, c_d$ *are arbitrary complex numbers.*

Lemma 2 *Assume the homogeneous system* (48) *does not have any* p*-periodic solution, then the inhomogeneous system* (47) *has one and only one* p*-periodic solution.*

Please observe that when we say that the system (48) has d particular solutions, we mean that there exist d particular initial conditions $\boldsymbol{y}_0$, moreover the expression p-periodic stands for periodic with period equal to p.

References

1. A. J. Barlow, J. J. Ramskov-Hansen, and D. N. Payne, Appl. Opt., **20**, 2962–2968 (1981).
2. A. J. Barlow, D. N. Payne, M. R. Hadley, and R. J. Mansfield, Electron. Lett., **17**, 725–726 (1981).
3. R. I. Laming and D. N. Payne, J. Lightwave Technol., **7**, 2084–2094 (1989).
4. I. G. Clarke, Opt. Lett., **18**, 158–160 (1993).
5. A. Hart, R. G. Huff, and K. L. Walker, U.S. Patent no. 5298047, 1994.
6. A. Hart, R. G. Huff, and K. L. Walker, U.S. Patent no. 5418881, 1995.
7. R. E. Schuh, X. Shan, A. S. Siddiqui, and E. S. R. Sikora, Proc. OFMC'97 122–125 (1997).
8. R. E. Schuh, X. Shan, and A. S. Siddiqui, J. Lightwave Technol., **16**, 1583–1588 (1998).
9. M. J. Li and D. A. Nolan, Opt. Lett., **23**, 1659–1661 (1998).
10. K. Walker, Tech. Digest OFC (paper WJ4), Atlanta (GA), 2003.
11. J. G. Ellison and A. S. Siddiqui, IEE Proc.-Optoelectron., **146**, 137–141 (1999).
12. J. G. Ellison and A. S. Siddiqui, IEE Proc.-Optoelectron., **148**, 176–182 (2001).
13. M. Ferrario, P. Martelli, S. Pietralunga, and M. Martinelli, Tech. Digest. OFC (paper WJ3), Atlanta (GA), 2003.
14. R. Ulrich and A. Simon, Appl. Opt., **18**, 2241–2251 (1979).
15. S. C. Rashleigh, J. Lightwave Technol., **1**, 312–331 (1983).
16. H. Kogelnik, R. M. Jopson, and L. E. Nelson, *Polarization-Mode Dispersion* I. P. Kaminov and T. Li, editors (Academic Press, San Diego, 2002).
17. P. K. A. Wai and C. R. Menyuk, J. Lightwave Technol., **14**, 148–157 (1996).
18. A. J. Barlow and D. N. Payne, J. Quantum Electron., **19**, 834–839 (1983).
19. D. Sarchi and G. Roba, Tech. Digest OFC (paper WJ2), Atlanta (GA), 2003.
20. R. E. Schuh, E. S. R. Sikora, N. G. Walker, A. S. Siddiqui, L. M. Gleeson, and D. H. O. Bebbington, Elett. Lett., **28**, 1772–1773 (1995).
21. A. Galtarossa, P. Griggio, A. Pizzinat, and L. Palmieri, Opt. Lett., **28**, 1639–1641 (2003).

22. A. Galtarossa, L. Palmieri, and A. Pizzinat, J. Lightwave Technol., **19**, 1502–1512 (2001).
23. C. D. Poole, J. H. Winters, and J. A. Nagel, Opt. Lett., **6**, 372–374 (1991).
24. G. J. Foschini and C. D. Poole, J. Lightwave Technol., **9**, 1439–1456 (1991).
25. V. A. Yakubovich and V. M. Starzhinskii, *Linear differential equations with periodic coefficients* (John Wiley & Sons, New York, 1975).
26. A. Galtarossa, L. Palmieri, A. Pizzinat, M. Schiano, and T. Tambosso, J. Lightwave Technol., **18**, 1389–1394 (2000).
27. A. Pizzinat, B. Marks, L. Palmieri, C. Menyuk, and A. Galtarossa, Opt. Lett., **28**, 390–392 (2003).
28. J. Kevorkian and J. D. Cole, *Perturbation methods in applied mathematics* (Springer-Verlag, New York, 1981).
29. M. Wang, T. Li, and S. Jian, Opt. Express, **11**, 2403–2410 (2003).
30. F. Curti, B. Daino, G. De Marchis, and F. Matera, J. Lightwave Technol., **8**, 1162–1166 (1990).
31. A. Galtarossa and L. Palmieri, Tech. Digest. OFC (paper WA4), Anaheim (CA), 2002.
32. A. Galtarossa, L. Palmieri, M. Schiano, and T. Tambosso, OSA Opt. Lett., **25**, 384–386 (2000).
33. A. Galtarossa, L. Palmieri, M. Schiano, and T. Tambosso, OSA Opt. Lett., **26**, 962–964 (2001).
34. A. Pizzinat, L. Palmieri, B. S. Marks, C. R. Menyuk, and A. Galtarossa, J. Lightwave Technol., **21**, 3355–3363 (2003).
35. B. Øksendal, *Stochastic Differential Equations*, (Springer-Verlag, Berlin, 2000).
36. L. Cesari, *Asymptotic Behavior and Stability Problems in Ordinary Differential Equations* (Springer-Verlag, New York, 1971).
37. A. Galtarossa, P. Griggio, A. Pizzinat, and L. Palmieri, Opt. Lett., **27**, 692–694 (2002).
38. A. Galtarossa, P. Griggio, L. Palmieri, and A. Pizzinat, J. Lightwave Technol., **22**, 1127-1136 (2004).
39. A. Galtarossa, L. Palmieri, A. Pizzinat, B. S. Marks, and C. R. Menyuk, J. Lightwave Technol., **21**, 1635–1643 (2003).
40. W. H. Press, S. A. Teukolsky, W. T. Vetterling, and B. P. Flannery, *Numerical recipes in C: the art of scientific computing* (Cambridge University Press, Cambridge, 1992).
41. Y. Tan, J. Yang, W. L. Kath, and C. R. Menyuk, J. Opt. Soc. Am. A, **19**, 992–1000 (2002).
42. A. Pizzinat, B. S. Marks, L. Palmieri, C. R. Menyuk, and A. Galtarossa, IEEE Ph. Technol Lett., **15**, 819–821 (2003).
43. C. R. Menyuk and P. K. A. Wai, J. Opt. Soc. Am. B, **11**, 1288–1296 (1994).

DOI: 10.1007/s10297-004-0017-x
Originally published in J. Opt. Fiber. Commun. Rep. 1, 181–200 (2004)

PMD emulation

A.E. Willner[1] and M.C. Hauer[2]

[1]Department of Electrical Engineering-Systems
University of Southern California
Los Angeles, CA 90089-2565 USA
Email: willner@csi.usc.edu
[2]Department of Electrical Engineering-Electrophysics
University of Southern California
Los Angeles, CA 90089-0483 USA
Email: hauer@usc.edu

Abstract. As PMD has become an increasingly significant issue in high-bit-rate fiber optic systems, a need has developed for laboratory instruments and software tools capable of rapidly exploring the effects of PMD on various test items. The random nature of PMD dictates that to characterize its effects on transmitter/receiver pairs and on PMD compensation systems, one must repeatedly measure the system performance over a wide sample space of PMD states. This need has spurred the development of several methods for accurately and rapidly emulating the random variations of PMD in real fibers, as well as techniques for generating specific components and combinations of first- and higher-order PMD in a predictable and repeatable way. This chapter reviews several of these methods for both statistical and deterministic PMD emulation. The underlying concepts and rationales for various design architectures are discussed. A common analytical model for describing multisection all-order emulators is presented and a simple design example is used to further illustrate the concepts.

1. Introduction

Polarization-mode dispersion (PMD) has emerged as a critical issue for achieving high-performance, $\geq$ 10-Gbit/s/channel optical fiber transmission systems. PMD arises in an optical fiber from asymmetries in the fiber core that induce a small amount of birefringence that randomly varies along the length of the fiber. This birefringence causes the power in each optical pulse to split between the two polarization modes of the fiber and travel at different speeds, creating a differential group delay (DGD) between the two modes that can result in pulse spreading and intersymbol interference. PMD may be represented as a vector on the Poincaré sphere whose magnitude is the DGD and whose direction represents the polarization state of one of the two orthogonal principal states of polarization, or PSPs, of the fiber (at either the fiber input or output) [1]. Owing to random geometric and stress variations along the fiber core and environmental perturbations, the DGD and PSPs vary randomly with time and frequency and are therefore characterized by statistical parameters [2]. The probability distributions of the first and higher order components of PMD in long single-mode fibers are well known [3]. Analytical and experimental results show that the DGD follows a Maxwellian probability distribution and that the high-DGD points in the tail of this distribution are likely to cause system outages [4]. Additionally, owing to the effects of higher-order PMD, system outages can also occur at low DGD values, or after the DGD has been compensated [5]. These outages may happen for only a few minutes per year and will depend on the average DGD value and environmental changes. For example, in a 25-day field trial, 10-Gbit/s data was transmitted over a 110-km route containing fiber manufactured in 1994 with an average PMD of $\sim$ 25 ps and the system receiver (without PMD compensation) lost synchronization due to PMD for several seconds on three separate instances [6]. Thus, in order to characterize the effects of PMD on transmitter/receiver pairs and PMD compensation systems, one must repeatedly measure the system performance over a wide sample space of PMD states. This is difficult to accomplish in the laboratory because high-PMD transmission fibers are not commercially available and, even if they were, testing times would be prohibitively long due to the large number of different fiber ensembles that must be explored. These issues have spurred the development of several methods for accurately and rapidly emulating the random variations of PMD in real fibers, as well as techniques for generating specific components and combinations of first- and higher-order PMD in a predictable and repeatable way. These efforts define the field of PMD emulation.

2. Statistical PMD Emulators

A device intended to accurately and rapidly reproduce the statistical variations of PMD in real fibers can be considered a statistical emulator. Such devices are also frequently called "all-order emulators." To mimic the statistics of a real fiber, any such emulator should therefore meet the following key performance metrics:

1. The DGD should be Maxwellian-distributed over an ensemble of fiber realizations at any fixed optical frequency.

2. The emulator should produce accurate higher-order PMD statistics and should be able to reach any combination of first- and higher-order PMD values.

3. When averaged over an ensemble of fiber realizations, the frequency autocorrelation function of the PMD emulator should tend toward zero outside a limited frequency range to provide accurate PMD emulation for WDM channels ([7, 8]).

In addition to these performance requirements, the following features are desirable to form a practical measurement tool:

1. Stability—the PMD of the emulator should remain stable over the measurement period, which may last minutes to hours.

2. Low loss—the emulator should ideally have low insertion loss and exhibit negligible polarization dependent loss (PDL).

3. Simplicity—the implementation of the emulator should allow easy control from one emulator state to the next.

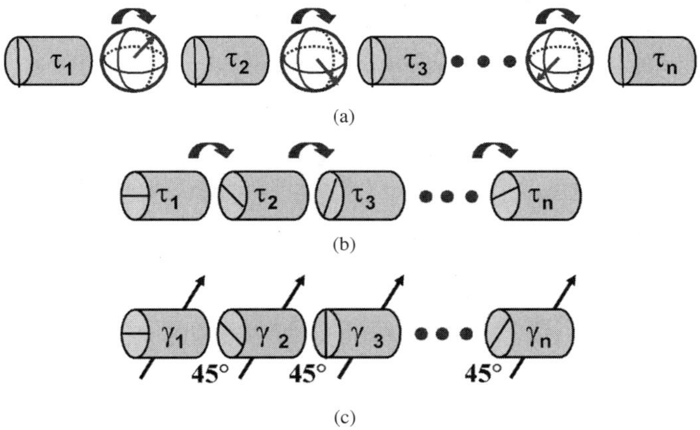

Fig. 1. Three common statistical emulator configurations constructed from a concatenation of many birefringent elements (a) with uniformly-distributed polarization transformations (pol. controllers) between sections, (b) with rotatable sections, and (c) with fixed 45° angles between sections and variable birefringence elements. τ is the DGD and γ is the birefringence.

The first requirement, stability, is especially difficult to achieve since most statistical emulators are constructed of several birefringent elements, which tend to be extremely sensitive to environmental changes, causing the polarization state at the emulator output to vary over short periods of time even though the control parameters are held constant (polarization controller settings, crystal rotations, etc.).

Statistical PMD emulators are typically constructed from a concatenation of several linear birefringent elements. These elements may be sections of polarization-maintaining (PM) fiber, birefringent crystals, or any other device that provides a differential group delay between the two orthogonal polarization axes (e.g., a polarization beam splitter followed by two paths of differing lengths and a polarization beam combiner). To achieve different PMD states, some property of the emulator must be varied between samples, such as the polarization coupling between sections, the wavelength, or the birefringence of each section. If enough sections are used, the PMD statistics of the emulator will converge toward those of a real fiber. Most of these emulators can be classified into one of four main categories:

1. Emulators with fixed orientations between sections—this type of emulator is typically constructed by splicing equal or unequal-length sections of PM fiber at either random or 45° angles [4]. As explained in [9], the use of 45° angles leads to the most rapid frequency decorrelation of the resultant PMD vector and the use

of unequal sections avoids an undesired periodicity in the frequency autocorrelation function. However, this emulator configuration has limited use since different PMD states can only be obtained by widely varying the wavelength or by cycling through environmental changes.

2. Emulators with uniformly-distributed polarization transformations between sections [Fig. 1(a)]—these emulators are obtained by placing polarization controllers between the birefringent sections to transform the polarization state according to a uniform distribution over the Poincaré sphere ([9, 10]).

3. Emulators with rotatable sections [Fig. 1(b)]—the birefringent sections are randomly rotated relative to each other to obtain different PMD states. Examples include birefringent crystals mounted on rotation stages [11] or separated by rotatable polarization mode-mixers (thin waveplates) [12], PM fibers connected by rotatable connectors [13], and a long strand of PM fiber with fiber-twisters placed periodically along its length to vary the polarization coupling between sections [14].

4. Emulators with tunable-birefringence sections connected at fixed 45° angles [Fig. 1 (c)]—for this type of emulator, the orientation between sections remains fixed at 45° (to obtain rapid frequency decorrelation), while the birefringence of each section is varied to obtain different PMD states. Examples of variable birefringence elements include voltage-controlled lithium-niobate crystals and PM fiber sections with small metallic heaters deposited on the fiber surface to temperature tune the birefringence [15].

For the latter three cases, statistical PMD samples are obtained at a given wavelength by randomly varying the polarization coupling between each section between measurements. Alternatively, the statistics of the emulator may be obtained by measuring the PMD at many different wavelengths while the coupling between sections remains constant. However, this wavelength scanning technique is likely to produce undesirable statistics (i.e., a non-Maxwellian DGD distribution) if the polarization coupling between sections is not equal to 45°, the case that leads to the most rapid frequency decorrelation of the PMD vector.

Early studies of PMD emulators focused on the number of sections required to produce accurate first- and higher-order statistics, on the differences between using polarization controllers and polarization rotators between sections, and on obtaining a frequency autocorrelation function that quadratically falls to zero outside the PMD vector's correlation bandwidth [1, 8–11, 13].

For example, in [13], Khosravani et al. compare the performance of two experimental emulators using 3 and 15 sections of PM fiber connected by rotatable-key connectors. The lengths of the PM fiber sections were chosen randomly, with an average of $\sim$ 7 m and a 20% Gaussian deviation [4]. The beat length of the PM fiber was measured to be $\sim$ 3.1 mm at 1550 nm, so the average DGD of the 15-section emulator was $\sim$ 40 ps. The total loss of the emulator was reported to vary from 6 to 10 dB, depending on the rotation angles between the PM fiber sections. The polarization dependent loss was less than 0.2 dB, and the DGD values were measured using the Jones matrix method [16].

Figures 2(a) and (b) show the DGD probability density functions (pdf) for the 3- and 15-section emulators at a fixed wavelength of 1555 nm. 1000 samples were taken by randomly rotating the physical angles between the fibers. The wavelength was swept over 1 nm with 0.02-nm steps for each set of coupling angles to obtain a DGD pdf at 50 different wavelengths. Good distributions were obtained at other wavelengths as well, and the average DGD was very close to 40 ps at all wavelengths. Using numerical simulations, the authors verified that with 15 sections of PM fiber, a Maxwellian pdf is achieved out to three times the average DGD in the tail of the pdf. As shown in Fig. 2(c), the 15-section emulator also exhibits good second-order PMD statistics that closely match the ideal theoretical distribution. The results also show that three sections are clearly not sufficient to reproduce realistic PMD statistics.

In WDM systems, not only is the DGD pdf of each channel Maxwellian distributed, but the PMD characteristics of channels separated by large frequencies are statistically independent, i.e., the PMD vectors are uncorrelated. For 40 ps of PMD, a real fiber shows negligible correlation between PMD vectors when the channel spacing is more than 0.2 nm [1, 7]. As mentioned previously, to avoid periodicity in the autocorrelation function and to decrease the residual correlation, it is preferable to employ unequal length sections in PMD emulators [8]. Figure 3 shows the autocorrelation function of the PMD vector for the emulators in [13] with 3, 10, and 15 sections. For the 15-section, unequal-length PMD emulator, an average level of 10% correlation remains between well spaced wavelengths. The authors also show that the use of polarization controllers rather than rotators between sections only slightly reduces the residual correlation.

A detailed performance comparison of emulator types 1 through 3, for various numbers of unequal-length sections, may be found in [9]. The authors ultimately conclude that a 15-section emulator with rotatable sections presents a good compromise between meeting the theoretical design criteria and practical implementation. A number of other statistical emulator designs have also been demonstrated in the recent literature [11, 12, 14, 15]. These emulators demonstrate the advantages of electrical control, low-loss, and a compact design that is easily expandable to many sections. For example, Fig. 4 shows a design in which a single, long strand of PM fiber is separated into 64 segments by affixing fiber twisters between sections [14]. The fiber twisters are rotated using individual step motors to randomly couple the polarization state between sections. This architecture accommodates numerous sections with low insertion loss since there are no fiber splices or connectors. A type 4 statistical emulator is shown in Fig. 5(a) [15]. In this case, 30 unequal length sections of PM fiber (average length $= 5.7$ m) are fusion spliced together at fixed 45° angles. A 2.5-cm-long thin-film metal heater is deposited near the center of each fiber section to form 30 electrically tunable waveplates. Individual voltages are applied to each heater to thermally tune their birefringence, which, in turn, varies the output polarization state of each section. The fiber heaters are mounted in a silicon v-groove array to achieve a compact device. Figure 5(b) shows the measured DGD distribution of the device in comparison with the theoretical Maxwellian for a fiber with 42-ps average DGD. While many designs have been proposed, each with its own set of advantages and tradeoffs, it is apparent that all of the reported configurations for statistical emulators will meet the desired requirements if enough unequal-length sections are used. The choice of which to employ will therefore depend on the user's application and resources.

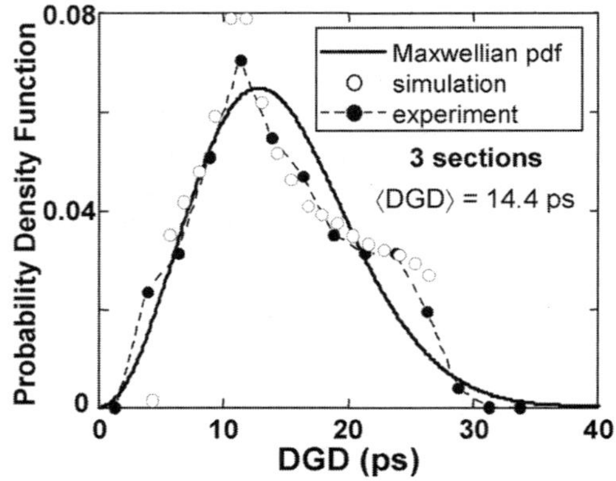

(a)

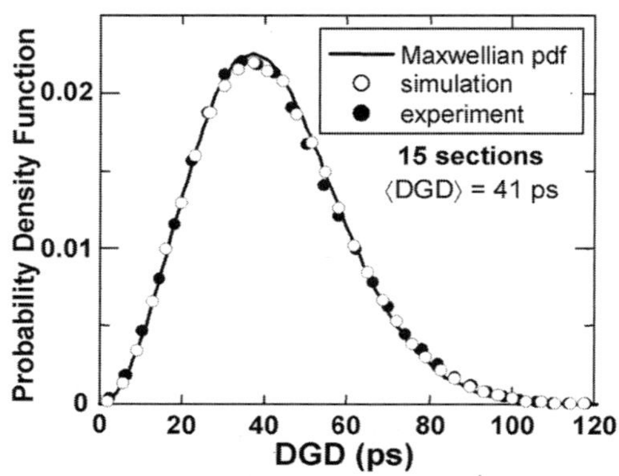

(b)

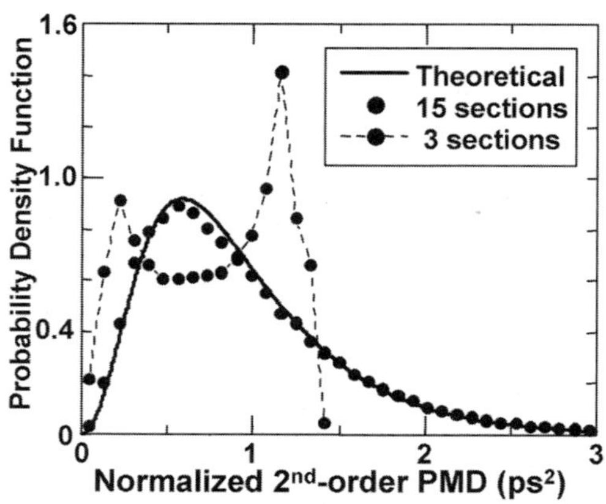

(c)

Fig. 2. Measured DGD distributions at 1555 nm for two statistical emulators with (a) 3 sections and (b) 15 sections separated by rotatable connectors. (c) Simulation results of the normalized second-order PMD magnitude for the 3- and 15-section emulators. From [13].

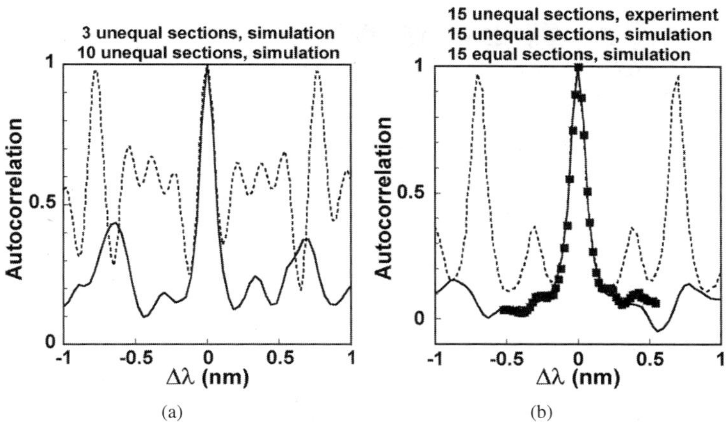

Fig. 3. Frequency autocorrelation function of the PMD vector for (a) 3- and 10-section PMD emulators with unequal lengths of PM fibers (simulation), and (b) a 15-section emulator with equal and unequal lengths of PM fibers (simulation and experiment). Note the undesired periodicity that results when equal-length sections are used. From [13].

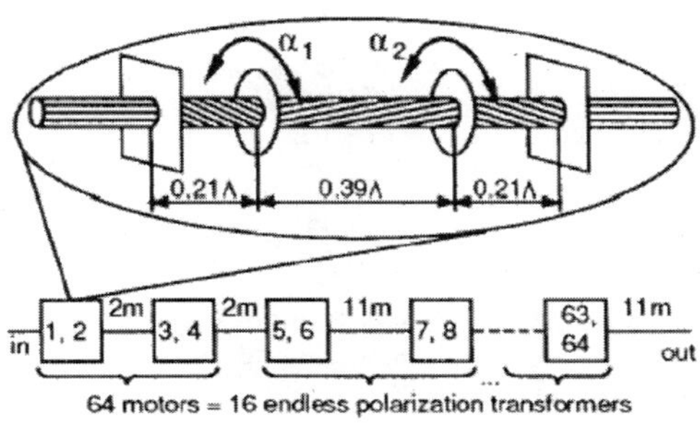

Fig. 4. A 64-section emulator constructed from a single strand of PM fiber. Each section is separated by a fiber twister that is affixed to the fiber and controlled with individual step motors. From [14].

3. Design Example

An example will best illustrate how an emulator is typically designed to meet the above requirements. The application of our example emulator will be to characterize the performance of a PMD compensator for different WDM channels at 10 Gbit/s. Thus, in our experimental setup, the emulator will be followed by the compensator under test and a receiver. A bit-error-rate test set will be used to measure the signal impairment before and after the compensator for different PMD states of the emulator.

To obtain statistics similar to those shown in Figs. 2 and 3, a 15-section emulator with randomly chosen unequal lengths will be used. The emulator will be designed to have a mean DGD of 40 ps. This is a typical value for emulators that are used to test 10 Gbit/s NRZ systems, for which severe outage probabilities will occur when the

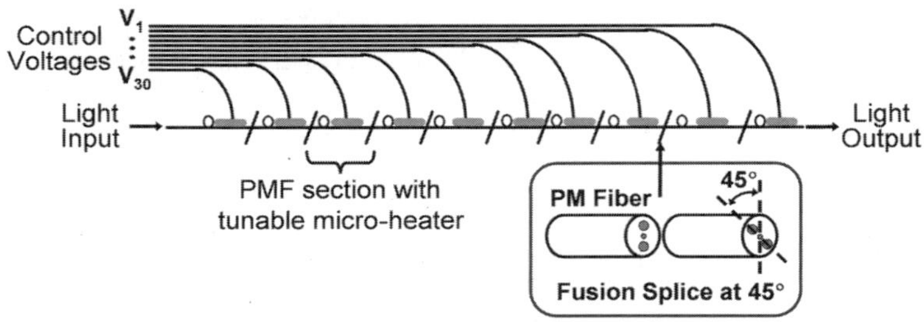

(a)

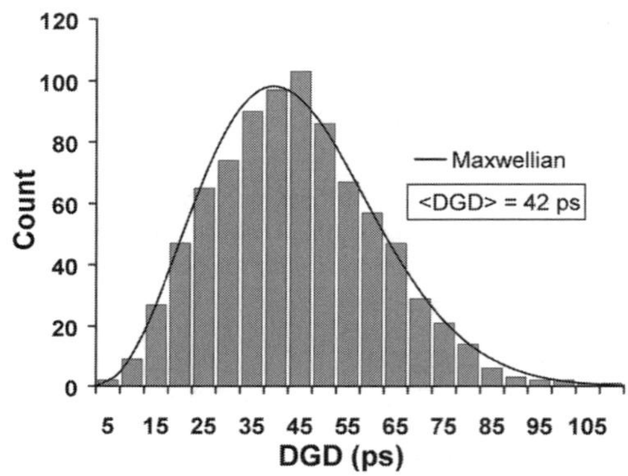

(b)

Fig. 5. (a) Schematic diagram of a type 4 emulator with evaporated micro-heaters deposited on 30 PM fiber sections spliced together at 45°. (b) Measured DGD distribution of 850 experimental measurements that were taken by randomly varying the 30 heater voltages between samples. The solid curve is the Maxwellian distribution that would be expected from a real fiber with the same average DGD. From [15].

DGD is greater than or equal to a bit-time (100 ps at 10 Gbit/s). The mean DGD is chosen to be 30 to 40% of a bit-time to ensure that several high-DGD points in the tail of the distribution (at two to three times the average DGD) will be encountered during a test of only 1000 or so samples (to allow for reasonable testing times). Additionally, as demonstrated in [17], an emulator with increased mean PMD (compared with the installed fiber), can be used to perform accelerated outage probability testing to predict the long-term performance of a system (over years) in a short time. Given this desired mean DGD and the number of sections, Eq. (1) may be used to determine the DGD needed for each section (initially assuming equal length sections):

$$\text{DGD/section} = \frac{\text{DGD}_{\text{rms}}}{\sqrt{N}} = \frac{\sqrt{\frac{3\pi}{8}}\langle\text{DGD}\rangle}{\sqrt{N}} = \frac{\sqrt{\frac{3\pi}{8}}(40\text{ ps})}{\sqrt{15}} = 11.21\text{ ps/section}. \tag{1}$$

Knowing that the DGD of a fiber is Maxwellian distributed, this formula is derived from the known relationship between the rms and the mean DGD [1]. To obtain unequal length sections, 15 Gaussian-distributed random numbers are chosen with a mean of 11.21 ps/section and a standard deviation that is 20% of the mean value. This standard deviation is known to provide sufficient randomness of the DGD values per section so as to avoid undesired periodicities in the frequency autocorrelation function (which do not occur in real fibers) [7–9]. The resulting DGDs for each section are: 8.16, 10.88, 13.76, 11.84, 10.08, 10.72, 16.00, 13.76, 8.64, 11.52, 11.04, 11.36, 9.76, 11.84, and 7.36 ps. Using $\langle \mathrm{DGD} \rangle = \sqrt{\tau_1^2 + \tau_2^2 + \cdots + \tau_{15}^2}$, where τ_i is the DGD of the ith section, the computed average DGD of the emulator is 43.9 ps, a little higher than our design value.

Assume that PM fiber will be used to form the DGD sections of the emulator and that a measurement of 1 m of the PM fiber yields a DGD of 1.6 ps/m. This means that the lengths of the 15 sections should be: 5.1, 6.8, 8.6, 7.4, 6.3, 6.7, 10.0, 8.6, 5.4, 7.2, 6.9, 7.1, 6.1, 7.4, and 4.6 m. This concludes the design of the emulator sections. Any of the methods described above for type 2 through 4 emulators may now be implemented to adjust the emulator's PMD state between measurements.

4. Analytical Emulator Model

To theoretically investigate and simulate the behavior of PMD emulators, the typical PMD concatenation equations are applied [1]. Presented here is a brief overview of the derivation of the recursion relation often used to model PMD emulators in Stokes space (for further details, see [1], [9], and [20]).

The polarization mode dispersion of a fiber is represented as a vector $\boldsymbol{\Omega}$ on the Poincaré sphere and is typically defined by the key equation:

$$\frac{d\mathbf{s}}{d\omega} = \boldsymbol{\Omega} \times \mathbf{s} \tag{2}$$

that relates the rate of change with frequency of the input or output state of polarization to the PMD vector. The vector $\mathbf{s}$ is the three-dimensional stokes vector for either the input or output polarization state. The magnitude $\Delta\tau$ of the dispersion vector is the DGD:

$$\Delta\tau = \mid \boldsymbol{\Omega} \mid = \sqrt{\boldsymbol{\Omega}_x^2 + \boldsymbol{\Omega}_y^2 + \boldsymbol{\Omega}_z^2}. \tag{3}$$

Each birefringent section of the emulator can be represented by a 3×3 Müller matrix $\mathbf{B}$ that transforms the polarization state from the input to the output of the $n+1$th section. The following recursion relation can therefore be used to determine the polarization dispersion vector at the output of the $n+1$th section:

$$\boldsymbol{\Omega}_{n+1} = \Delta\boldsymbol{\Omega}_{n+1} + \mathbf{B}\boldsymbol{\Omega}_n. \tag{4}$$

where $\boldsymbol{\Omega}_n$ is the PMD vector at the output of the previous n sections and $\Delta\boldsymbol{\Omega}_{n+1}$ is the PMD vector of the $n+1$th section. Similar recursion relations may be found for the magnitude of the second and third orders of PMD and are given in the appendix of [20].

Each birefringent section of the emulator is simply modeled as a waveplate. On the Poincaré sphere, a waveplate is represented as a vector in the x-y plane, rotated by some angle θ relative to the x axis. We will define $\theta = 0$ as the case when the fast axis of the waveplate is aligned with the x axis of the sphere, corresponding to the horizontal axis of the laboratory. To propagate through the waveplate, the input state of polarization is first multiplied on the left by a rotation matrix, $R_z(\theta)$, that rotates the coordinate system about the z axis to align it with the fast axis of the waveplate. This result is then multiplied by the Müller propagation matrix, $R_x(\gamma)$, where the phase γ accounts for the birefringence-induced phase retardation (often called the "birefringent phase") between the two polarization modes that causes the input polarization state to precess about the waveplate-vector (oriented along the local x axis, which explains the 'x' subscript of R). At this point, the product is typically multiplied by $R_z(-\theta)$, the inverse of the original rotation matrix, to rotate back into the original coordinate system. However, for the case of PM fibers, this final step may be omitted. Since PM fiber sections are long and flexible, it is possible for the fast axis at the fiber's input to be oriented differently, relative to the x axis of the laboratory, than at the fiber's output. So the input of each section may be rotated by the desired angle θ while the output remains fixed at $\theta = 0$, so that the final rotation matrix is simply $R_z(\theta = 0) = I$, the identity matrix. For birefringent crystals, the final rotation should be included, but even if it is omitted, the PMD statistics will remain the same, and the output PMD vector will simply be off by some unitary transformation. Thus the complete propagation matrix **B** for each birefringent section where the axes at the input are rotated through an angle θ while the output axes remain fixed at $\theta = 0$ is

$$\mathbf{B} = \mathbf{R}_x\mathbf{R}_z = \begin{pmatrix} 0 & -\Omega_z & \Omega_y \\ \Omega_z & 0 & -\Omega_x \\ -\Omega_y & -\Omega_x & 0 \end{pmatrix}, \tag{5}$$

which relates the propagation matrix to the elements of the PMD vector. For emulators in which each birefringent section is simply rotated relative to the next, the rotation matrix is

$$\mathbf{R}_z(\theta) = \begin{pmatrix} \cos 2\theta & \sin 2\theta & 0 \\ -\sin 2\theta & \cos 2\theta & 0 \\ 0 & 0 & 1 \end{pmatrix}, \tag{6}$$

where θ represents the physical rotation angle between sections. For the case with polarization controllers (or some polarization scrambling device) between sections, the PMD vector is randomly distributed over the Poincaré sphere between sections. In this case, the rotation matrix can be described as

$$\mathbf{R}_z(\theta,\phi) = \begin{pmatrix} \phi & \sqrt{1-\phi^2}\cos\theta & \sqrt{1-\phi^2}\sin\theta \\ -\sqrt{1-\phi^2} & \phi\cos\theta & \phi\sin\theta \\ 0 & -\sin\theta & \cos\theta \end{pmatrix}, \tag{7}$$

where θ and ϕ are uniformly distributed in the intervals $[0, 2\pi]$ and $[-1, 1]$, respectively [10].

The propagation matrix $\mathbf{R}_x(\gamma)$ for each birefringent section is

$$\mathbf{R}_x(\gamma) = \begin{pmatrix} 1 & 0 & 0 \\ 0 & \cos\gamma & -\sin\gamma \\ 0 & \sin\gamma & \cos\gamma \end{pmatrix}, \tag{8}$$

where

$$\gamma = \omega \cdot \Delta\tau_n. \tag{9}$$

The phase γ is the accumulated birefringent phase retardation between the two polarization modes. The optical carrier frequency is ω and $\Delta\tau_n$ is the DGD of the nth section. From Eqs. (5) through (8), the PMD vector for the nth emulator section is

$$\Delta \boldsymbol{\Omega}_n = \begin{pmatrix} \Delta\Omega_{x_n} \\ \Delta\Omega_{y_n} \\ \Delta\Omega_{z_n} \end{pmatrix} = \begin{pmatrix} \Delta\tau_n \\ 0 \\ 0 \end{pmatrix}. \tag{10}$$

Using Eq. (4), the following recursion relation for the polarization dispersion vector for the case of sections connected by polarization rotators is obtained:

$$\begin{pmatrix} \Omega_{x_{n+1}} \\ \Omega_{y_{n+1}} \\ \Omega_{z_{n+1}} \end{pmatrix} = \begin{pmatrix} \Delta\tau_{n+1} \\ 0 \\ 0 \end{pmatrix} + \begin{pmatrix} \cos 2\theta & \sin 2\theta & 0 \\ -\cos 2\gamma \sin 2\theta & \cos 2\gamma \cos 2\theta & -\sin 2\gamma \\ -\sin 2\gamma \sin 2\theta & \sin 2\gamma \cos 2\theta & \cos 2\gamma \end{pmatrix} \begin{pmatrix} \Omega_{x_n} \\ \Omega_{y_n} \\ \Omega_{z_n} \end{pmatrix}. \tag{11}$$

This equation can be used with Monte Carlo methods to choose random values for θ or γ or both to model the various types of emulators described above.

5. Limitations

The major limitation of statistical PMD emulators is that the finite number of sections puts a cap on the maximum DGD that can be obtained by the emulator. This peak value corresponds to the case when the birefringent axes of all of the emulator sections become aligned. The peak DGD value is approximately the product of the rms DGD of the emulator and the square root of the number of sections. A real single-mode fiber contains hundreds to thousands of birefringent sections [18], so emulators will always tend to have a maximum DGD that is much less than in a real fiber.

In the paper by G. Biondini et al. [19], the authors show that the tail of the DGD distribution diverges from the ideal Maxwellian for values greater than about 3.5 times the average DGD. They suggest that for applications requiring accurate statistics in the tail of the distribution, the emulators should be designed to have a DGD that is much larger than the range expected in practice.

6. Deterministic Emulators

Following the development of statistical emulators, it became evident that the ability to deterministically generate particular values of DGD and higher-order PMD would be highly desirable. The goal of a deterministic PMD emulator is therefore not to mimic the statistical variations of a real fiber, but rather to predictably and repeatably generate desired values of PMD at some wavelength or span of wavelengths. This allows the tester to quantify the performance of a device over a well-defined portion of the PMD sample space and to focus on PMD states that are particularly significant or problematic. This is a critical requirement for the development of standardized testing methods and to enable designers to make apples-to-apples comparisons of transmitter–receiver pairs. Deterministic emulators are usually configured to generate only certain components or combinations of first- and higher-order PMD in order to isolate the effects due to each ([12, 20–28]). For example, it may be desirable to examine the effects of second-order PMD on the performance of a first-order compensator by using an emulator for which the DGD remains fixed and the second-order PMD is variable [21]. This level of programmability and repeatability usually comes at the cost of having a limited set of PMD coordinates that can be explored for a given source configuration. Thus, as a final testing stage, it is still often necessary and desirable to test the system using an all-order statistical emulator to confirm the system performance under conditions that closely mimic the full random variations of real fibers.

It is sometimes the case that generating a variable amount of first-order PMD is sufficient, or even required, for certain experiments. This is the case, for instance, when there is a desire to isolate the effects of DGD on a system. An example application is the testing of first-order compensators and pure DGD monitors [30]. The simplest construction employs a polarization beam splitter to separate the two polarization modes, one of which is then delayed relative to the other by inserting a variable time delay element in one branch before recombining them with a polarization beam combiner, as illustrated in Fig. 6. However, this mechanical configuration is difficult to control precisely and is known to suffer from poor stability of the output polarization state.

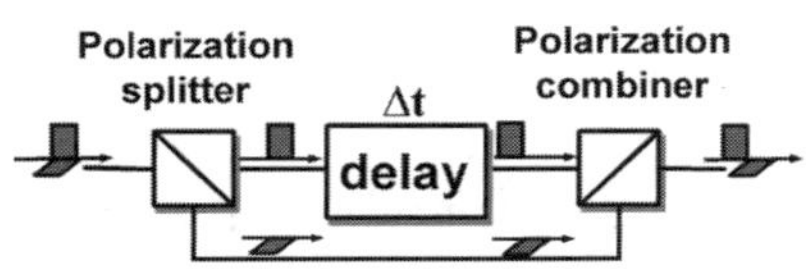

Fig. 6. Basic variable DGD element.

More recent approaches overcome these issues [22–24]. In [22], each variable DGD element consists of several birefringent crystals whose lengths increase in a binary series and are separated by electrically driven magneto-optic polarization switches, as illustrated in Fig. 7(a). The polarization switches are driven between the 0° and 90° states [Fig. 7(b)] to precisely align the fast and slow axes of the birefringent crystals so as to either add or subtract the DGD of the following segment without adding higher-order PMD. This variable DGD element can be digitally programmed to generate any DGD value from -45 ps to $+45$ ps with a tuning speed of < 1 ms and a resolution of 1.40 ps. Figure 7(c) shows a plot of the measured versus designed DGD values, as well as the small amount of second-order PMD of the device (approximately equivalent to that of a PM fiber). Only a single DGD element is needed to programmatically generate any desired statistical distribution of pure first-order PMD.

With a programmable DGD element, it is possible to map out a Maxwellian probability distribution by simply using a microprocessor to control the element to provide the proper counts of each DGD value to generate an ideal histogram, even in the tail of the distribution (there will still exist a peak achievable DGD, but it can be designed to be a large value). One drawback of the statistical emulators described above is that their statistics (i.e., average DGD) are fixed at the time of fabrication and cannot easily be reconfigured to emulate different fiber plants. Also, at least two or three PMD emulators with different average DGD values are needed to perform accelerated outage probability testing, in which data is collected using multiple emulators with higher than expected average PMD to rapidly induce a large number of outages. This data is then used to extrapolate to lower average PMD values in analogy to accelerated lifetime tests using elevated temperatures [17]. To obtain higher-order PMD, multiple programmable DGD elements can be concatenated as shown in Fig. 8(a), where three

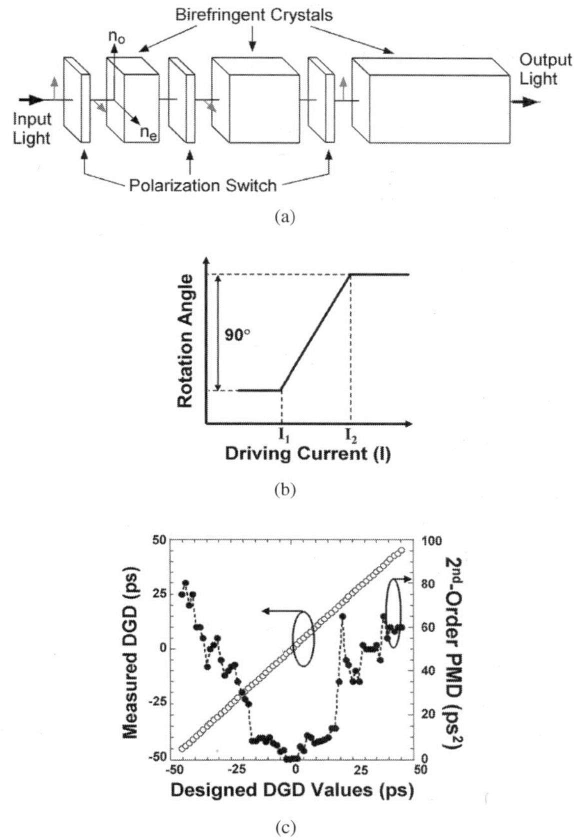

Fig. 7. (a) Digitally programmable DGD element. (b) The magneto-optic polarization switches are current driven to precisely switch the polarization state between the two orthogonal axes of each birefringent crystal. (c) Measured versus designed DGD values and second-order PMD. From [22].

elements are connected by two fiber-squeezer-based polarization controllers [23]. To obtain a Maxwellian DGD distribution at the output of all three sections, the DGD values of each element are varied according to a Maxwellian with average $\Delta\tau$. This yields an average DGD of $3^{1/2}(\Delta\tau)$ for the total emulator and an average second-order PMD distribution that has the correct shape but falls slightly short of that expected for a real fiber because only three sections are used. The tunability of the PMD statistics can be seen in Fig. 8(b), which shows three different distributions generated by the emulator with $\langle \text{DGD} \rangle = 10$, 25 and 35 ps. As expected, the DGD values closely match the expected Maxwellian distribution. The corresponding second-order PMD distributions (not shown) had averages of 38, 268, and 471 ps^2, which are $\sim$ 30% lower than expected for a real fiber.

The authors in [24] propose an alternative design for a variable DGD element that produces zero second-order PMD using four identical, fixed DGD sections (e.g.,

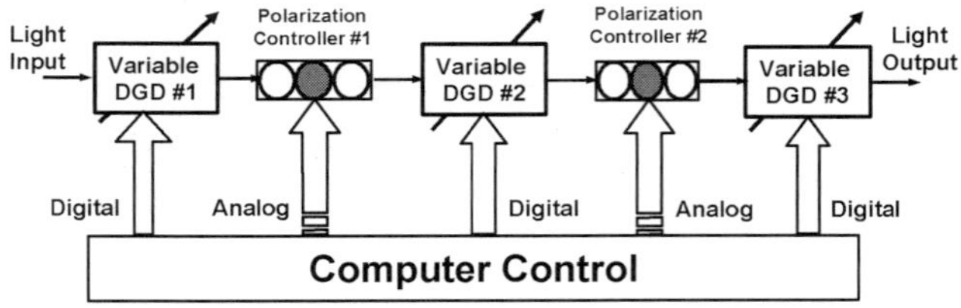

(a)

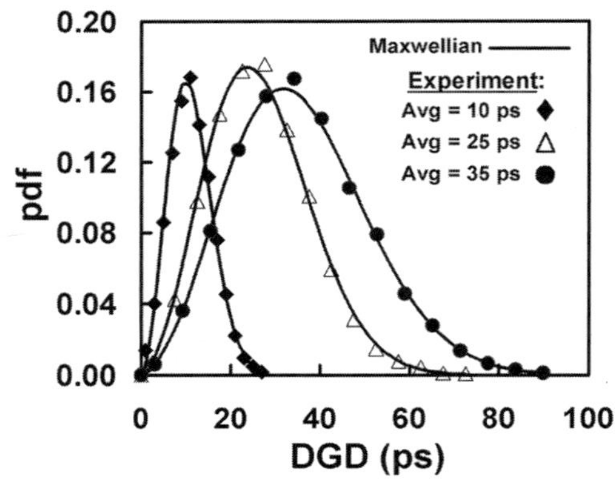

(b)

Fig. 8. Three programmable DGD elements concatenated with two electrically driven polarization controllers to form an emulator with adjustable statistics. (b) Three output DGD distributions with different mean values, showing a good fit to the Maxwellian pdf for each case. From [23].

birefringent crystals) as shown in Fig. 9(a). The sections are connected via tunable phase plates that could be implemented using electro-optic or magneto-optic materials, or the entire module could be integrated using MEMS technology. The produced versus designed DGD values from 0 to 50 ps are shown in Fig. 9(b). The zero second-order PMD produced by the device is shown in Fig. 9(c), where it is compared with that produced by two 25-ps PM fiber sections connected via a polarization controller. The authors also find that the third-order PMD is low, in that it is half the value generated by the two PM fiber segments. This symmetric architecture of four identical DGD elements grouped into two blocks thus produces pure variable DGD with no second-order PMD and low third-order PMD and has the potential for high-speed tuning and a compact design.

Another recent publication presented first and second-order PMD emulation using only two cascaded, equal-length, DGD sections to obtain rare events of DGD and second-order PMD to concentrate the measured results on those cases most likely to result in system outages [25].

The goal of a deterministic multiorder PMD emulator is to predictably generate any desired combination of first-, second- and higher-order PMD. However, this has been a difficult task to achieve as it requires total control over the three key parameters of each birefringent section: the DGD (delay), the orientation of the birefringent axes

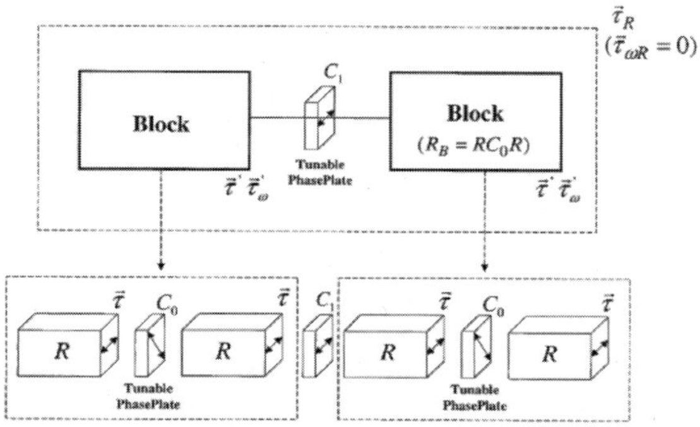

(a)

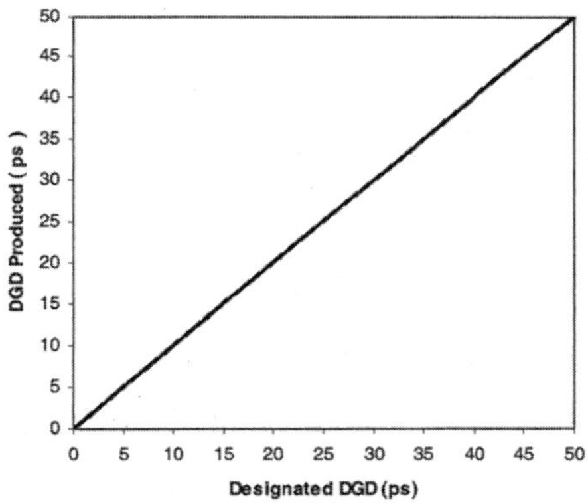

(b)

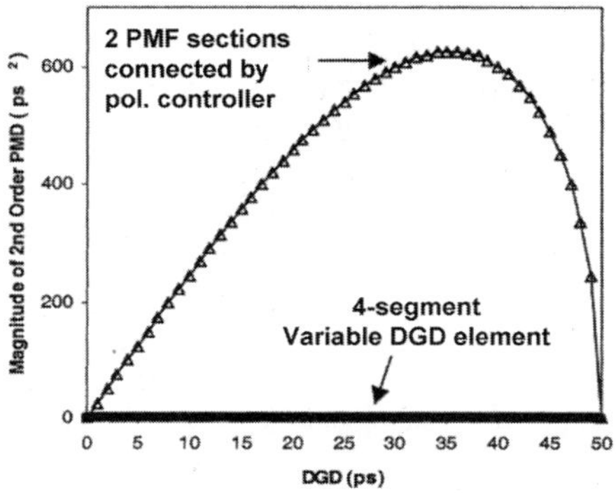

(c)

Fig. 9. (a) Variable DGD module with no second-order PMD consisting of four identical birefringent elements connected by tunable phase plates. (b) DGD generated for each designated value over the 0 to 50 ps tuning range. (c) second-order PMD generated by the module is zero, in comparison to 2 sections of 25-ps PM fiber connected by a polarization controller. From [24].

(polarization mode mixing), and the birefringent phase, as illustrated in Fig. 10(a). Owing to fabrication tolerances or environmental sensitivity, lack of control over the phase of the sections has been the prohibiting factor. To resolve this issue, J. Damask et al. have employed a Fourier-based approach to PMD generation and introduced the concept of a "coherent PMD source." They use a concatenation of four birefringent crystals separated by rotatable waveplates, as shown in Figs. 10(b) and (c), that they call "ECHO" for "enhanced coherent higher order" PMD instrument [26]. The key component is the Evans phase shifter [29], an element that is added after each crystal (excluding the two end sections) to bring the phase of each segment to within a small fraction of the birefringent beat length. These phase shifters are used to drive the emulator into a "coherent" state, meaning that the Fourier components of the square of the DGD spectrum are integral multiples of a unit frequency and have the same Fourier phase, as shown in Fig. 10(d). Once the phase shifters are adjusted for coherence, they are rotated together to frequency shift the periodic DGD and/or higher-order PMD spectra in a predictable manner. With such a source, completely predictable and flexible generation of PMD including first-, second- [both depolarization and polarization-dependent chromatic dispersion (PDCD)], and higher-order PMD can be generated. The data plotted in Fig. 11 illustrates the advantage of the 4-stage coherent PMD source [Fig. 11(b)] over a previously presented 12-stage incoherent programmable source [Fig. 11(a)], in which the birefringent phase of each section was not controlled [31]. The light contour lines in the plots show the joint pdf of the first and second-order PMD of a real fiber. The incoherent emulator, while programmable, is unable to cover the entire state-space, whereas the coherent source can deterministically reach any state of interest expected of a real fiber.

7. Simulation Tools

As a final note, an important component of the emulation field has been the computer modeling of PMD emulators for use in simulations, often for evaluating different compensator configurations. In the past, these simulations simply used Monte Carlo techniques to generate random PMD samples. However, even with a computer, it is prohibitively time-consuming to perform enough trials to evaluate the effects of extremely rare PMD events that cause outage probabilities of 10^{-6} (< 1 min/yr) or less. The tool of importance sampling (IS) has thus attracted much recent attention [32–36]. Importance sampling (IS) is a powerful tool for obtaining very low probability events with relatively few sample points [32]. This is accomplished by altering the method of obtaining the random samples to concentrate the measured results in the area of interest in the sample space. This distorts the probability distribution of the measured results, so each sample must then be appropriately weighted to map the measured values back onto the proper distribution function. Although initially confined to computer simulation studies, the recent developments of more programmable PMD sources, such as the variable DGD element and the coherent PMD source described above, are making it feasible to experimentally employ importance sampling techniques [36].

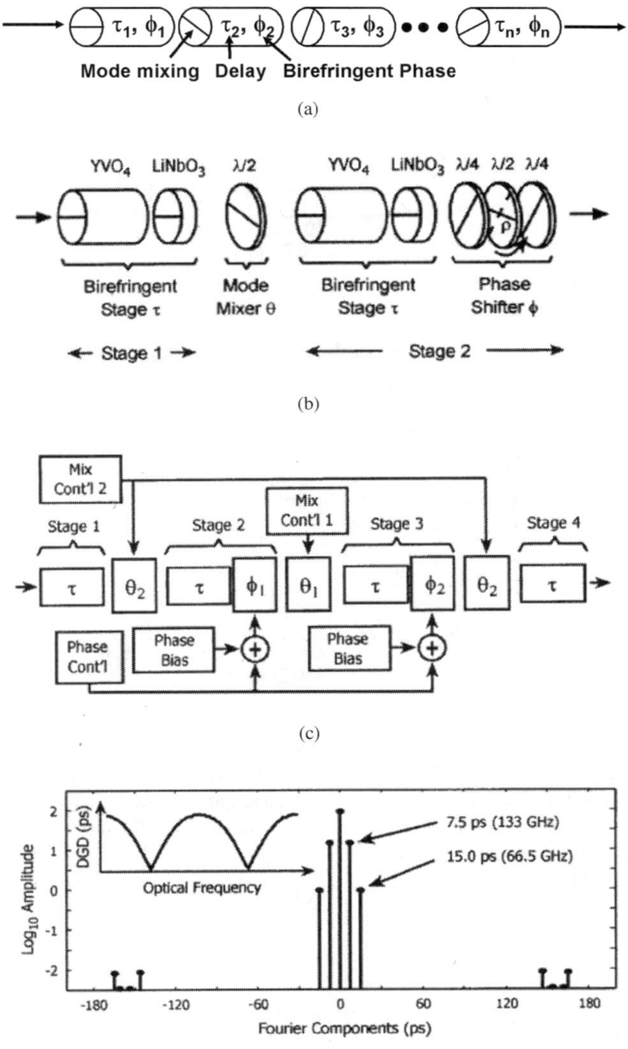

Fig. 10. (a) Three key control parameters for each segment of a multistage deterministic emulator are the DGD (delay), the orientation of the birefringent axes (polarization mode mixing), and the birefringent phase (accumulated phase delay between the fast and slow axes). (b) and (c) Architecture of a 4-stage coherent programmable PMD source (ECHO), from [26]. (d) The Fourier components of the measured square of the DGD spectrum (inset) for the ECHO with 7.5 ps DGD per stage, from [26].

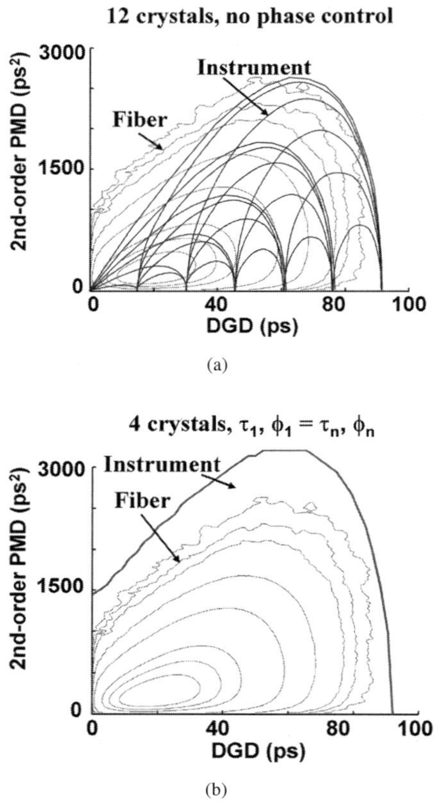

Fig. 11. Comparison of the addressable PMD space of (a) a programmable 12-stage incoherent (no phase control) PMD source and (b) the 4-stage coherent PMD source from Fig. 10. The light contour lines show the joint pdf of the first and second-order PMD states of a real fiber. The darker lines in (a) and the region bounded by the outer dark line in (b) indicate the addressable regions. From [31].

References

[1] H. Kogelnik, R. M. Jopson, and L. E. Nelson, Optical Fiber Telecommunications IV B, San Diego, chapter 15, pp. 725–861, 2002.

[2] G. J. Foschini and C. D. Poole, "Statistical theory of polarization dispersion in single mode fibers," IEEE J. Lightwave Technol., **9**, 1439–1456 (Nov. 1991).

[3] G. J. Foschini, L. E. Nelson, R. M. Jopson, and H. Kogelnik, "Probability densities of second-order polarization mode dispersion including polarization dependent chromatic fiber dispersion," IEEE Photon. Technol. Lett., **12**, 293–295 (March 2000).

[4] C. H. Prola, Jr., J. A. Pereira da Silva, A. O. Dal Forno, R. Passy, J. P. von der Weid, and N. Gisin, "PMD emulators and signal distortion in 2.48-Gb/s IM-DD lightwave systems," IEEE Photon. Technol. Lett., **9**, 842–844 (June 1997).

[5] F. Bruyère, L. Pierre, J. P. Thiery, and B. Clesca, "Penalties induced by higher-order PMD at 10 Gbit/s in nondispersion-shifted fibers," in Tech. Digest Optical Fiber Communications Conf., paper WC1, Feb. 1997.

[6] D. L. Peterson Jr., B. C. Ward, K. B. Rochford, P. J. Leo and G. Simer, "Polarization mode dispersion compensator field trial and field fiber characterization," Optics Express, **10**, 614–621 (July 2002).

[7] M. Karlsson and J. Brentel, "Autocorrelation function of the polarization-mode-dispersion vector," Opt. Lett., **24**, 939–941 (July 1999).

[8] B. S. Marks, I. T. Lima, and C. R. Menyuk, "Autocorrelation function for polarization mode dispersion emulators with rotators," Opt. Lett., **27**, 1150–1152 (July 2002).

[9] I. T. Lima, R. Khosravani, P. Ebrahimi, E. Ibragimov, C. R. Menyuk, and A. E. Willner, "Comparison of polarization mode dispersion emulators," IEEE J. Lightwave Technol., **19**, 1872–1881 (Dec. 2001).

[10] A. Djupsjöbacka, "On differential group-delay statistics for polarization-mode dispersion emulators," IEEE J. Lightwave Technol., **19**, 285–290 (Feb. 2001).

[11] J. N. Damask, "A Programmable Polarization-Mode Dispersion Emulator for Systematic Testing of 10 Gb/s PMD Compensators," in Tech. Digest Optical Fiber Communications Conf., paper ThB3, March 2000.

[12] J. N. Damask, G. J. Simer, K. B. Rochford, and P. R. Myers, "Demonstration of a programmable PMD source," IEEE Photon. Technol. Lett., **15**, 296–298 (Feb. 2003).

[13] R. Khosravani, I. T. Lima, P. Ebrahimi, E. Ibragimov, A. E. Willner, and C. R. Menyuk, "Time and frequency domain characteristics of polarization-mode dispersion emulators," IEEE Photon. Technol. Lett., **13**, 127–129 (Feb. 2001).

[14] R. Noé, D. Sandel, M. Yoshida-Dierolf, S. Hinz, V. Mirvoda, A. Schöpflin, C. Glingener, E. Gottwald, C. Scheerer, G. Fischer, T. Weyrauch, and W. Haase, "Polarization mode dispersion compensation at 10, 20, and 40 Gb/s with various optical equalizers," IEEE J. Lightwave Technol., **17**, 1602–1616 (Sept. 1999).

[15] M. C. Hauer, Q. Yu, E. R. Lyons, C. H. Lin, A. A. Au, H. P. Lee, and A. E. Willner, "Electrically controllable all-fiber PMD emulator using a compact array of thin-film microheaters," IEEE J. Lightwave Technol., **22**, 1059–1065 (April 2004).

[16] B. L. Heffner, "Accurate, automated measurement of differential group delay dispersion and principal state variation using Jones matrix eigenanalysis," IEEE Photon. Technol. Lett., **5**, 814–817 (July 1993).

[17] W. Shieh, "Accelerated outage probability testing for PMD induced impairment," Photon. Technol. Lett., **12**, 1364–1366 (Oct. 2000).

[18] A. Galtarossa, L. Palmieri, A. Pizzinat, M. Schiano, and T. Tambosso, "Measurement of local beat length and differential group delay in installed single-mode fibers," IEEE J. Lightwave Tec., **18**, 1389–1394 (Oct. 2000).

[19] G. Biondini, W. L. Kath, and C. R. Menyuk, "Non-Maxwellian DGD distributions of PMD emulators," in Tech. Digest Optical Fiber Communications Conf., paper ThA5, March 2001.

[20] P. B. Phua and H. A. Haus, "A deterministically controlled four-segment polarization-mode dispersion emulator, IEEE J. Lightwave Technol., **20**, 1132–1140 (July 2002).

[21] P. B. Phua and H. A. Haus, "Variable second-order PMD module without first-order PMD," IEEE J. Lightwave Technol., **20**, 1951–1956 (Nov. 2002).

[22] L.-S. Yan, C. Yeh, G. Yang, L. Lin, Z. Chen, Y. Q. Shi, A. E. Willner and X. Steve Yao, "Programmable group delay module using binary polarization switching," IEEE J. Lightwave Technol., **21**, 1676–1684 (July 2003).

[23] L.-S. Yan, M.C. Hauer, C. Yeh, G. Yang, L. Lin, Z. Chen, Y. Q. Shi, X. Steve Yao, A. E. Willner, and W. L. Kath, "High-Speed, Stable and Repeatable PMD Emulator with Tunable Statistics," in Tech. Digest Optical Fiber Communications Conf., paper MF6, March 2003.

[24] P. B. Phua and H. A. Haus, "Variable differential-group-delay module without second-order PMD," IEEE J. Lightwave Technol., **20** (9), 1788–1794 (Sept. 2002).

[25] M. Wegmuller, S. Demma, C. Vinegoni, and N. Gisin, "Emulator of first- and second-order polarization-mode dispersion," IEEE Photon. Technol. Lett., **14**, 630–632 (May 2002).

[26] J. N. Damask, P. R. Myers, A. Boschi, and G. J. Simer, "Demonstration of a Coherent PMD Source," in IEEE Photon. Technol. Lett., **15**, 1612–1614 (Nov.2003).

[27] L. Möller, J. H. Sinsky, S. Chandrasekhar, L. L. Buhl, S. X. Yao, C. Yeh, and Y. Shi, "A tunable interferometrically stable three-section higher-order PMD emulator," IEEE Photon. Technol. Lett., **15**, 230–232 (Feb. 2003).

[28] H. Kogelnik, L. E. Nelson and J. P. Gordon, "Emulation and inversion of polarization-mode dispersion," IEEE J. Lightwave Technol., **21**, 482–495 (Feb. 2003).

[29] J. Evans, "The birefringent filter," J. Opt. Soc. Amer., **39**, 229–242 (1949).

[30] S.M.R. Motaghian Nezam, L.-S. Yan, J. McGeehan, Y. Q. Shi, A.E. Willner, and Steve Yao, "Wide-dynamic-range DGD monitoring by partial optical signal spectrum DOP measurement" in Tech. Digest Optical Fiber Communications Conf., post-deadline paper FD8, March 2002.

[31] J. N. Damask, "Construction and Applications of Programmable PMD Sources," Presentation given at the University of Southern California, July 1, 2003.

[32] S. L. Fogal, G. Biondini, and W.L. Kath, "Multiple importance sampling for first- and second-order polarization-mode dispersion," IEEE Photon. Technol. Lett., **14**, 1273–1275 (2002).

[33] W.L. Kath and G. Biondini, "Calculating PMD statistics and outage probabilities with importance sampling," Conference on Optical Fiber Communication, paper THA1, 2003.

[34] A. Lima, I.T Lima, B. Marks, C. Menyuk, "Performance analysis of single-section PMD compensators using multiple importance sampling," Conference on Optical Fiber Communication, paper THA3, 2003.

[35] G. Biondini and W.L. Kath, "PMD emulation with importance sampling and Maxwellian length sections," Conference on Optical Fiber Communication, paper THA2, 2003.

[36] L.-S. Yan, M.C. Hauer, Y. Q. Shi, S. Yao, A.E. Willner, and W.L. Kath, "Experimental importance sampling using a 3-Section PMD emulator with programmable DGD elements," Conference on Optical Fiber Communication, paper THA4, 2003.